Progress in Inflammation Research

Series editors
Michael J. Parnham
Fraunhofer IME & Goethe University
Frankfurt, Germany

Achim Schmidtko
Goethe University
Frankfurt, Germany

More information about this series at http://www.springer.com/series/4983

Dieter Steinhilber
Editor

Lipoxygenases in Inflammation

Editor
Dieter Steinhilber
Institute of Pharmaceutical Chemistry
Goethe University Frankfurt
Frankfurt am Main
Germany

Series Editors
Michael J. Parnham
Fraunhofer IME & Goethe University
Frankfurt
Germany

Achim Schmidtko
Goethe University
Frankfurt
Germany

Progress in Inflammation Research
ISBN 978-3-319-80212-1   ISBN 978-3-319-27766-0   (eBook)
DOI 10.1007/978-3-319-27766-0

© Springer International Publishing Switzerland 2016
Softcover reprint of the hardcover 1st edition 2016
This work is subject to copyright. All rights are reserved by the Publisher, whether the whole or part of the material is concerned, specifically the rights of translation, reprinting, reuse of illustrations, recitation, broadcasting, reproduction on microfilms or in any other physical way, and transmission or information storage and retrieval, electronic adaptation, computer software, or by similar or dissimilar methodology now known or hereafter developed.
The use of general descriptive names, registered names, trademarks, service marks, etc. in this publication does not imply, even in the absence of a specific statement, that such names are exempt from the relevant protective laws and regulations and therefore free for general use.
The publisher, the authors and the editors are safe to assume that the advice and information in this book are believed to be true and accurate at the date of publication. Neither the publisher nor the authors or the editors give a warranty, express or implied, with respect to the material contained herein or for any errors or omissions that may have been made.

Printed on acid-free paper

This Springer imprint is published by Springer Nature
The registered company is Springer International Publishing AG Switzerland

Preface

Lipoxygenases oxidize polyunsaturated fatty acids and can generate lipid mediators which have a broad spectrum of physiological and pathophysiological functions. Data gathered during the last decades suggest that lipoxygenase metabolites are involved in host defence reactions and the cardiovascular system and contribute to the development of inflammatory and allergic diseases, cardiovascular disease, and cancer but are also involved in the resolution of inflammation. This *Progress in Inflammation Research* volume summarizes the physiological and pathophysiological functions of lipoxygenases.

Contents

Lipoxygenases: An Introduction 1
Dieter Steinhilber

5-Lipoxygenase 7
Oliver Werz and Olof Rådmark

Leukotriene A_4 Hydrolase and Leukotriene C_4 Synthase 31
Agnes Rinaldo-Matthis and Jesper Z. Haeggström

Catalytic Multiplicity of 15-Lipoxygenase-1 Orthologs (ALOX15) of Different Species 47
Hartmut Kühn, Felix Karst, and Dagmar Heydeck

Platelets and Lipoxygenases 83
Michael Holinstat, Katrin Niisuke, and Benjamin E. Tourdot

Lipoxygenases and Cardiovascular Diseases 101
Andrés Laguna-Fernández, Marcelo H. Petri,
Silke Thul, and Magnus Bäck

Role of Lipoxygenases in Pathogenesis of Cancer 131
J. Roos, B. Kühn, J. Fettel, I.V. Maucher, M. Ruthardt, A. Kahnt,
T. Vorup-Jensen, C. Matrone, D. Steinhilber, and T.J. Maier

The Physiology and Pathophysiology of Lipoxygenases in the Skin 159
Peter Krieg and Gerhard Fürstenberger

5-Oxo-ETE and Inflammation 185
William S. Powell and Joshua Rokach

Lipoxins, Resolvins, and the Resolution of Inflammation 211
Antonio Recchiuti, Eleonora Cianci, Felice Simiele, and Mario Romano

Index 241

List of Contributors

Magnus Bäck Translational Cardiology, Center for Molecular Medicine, L8:03, Karolinska University Hospital, Stockholm, Sweden

Department of Medicine, Karolinska Institutet, Stockholm, Sweden

Department of Cardiology, Karolinska University Hospital, Stockholm, Sweden

Eleonora Cianci Department of Medical, Oral and Biotechnological Sciences, G. D'Annunzio University of Chieti-Pescara, Chieti, Italy

Department of Medicine and Aging Sciences, G. D'Annunzio University of Chieti-Pescara, Chieti, Italy

Center of Excellence on Aging, G. D'Annunzio University of Chieti-Pescara, Chieti, Italy

Gerhard Fürstenberger Molecular Diagnostics of Oncogenic Infections, German Cancer Research Center, Heidelberg, Germany

J. Fettel Institute of Pharmaceutical Chemistry/ZAFES, Goethe-University, Frankfurt/Main, Germany

Jesper Z. Haeggström Department of Medical Biochemistry and Biophysics, Karolinska Institutet, Stockholm, Sweden

Dagmar Heydeck Institute of Biochemistry, Charité—University Medicine Berlin, Berlin, Germany

Michael Holinstat University of Michigan Medical School, Ann Arbor, MI, USA

B. Kühn Institute of Pharmaceutical Chemistry/ZAFES, Goethe-University, Frankfurt/Main, Germany

A. Kahnt Institute of Pharmaceutical Chemistry/ZAFES, Goethe-University, Frankfurt/Main, Germany

Felix Karst Institute of Biochemistry, Charité—University Medicine Berlin, Berlin, Germany

Peter Krieg Molecular Diagnostics of Oncogenic Infections, German Cancer Research Center, Heidelberg, Germany

Hartmut Kühn Institute of Biochemistry, Charité—University Medicine Berlin, Berlin, Germany

Andrés Laguna-Fernández Translational Cardiology, Center for Molecular Medicine, L8:03, Karolinska University Hospital, Stockholm, Sweden

Department of Medicine, Karolinska Institutet, Stockholm, Sweden

T.J. Maier Department of Biomedicine, Aarhus University, Aarhus C, Denmark

C. Matrone Department of Biomedicine, Aarhus University, Aarhus C, Denmark

I.V. Maucher Institute of Pharmaceutical Chemistry/ZAFES, Goethe-University, Frankfurt/Main, Germany

Katrin Niisuke University of Michigan Medical School, Ann Arbor, MI, USA

Marcelo H. Petri Translational Cardiology, Center for Molecular Medicine, L8:03, Karolinska University Hospital, Stockholm, Sweden

Department of Medicine, Karolinska Institutet, Stockholm, Sweden

William S. Powell Meakins-Christie Laboratories, McGill University Health Centre Research Institute, Centre for Translational Biology, Montreal, QC, Canada

Olof Rådmark Division of Chemistry II, Department of Medical Biochemistry and Biophysics, Karolinska Institutet, Solna, Sweden

Antonio Recchiuti Department of Medical, Oral and Biotechnological Sciences, G. D'Annunzio University of Chieti-Pescara, Chieti, Italy

Center of Excellence on Aging, G. D'Annunzio University of Chieti-Pescara, Chieti, Italy

Agnes Rinaldo-Matthis Department of Medical Biochemistry and Biophysics, Karolinska Institutet, Stockholm, Sweden

Joshua Rokach Department of Chemistry, Claude Pepper Institute, Florida Institute of Technology, Melbourne, FL, USA

Mario Romano Department of Medical, Oral and Biotechnological Sciences, G. D'Annunzio University of Chieti-Pescara, Chieti, Italy

Center of Excellence on Aging, G. D'Annunzio University of Chieti-Pescara, Chieti, Italy

J. Roos Institute of Pharmaceutical Chemistry/ZAFES, Goethe-University, Frankfurt/Main, Germany

List of Contributors

M. Ruthardt Department of Hematology, Goethe University, Frankfurt, Germany

Felice Simiele Department of Medical, Oral and Biotechnological Sciences, G. D'Annunzio University of Chieti-Pescara, Chieti, Italy

Center of Excellence on Aging, G. D'Annunzio University of Chieti-Pescara, Chieti, Italy

Dieter Steinhilber, Ph.D. Institute of Pharmaceutical Chemistry/ZAFES, Goethe University, Frankfurt/Main, Germany

Silke Thul Translational Cardiology, Center for Molecular Medicine, L8:03, Karolinska University Hospital, Stockholm, Sweden

Department of Medicine, Karolinska Institutet, Stockholm, Sweden

Benjamin E. Tourdot University of Michigan Medical School, Ann Arbor, MI, USA

T. Vorup-Jensen Department of Biomedicine, Aarhus University, Aarhus C, Denmark

Oliver Werz Chair of Pharmaceutical/Medicinal Chemistry, Institute of Pharmacy, University Jena, Jena, Germany

Lipoxygenases: An Introduction

Dieter Steinhilber

Abstract Oxidation of polyunsaturated fatty acids by lipoxygenases (LO) leads to a variety of fatty acid metabolites which play important roles in physiology but also in pathophysiology. Data accumulated during the last decades point to the fact that lipoxygenase metabolites are involved in host defence reactions, the cardiovascular system and contribute to the development of inflammatory and allergic diseases, cardiovascular disease and cancer. This *Progress in Inflammation Research* volume summarizes the physiological and pathophysiological functions of lipoxygenases.

1 Catalytic Properties of Lipoxygenases

Lipoxygenases (LOs) are a family of non-heme iron containing enzymes which catalyze the oxygenation of polyunsaturated fatty acids (PUFA) [1]. The first step of the reaction cycle consists of the abstraction of a hydrogen radical (Fig. 1). A prerequisite for this reaction step is the presence of two double bonds in allylic position. This implicates that fatty acids have to carry at least two double bounds spaced by a methylene group in order to serve as lipoxygenase substrate. The second reaction step consists of the rearrangement of a double bond which leads to a trans-configured double bond and the generation of a conjugated diene system. The final steps of the reaction are the insertion of molecular oxygen and hydrogen, respectively. In arachidonic acid which is considered to be one of the most important lipoxygenase substrates, the hydrogens at C(7), C(10) and C(13) fulfill the requirements for lipoxygenase-mediated abstraction which leads to six possible sites for oxygen insertion (Fig. 2). In mammalian cells, lipoxygenases that insert oxygen at positions 5, 8, 12 and 15 have been discovered (Fig. 3). The different lipoxygenases are named according to the position of the oxygen insertion at the arachidonic acid although this nomenclature has turned out to be difficult since e.g. 15-LO1 orthologs from different species show different positional specificities for oxygen insertion. The human lipoxygenases are listed in Table 1.

D. Steinhilber, Ph.D. (✉)
Institute of Pharmaceutical Chemistry/ZAFES, Goethe University, Max-von-Laue-Str. 9, 60438 Frankfurt/Main, Germany
e-mail: steinhilber@em.uni-frankfurt.de

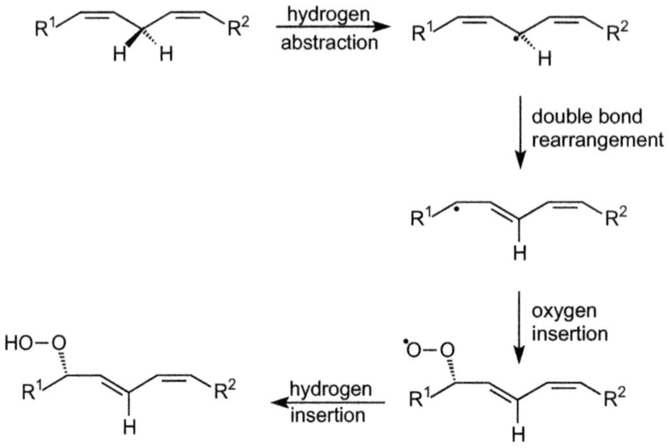

Fig. 1 Mechanism of oxidation of polyunsaturated fatty acids by lipoxygenases

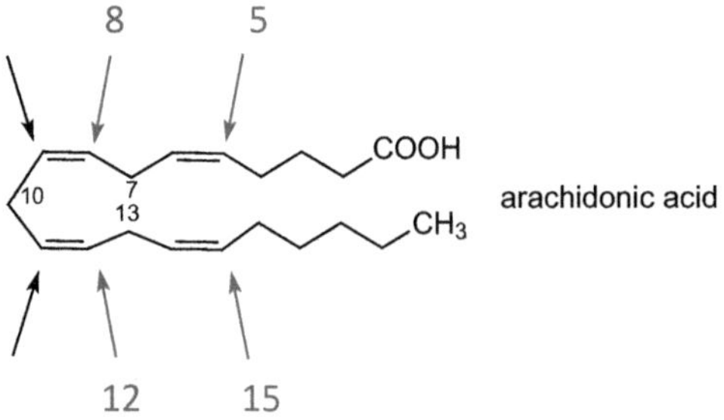

Fig. 2 Arachidonic acid and possible sites of oxygen insertion, indicated by *arrows*. The positions of hydrogen abstractions are C(7), C(10) and C(13)

It has to be emphasized that other polyunsaturated fatty acids such as linoleic acid, dihomo-γ-linoleic acid, eicosapentaenoic acid or docosahexaenoic acid can also serve as substrates for lipoxygenases and that the various lipoxygenases display different substrate specificities [2–4]. Most of the lipoxygenases can only convert free fatty acids whereas 15-LO1 is able to oxidize esterified fatty acids in phospholipids [5]. The mechanism of this reaction is unclear since phospholipids do not fit into the catalytic site of 15-LO without considerable rearrangement.

2 Regulation of Lipoxygenase Activity

Lipoxygenases consist of two domains, the catalytic domain and the regulatory C2-like domain. The catalytic center contains a nonheme iron in a cavity that accommodates the substrate and accomplishes the stereospecific insertion of

Lipoxygenases: An Introduction

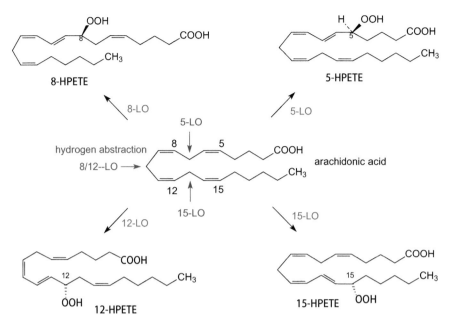

Fig. 3 Formation of 5-, 8-, 12- and 15-HPETE from arachidonic acid by the respective lipoxygenases (LOs). The positions of hydrogen abstraction are indicated by *blue arrows*

molecular oxygen. As mentioned above, only PUFAs can serve as substrates for lipoxygenases.

In general, lipoxygenases are enzymes with tightly regulated cellular activities. Intracellular calcium levels, membrane lipids, ATP, diacylglycerols, phosphatidylinositol derivatives, enzyme phosphorylations and various lipoxygenase interacting proteins have been shown to be involved in the regulation of lipoxygenase activities. In vitro, catalysis of LOs is characterized by a lag phase, a catalytic phase and an inactivation phase which occurs after a few minutes. The molecular mechanism for the rapid inactivation of lipoxygenases is still unclear. During the initial lag phase the iron of the enzymes is oxidized. Thus, an important parameter for the onset of a lipoxygenase reaction is the redox tonus. Activation of resting LOs requires the oxidation of the active site iron from the ferrous to the ferric state (Fig. 4). This oxidation is necessary in order to enter the catalytic cycle. In-vitro, lipid hydroperoxides can shorten the lag phase of LOs by converting the enzyme into the active form. During the catalytic cycle, the active site iron switches between the ferric and the ferrous state. Among various lipid hydroperoxides (ROOH), 5- and 12-hydro(pero)xyeicosatetraenoic acid (HETE), and 13-hydroperoxyoctadecadienoic acid can activate lipoxygenases, whereas hydrogen peroxide and several other organic hydroperoxides failed in this respect suggesting that distinct structural features are required for the hydroperoxides in order to serve as lipoxygenase activator.

Table 1 Human lipoxygenases

Lipoxygenase	Gene ID	Substrates	Physiological role	Expression profile
5-LO	ALOX5	Arachidonic acid Eicosapentaenoic acid Docosahexaenoic acid	Innate immune response Inflammation Resolution of inflammation Cell growth	Mainly leukocytes
12-LO	ALOX12	Arachidonic acid	Regulation of platelet function, cell growth and metastasis	Mainly platelets
12R-LO	ALOX12B	Ceramides	Epidermal differentiation	Skin
eLO3	ALOXE3	12R-HPETE	Epidermal differentiation	Skin
15-LO1	ALOX15	Linoleic acid, arachidonic acid Eicosapentaenoic acid Docosahexaenoic acid Esterified PUFAs	Erythropoiesis, inflammation Cell growth and apoptosis	Eosinophils, stimulated monocytes
15-LO2	ALOX15B	Linoleic acid, arachidonic acid Eicosapentaenoic acid Docosahexaenoic acid	Inflammation Resolution of inflammation	Monocytes, hair roots, skin, prostate

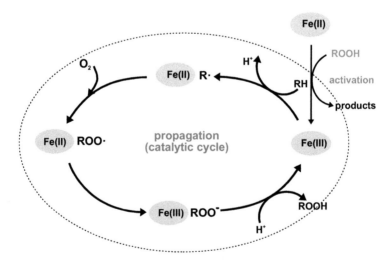

Fig. 4 Catalytic redox cycle of lipoxygenases

3 Physiological and Pathophysiological Roles of Lipoxygenases

Although mammalian lipoxygenases have been discovered more than four decades ago, their physiological and pathophysiological functions are only partially known. 5-LO is one of the best characterized enzymes. It converts arachidonic acid into 5-HPETE and generates leukotriene A_4 which is an unstable intermediate in the biosynthesis of leukotrienes. It can then be further converted by LTA_4 hydrolase or LTC_4 synthase to biologically active leukotrienes [6]. The 5-lipoxygenase pathway is involved in the onset of host defence and inflammatory reactions. The paper by Werz et al. in this issue highlights the 5-LO enzyme and its regulation [4]. The 5-LO product 5-HPETE can be reduced to 5-HETE by peroxidases and then subsequently be converted to 5-oxo-ETE, a mediator with proinflammatory and antiapoptotic properties [7]. ALOX12 (platelet-type) 12-LO is mainly expressed in platelets and it seems to be involved in the regulation of platelet functions and cancer metastasis. Although 12-HETE is the main metabolite generated by 12-LO in platelets, it is at present unclear which 12-LO-derived oxylipins are the main players in the regulation of platelet activation. The current status of our understanding of 12-LO in platelets is summarized in this article [2]. Data from LO-knockout studies have provided deep insights into the roles of lipoxygenases in the skin [3]. 12R-LO and eLO3 have been shown to be involved in epidermal differentiation and skin development and are important for the maintenance of the water barrier of the skin.

Regarding its biological function, 15-LO seems to be the most complex one. Pro- and anti-inflammatory properties as well as mitogenic or proapoptotic properties have been reported in the context of 15-LO activity. Human 15-LOs preferentially generate 15-HPETE from arachidonic acid whereas the mouse orthologs as well as orthologs from many other species preferentially form 12-HPETE. From this observation, it can be concluded that either the position of oxygen insertion [C(12) or C(15)] is not critical for the biological function or that arachidonic acid is not the crucial substrate for 15-LO. The paper by Kühn summarizes the properties of the 15-LO orthologs [5].

In order to link certain lipoxygenase metabolites and activities to distinct physiological functions, lipidomics will be a helpful tool to identify lipid mediators which play a key role. It has to be considered that lipoxygenases can use different substrates and that various lipoxygenases can interact to produce complex oxylipins. 5-LO is known since many years to produce proinflammatory leukotrienes. However in concert with 12-LO or 15-LOs and by the usage of alternative substrates such as eicosapentaenoic acid or docosahexaenoic acid, it can produce lipoxins, resolvins or maresins which possess potent anti-inflammatory and proresolving activities [8]. The proresolving properties of these mediators are well documented and GPCRs such as FPR2 (ALX) and GPR32 have been suggested to be involved in the transduction of proresolving signals. However, the signalling via these receptors is less clear and does not seem to follow canonical

GPCR signal transduction. Thus, more work will be necessary to unravel the mechanisms that contribute to the proresolving properties of these lipid mediators.

Due to their functions in the onset and resolution of inflammation, LOs were suggested to play also a role in atherosclerosis development. The recent achievements in our understanding of the role of LOs in atherosclerosis development are summarized here [9].

Besides regulation of inflammatory responses, LOs and their products have been shown to be involved in the regulation of cell proliferation, apoptosis and cancer metastasis. Although a considerable number of studies were performed with lipoxygenase inhibitors where off-target effects were not thoroughly addressed e.g. by add-back experiments with lipoxygenase products such as HETEs, oxo-ETEs or leukotrienes, there is accumulating evidence from LO knockout or knockdown experiments that LOs and its products have growth-regulatory properties. These aspects are summarized here [10].

References

1. Kühn H, Banthiya S, van Leyen K (2015) Mammalian lipoxygenases and their biological relevance. Biochim Biophys Acta 1851(4):308–330. doi:10.1016/j.bbalip.2014.10.002
2. Holinstat M (2015) Platelets and lipoxygenase. In: Steinhilber D (ed) Lipoxygenases in inflammation, Progress in inflammation research. Springer, Cham
3. Krieg P, Fürstenberger G (2015) The physiology and pathophysiology of lipoxygenases in the skin. In: Steinhilber D (ed) Lipoxygenases in inflammation, Progress in inflammation research. Springer, Cham
4. Werz O, Rådmark O (2015) 5-Lipoxygenase. In: Steinhilber D (ed) Lipoxygenases in inflammation, Progress in inflammation research. Springer, Cham
5. Kühn H, Karst F, Heydeck D (2015) Catalytic multiplicity of 15-lipoxygenase-1 orthologs (ALOX15) of different species. In: Steinhilber D (ed) Lipoxygenases in inflammation, Progress in inflammation research. Springer, Cham
6. Rinaldo-Matthis A, Haeggström J (2015) Leukotriene A4 hydrolase and LTC4 synthase. In: Steinhilber D (ed) Lipoxygenases in inflammation, Progress in inflammation research. Springer, Cham
7. Powell WS, Rokach J (2015) 5-Oxo-ETE and inflammation. In: Steinhilber D (ed) Lipoxygenases in inflammation, Progress in inflammation research. Springer, Cham
8. Recchiuti A, Cianci E, Simiele F, Romano M (2015) Lipoxins, resolvins, and the resolution of inflammation. In: Steinhilber D (ed) Lipoxygenases in inflammation, Progress in inflammation research. Springer, Cham
9. Laguna-Fernández A, Petri MH, Müller S, Bäck M (2015) Lipoxygenases and cardiovascular disease. In: Steinhilber D (ed) Lipoxygenases in inflammation, Progress in inflammation research. Springer, Cham
10. Roos J, Kühn B, Fettel J, Maucher IV, Ruthardt M, Kahnt A, Vorup-Jensen T, Steinhilber D, Maier TJ (2015) Role of lipoxygenases in pathogenesis of cancer. In: Steinhilber D (ed) Lipoxygenases in inflammation, Progress in inflammation research. Springer, Cham

5-Lipoxygenase

Oliver Werz and Olof Rådmark

Abstract 5-Lipoxygenase (5-LO) catalyzes the first two steps in the biosynthesis of leukotrienes (LTs), lipid mediators of inflammation derived from arachidonic acid. The expression of 5-LO is essentially limited to myeloid cells but 5-LO can be expressed also in non-myeloid cancer cells. In contrast to other human LOs, 5-LO enzyme activity is tightly regulated by interacting proteins and several co-factors including Ca^{2+}/Mg^{2+}, ATP, phospholipids, diglycerides, and lipid hydroperoxides. 5-LO product synthesis in intact cells is modulated by phosphorylations at Ser271, Ser523, and Ser663, an elevated oxidative tone, and subcellular redistribution, apparently together with coactosin-like protein, from a soluble compartment to the nuclear membrane where it colocalizes with 5-LO-activating protein.

1 Introduction

The discovery of 5-lipoxygenase (5-LO) dates back to 1976, when novel transformations of polyunsaturated fatty acids were reported in glycogen-elicited rabbit peritoneal neutrophils. AA was transformed to 5(S)-hydroxy-6-trans-8,11,14-cis-eicosatetrenoic acid (5-HETE), and 8,11,14-eicosatrienoic acid to 8(S)-hydroxy-9-trans-11,14-cis-eicosatrienoic acid, these conversions involved the corresponding hydroperoxy acids [1]. Subsequently, it was found that 5-HPETE could be dehydrated to the allylic epoxide 5(S)-trans-7,9-trans-11,14-cis-eicosatetrenoic acid (leukotriene A_4, LTA_4), which was enzymatically hydrolysed to LTB_4. The novel pathway was also found in human leukocytes where the Ca^{2+}-ionophore A23187 had a major stimulatory [2], and slow reacting substance of anaphylaxis (LTC_4, LTD_4, LTE_4) was formed by reaction of LTA_4 with glutathione [3]. It was assumed that the conversion of AA to 5-HPETE was catalyzed by a LO and

---

O. Werz (✉)
Chair of Pharmaceutical/Medicinal Chemistry, Institute of Pharmacy, University Jena, Jena, Germany
e-mail: oliver.werz@uni-jena.de

O. Rådmark
Division of Chemistry II, Department of Medical Biochemistry and Biophysics, Karolinska Institutet, Solna, Sweden
e-mail: olof.radmark@ki.se

subsequent work demonstrated the existence of a special 5-LO. 5-LO is primarily expressed in various leukocytes, that is, polymorphonuclear leukocytes (neutrophils and eosinophils), monocytes and macrophages, dendritic cells, mast cells, B-lymphocytes, and in foam cells of human atherosclerotic tissue [4].

2 Reactions Catalyzed by 5-LO in Leukotriene Biosynthesis

In the biosynthesis of LTs (Fig. 1), the 5-LO enzyme has two catalytic functions: (I) insertion of molecular oxygen by the dioxygenase activity, and (II) epoxide formation by the LTA_4 synthase activity [5]. 5-HPETE is formed from AA by homolytic cleavage and abstraction of the pro-S hydrogen at C-7, radical rearrangement and antarafacial oxygen insertion at C-5, see [4] for review. The LTA_4 synthase activity of 5-LO involves abstraction of the pro-R hydrogen at C-10 of 5(S)-HPETE, with subsequent radical migration to C-6 and double bond rearrangement to form the unstable epoxide LTA_4 with a conjugated triene system (Fig. 1). Alternatively, 5(S)-HPETE can be reduced by glutathione peroxidases into the corresponding alcohol 5(S)-HETE. The ratio of LTA_4 versus 5(S)-HETE depends on the presence of modulating cofactors in the cell, and assay conditions *in vitro*. Thus, high AA substrate concentrations lead to lower ratios, while LTA_4 formation is promoted when 5-LO is membrane-associated, particularly in the presence of FLAP or coactosin-like protein (CLP) [6, 7]. In addition, 5(S)-HPETE can be further metabolized to 5-oxo-ETE [8], and by transcellular routes 5-LO together with 12-LO or 15-LO catalyses the formation of anti-inflammatory lipoxins [9].

Besides AA, some other PUFAs serve as substrates for the oxygenase activity of purified 5-LO, reviewed in [10]. These were 5,8,11,14,17-eicosapentaenoic acid, 5,8,11-eicosatrienoic acid, 5,8-eicosadienoic acid, 12-HPETE, and 15-HPETE. Apparently, fatty acids with 5,8-cis-double bonds are suitable substrates. The reaction with 8,11,14-eicosatrienoic acid involves abstraction of hydrogen at C-10, comparable to the LTA_4-synthase activity. While transformation of EPA as substrate was almost as efficient as with AA, the conversion of docosahexaenoic acid (DHA, 22:6) by 5-LO was 1 % of the activity with AA.

The fate of the unstable 5,6-epoxide LTA_4 is determined by the cell type where it is produced, in particular by the expression of terminal enzymes synthesizing LTs. Soluble LTA_4 hydrolase (LTA_4H) forms LTB_4 by stereospecific hydrolysis of LTA_4 [11]. Alternatively, LTA_4 can be converted to LTC_4 by addition of glutathione to the epoxide moiety [12]. In monocytes, macrophages, dendritic cells and mast cells, this is catalysed by LTC_4 synthase, while in endothelial cells MGST2 (another MAPEG family member) catalyses formation of LTC_4 from LTA_4 provided by transcellular metabolism. Removal of the γ-glutamyl residue of glutathione by a γ-glutamyltransferase leads to LTD_4 and elimination also of the glycine by

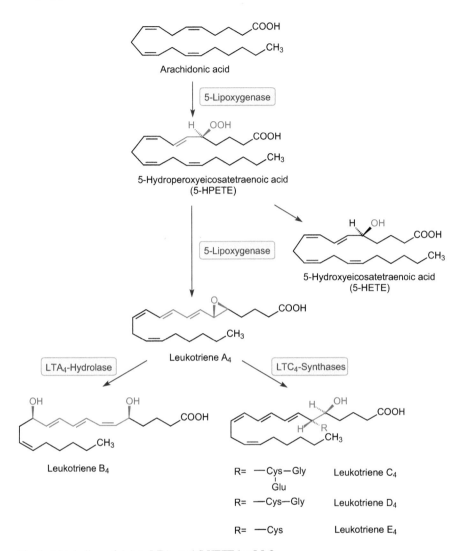

Fig. 1 Metabolism of AA to LTA$_4$ and 5-HETE by 5-LO

a dipeptidase yields LTE$_4$ (see Fig. 1). In preparations of purified 5-LO enzyme or in 5-LO transfected cell lines lacking LTA$_4$H or LTC$_4$S (e.g., HEK293 [13]), the LTA$_4$ decomposes nonenzymatically into the two isomers of LTB$_4$, that is, 6-trans LTB$_4$ and 6-trans-12-epi LTB$_4$.

3 Structure of 5-LO and Functions of 5-LO Domains

The mammalian 5-LOs contain 672 or 673 amino acids with a molecular weight of 78 kd. Based on the crystal structure of the rabbit reticulocyte 15-LO (a 12/15-LO) [14], it became evident that LOs from mammals consist of two domains: an N-terminal regulatory C2-like domain (residues 1–112 in 5-LO) which mainly consists of β-sheets, and a C-terminal catalytic domain (residues 126–673 in 5-LO) which is mainly helical in structure and contains iron. In the LO reaction mechanism, the active site iron functions as electron acceptor and donor, during hydrogen abstraction and peroxide formation. By electron paramagnetic resonance (EPR), the iron was ferrous, treatment with lipid hydroperoxides oxidized the iron to ferric form. EPR experiments also indicated a flexible iron ligand arrangement and selenide inhibited 5-LO by binding to iron, thus abolishing the EPR signal ($g = 6.2$) typical for active 5-LO. Attempts to crystallize 5-LO failed for many years but became possible after mutation of several destabilizing sequences [15]. Thus "stable 5-LO" lacks putative membrane insertion amino acids, and residues of the Ca^{2+}-binding site within the C2-like domain. Also a 5-LO specific destabilizing sequence (KKK) close to the C-terminus was exchanged [15].

As in other lipoxygenases, the 5-LO catalytic domain (residues 113–673) is composed of several α-helices. The conserved arched helix α8, found in all LOX structures, helps to shield the U-shaped cavity with the catalytic center at the bottom [15]. In stable 5-LO, a striking variation of helix α2 (defining one edge of the active site) was observed, when compared to the already known LO structures. In 5-LO this helix is short (three turns) and its unique orientation limits access to the catalytic iron. In particular the side chains of Phe177 and Tyr181 close off an access channel, thus nick-named the "FY cork" [15]. Helix α2 in 5-LO was also referred to as the "broken" helix [16]. Such a broken helix was found also in 11(R)-LOX [17], and it was proposed that helix α2 is part of a mobile lid controlling access to the active site [15, 17]. The more efficient catalysis following disruption of ionic association between the domains in 5-LO may support the concept of such a flexible lid in human 5-LO [18].

The N-terminal domain of 5-LO (residues 1–112) is a C2-like β-sandwich composed of two 4-stranded antiparallel β-sheets with high similarity to the C2-like domain of *C. perfringens* α-toxin (1QMD, a phospholipase C) [19]. Sequence and topology similarities between Polycystin-1, Lipoxygenase and α-Toxin led to the definition of the PLAT domain as a subset within the C2 family [20]. For 5-LO, residues in the N-terminal C2-like domain binds to Ca^{2+}, cellular membranes, glycerides, CLP, and Dicer, as described below.

4 5-LO Interacting Proteins

4.1 *5-Lipoxygenase-Activating Protein*

5-Lipoxygenase-Activating Protein (FLAP) was identified in studies with the antiinflammatory compound MK-886, which inhibited 5-LO in intact human leukocytes, but not in corresponding cell homogenates. By means of a photoaffinity analogue to MK-886 and an MK-886 affinity gel, FLAP could be purified from neutrophils, for review see [21]. The sequence of FLAP, a 18 kDA integral nuclear membrane protein, is similar (31 % identity) to microsomal LTC_4 synthase. In fact, both proteins are members of the MAPEG family (membrane associated proteins in eicosanoid and glutathione metabolism) [22].

In osteosarcoma cells transfected with cDNA for 5-LO and/or FLAP, both proteins were required for A23187-induced LT production from endogenous AA [21]. FLAP bound to a photoaffinity analog of AA, and cis-unsaturated fatty acids as AA competed with inhibitors (BAY X1005, MK-886) for binding to FLAP [24]. FLAP may function as an AA transfer protein, and FLAP is crucial for conversion of endogenous substrate by 5-LO [6]. Also the conversion of exogenous AA by 5-LO was stimulated by FLAP, and the utilization of another exogenous substrate (12(S)-HETE) was markedly (190-fold) stimulated [25]. FLAP promotes conversion of 5-HPETE to LTA_4 [6] which is a feature shared with CLP [7]. Otherwise there are no obvious similarities between FLAP and CLP.

In neutrophils, FLAP was present as mono- and dimers, and LT biosynthesis required the organization into a FLAP homodimer [26]. Using RBL-2H3 cell extracts, FLAP mono-, di-, and even trimers were found and mixed complexes of FLAP and LTC_4 synthase were also described [27]. Finally, the crystal structure revealed that inhibitor-bound FLAP is a homotrimer, and that each monomer has four transmembrane helices connected by two cytosolic and one lumenal loops [28]. The inhibitors bind to membrane-embedded pockets in FLAP, suggesting how they might prevent binding of AA. Association between FLAP and 5-LO could be demonstrated in mouse synovial PMN and in IgE-stimulated RBL-2H3 cells, using a membrane-permeant cross-linker. Also, antibody-based fluorescence lifetime imaging microscopy suggested different FLAP complexes on inner and outer nuclear membranes. It was concluded that FLAP may function as a 5-LO scaffold protein [29]. Binding between 5-LO and FLAP *in vitro* was demonstrated in pull-down experiments using immobilised GST-FLAP and GST-5-LO [30].

Most of the FLAP is associated with the nuclear membrane, but also with endoplasmatic reticulum [31]. Interestingly, in RBL-2H3 cells FLAP co-localized with the lipid-raft marker flotillin-1 and methyl-α-cyclodextrin, which depletes cells of cholesterol and disrupts membrane rafts, reduced LT biosynthesis [32]. In mantle cell lymphoma, also 5-LO was associated with lipid rafts [33]. Secretory vesicles from human neutrophils contain FLAP [34]. Also exosomes, isolated from human monocyte-derived macrophages or monocyte-derived dendritic cells, are quite rich in FLAP and also contain LTA_4 hydrolase and LTC_4 synthase [35].

4.2 Coactosin-Like Protein

Coactosin, first found in *Dictyostelium discoideum*, is a member of the ADF/Cofilin group of actin-binding proteins. It counteracted capping of actin filaments, thus promoting actin polymerization [36]. Mammalian CLP was identified as a 5-LO binding protein by a yeast two-hybrid system [37] and an interaction between CLP and 5-LO *in vitro* (1:1 molar stoichiometry) was shown by GST pull-down experiments, immunoprecipitation, cross-linking, and nondenaturing PAGE [38]. CLP can upregulate 5-LO activity *in vitro*. In absence of phosphatidylcholine (PC), CLP supported Ca^{2+}-induced 5-LO activity leading to 5-HETE, but not to LTA_4. On the other hand, in presence of PC, addition also of CLP upregulated LTA_4 formation three- to fourfold [7, 120]. Thus, CLP (as PC) functions as a scaffold for Ca^{2+}-induced 5-LO activity. In addition, CLP can bind to 5-LO in the absence of Ca^{2+} which stabilizes 5-LO and prevents non-turnover inactivation of the enzyme *in vitro*.

A role for CLP in cellular 5-LO activity was suggested by the similar migration patterns in neutrophils and Mono Mac 6 cells, both proteins translocated to a nuclear fraction upon cell activation [7, 120]. Hyperforin, a 5-LO inhibitor from St. John's wort interrupted binding of 5-LO to CLP *in vitro*, and impaired nuclear translocation of 5-LO in human neutrophils [39]. Recently, CLP and FLAP were knocked down in Mono Mac 6 cells, both proteins were required for 5-LO activity in LPS primed cells stimulated by fMLP (resembling pathophysiolological conditions) leading to LTC_4 formation [40]. Also, both CLP and FLAP were needed for translocation of 5-LO to the nuclear membrane in ionophore stimulated cells. It was concluded that formation of the 5-LO/CLP complex augments translocation from cytosol to nucleus, while FLAP stabilizes association of this complex with the perinuclear membrane.

Mutation of Trp102 in the 5-LO β-sandwich abolished binding of CLP to 5-LO [120]. Within the 5-LO/CLP complex, Trp102 which is partially hidden in the 5-LO interdomain cleft, seems not to bind CLP directly. Instead, Trp102 binds to Arg165 via a proposed π-cation interaction, and Arg165 binds to CLP-Lys131 via an hydrogen bond [120].

4.3 Dicer

Dicer is a multidomain RNA helicase/RNase III involved in the biogenesis of microRNAs (miRNAs) and small interfering RNA (siRNA) from dsRNA substrates, see [41] for review. Human dicer binds to trans-activation response RNA-binding protein (TRBP) and to protein activator of PKR (PACT). In a yeast two-hybrid screen for 5-LO binding proteins [37] the sequence giving the strongest interaction was the C-terminal third of Dicer. Full-length human Dicer (1912 amino acids) was cloned, and the ribonuclease activity and dsRNA binding properties were characterized [42]. Binding between 5-LO and Dicer was confirmed with purified proteins [43] and it was found that the N-terminal β-sandwich of 5-LO binds to the Dicer C-terminal 140 amino acids (the dsRNA-binding domain). Interestingly, presence of 5-LO changed the Dicer product pattern *in vitro*, leading to formation of 10–12 bp small RNAs [43]. *In vitro*, the 5-LO binding domain of Dicer can support Ca^{2+}-induced 5-LO activity, albeit not as efficiently as PC or CLP.

5 Regulation of 5-LO Enzyme Activity

The biosynthesis of 5-LO products is essentially limited to monocytes, macrophages, mast cells, neutrophils, dendritic cells and B-lymphocytes that possess the required enzyme machinery including PLA_2 enzyme(s), 5-LO, FLAP, LTA_4H and/or LTC_4 synthases [4]. Other cell types (platelets, endothelial cells, red blood cells) contain only LTA_4H, LTC_4 synthase, or MGST2, and can participate in transcellular LT biosynthesis, for review see [44]. The formation of 5-LO products in the cell is a tightly regulated process, and external stimuli capable of inducing signalling kinase pathways and second messengers (e.g., Ca^{2+} and diacylglycerides) are required [45]. Importantly, such activation and signalling events regulate not only 5-LO but also $cPLA_2$ allowing for coordinate provision of AA as substrate for 5-LO [46]. Leukocyte stimulation leads to translocation of both 5-LO and $cPLA_2$ from soluble compartments to the perinuclear region, in close proximity to nuclear membrane-associated FLAP (Fig. 2). $cPLA_2$ liberates AA from arachidonyl-PC, which is then thought to be transferred via FLAP to membrane-associated 5-LO for metabolism.

Ca^{2+}-mobilizing agents such as ionophores or thapsigargin are potent elicitors of LT biosynthesis, for review see [47]. Other agonists, including fMLP, PAF, C5a, and phagocytic particles like urate crystals or opsonized zymosan cause only minute LT formation, as compared to the ionophores. These agonists alone fail to substantially increase $[Ca^{2+}]_i$ and thus to activate $cPLA_2$ and to induce nuclear translocation of 5-LO. However, when fMLP, PAF, and C5a are combined with "priming" agents such as granulocyte macrophage colony-stimulating factor, tumor necrosis factor-α, phorbol esters, Epstein Barr virus, or lipopolysaccharides, formation of LTs is strongly enhanced due to multiple effects (reviewed in [47]). Various forms of cell-stress, e.g., osmotic shock, oxidative or genotoxic agents, lead to phosphorylation of 5-LO and subsequent LT synthesis which is independent on Ca^{2+} mobilization [48].

Full 5-LO activity under cell-free conditions depends on the presence of Ca^{2+}, sufficient amounts of hydroperoxides, phospholipids, ATP, and on optimal substrate concentration (see below). Also cellular 5-LO activity depends on these factors and on and additional determinants including subcellular localization of 5-LO and membrane interactions, dimerization of 5-LO, interaction with FLAP and CLP, phosphorylation events, and diacylglyceride formation [4]. Finally, systemic

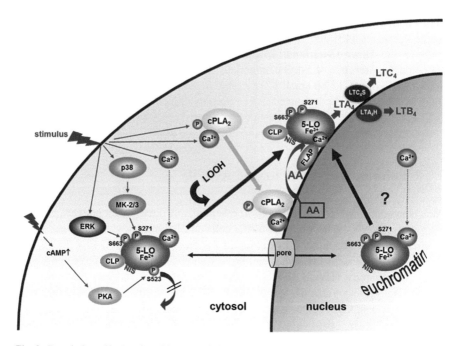

Fig. 2 Regulation of leukotriene biosynthesis in intact cells. 5-LO resides as soluble protein in the nucleoplasm or the cytosol, depending on the cell type and/or the phosphorylation status at serine residues. Upon cell stimulation soluble 5-LO binds Ca^{2+} and may become phosphorylated, and then translocates to the nuclear envelope where it binds to the outer or inner nuclear membrane via Ca^{2+} close to membrane-integrated FLAP. Cytosolic phopholipase A_2 is also activated by Ca^{2+} and phosphorylation, moves to the nuclear membrane as well and liberates AA that is transferred via FLAP to 5-LO. AA is then converted by 5-LO to LTA_4. The possibility that LTA_4 may be metabolized to LTC_4 or LTB_4, on the cytosolic/nuclear sides of the nuclear membrane respectively, has been discussed (29)

factors such as adenosine [49] and androgens [50, 51] regulate 5-LO product synthesis, probably via modulation of phosphorylation events.

5.1 Ca^{2+}

The bivalent ion Ca^{2+} activates purified 5-LO with EC_{50} in the range of 1–2 µM, and full activation is reached at 4–10 µM [47]. Also Mg^{2+} (at millimolar concentrations present in cells) can activate 5-LO [52]. 5-LO binds Ca^{2+} in a reversible manner, with a K_d close to 6 µM and a stoichiometry of maximum binding of two Ca^{2+} per 5-LO [53]. Similar results (two Ca^{2+} per 5-LO, K_{Ca} 7–9 µM) were obtained for the His-tagged C2-like domain (residues 1–115) [19]. Mutagenesis suggests that residues in ligand binding loop 2 of the 5-LO C2-like domain are important for both Ca^{2+} binding and for activation of 5-LO [54]. Note that 5-LO has some basal activity also in absence of Ca^{2+}/Mg^{2+} suggesting that the divalent cations are not part of catalysis.

The 5-LO stimulatory effect of Ca^{2+} activation of 5-LO requires the presence of PC or CLP. Ca^{2+} also lowers the K_m of 5-LO for AA and reaction kinetics are changed, leading to substrate inhibition. Binding of Ca^{2+} increases the hydrophobicity of 5-LO by neutralization of negatively charged amino acids, and Ca^{2+} seemingly mediates 5-LO binding to membranes involving Trp-13, Trp-75 and Trp-102 upon subcellular redistribution to the perinuclear region [19]. Apparently, Ca^{2+} can also modify the affinity of 5-LO for lipid hydroperoxide, since in presence of Ca^{2+}, GPX-1 was a less efficient 5-LO inhibitor [55]. This effect of Ca^{2+} was diminished for the 5-LO loop 2 mutant with decreased Ca^{2+} affinity [55]. Possibly, ferric 5-LO could be formed at a lower concentration of lipid hydroperoxide, when Ca^{2+} is present, and it was proposed that Ca^{2+} could induce lipid hydroperoxide binding to the β-sandwich of 5-LO [4].

5.2 ATP

5-LO catalytic activity is stimulated by ATP two- to sixfold (at 0.1–2 mM, Ka values = 30–100 μM) and to a lesser extent by other nucleotides, including ADP, AMP, cAMP, CTP, and UTP, for review see [47]. Stimulation of 5-LO by ATP required Ca^{2+} or Mg^{2+}, but was effective also in the absence of divalent cations [52, 56–58]. ATP hydrolysis appears not required for stimulation as the non-hydrolyzable analog γ-S-ATP causes comparable effects, and energy consumption or auto-phosphorylation have been excluded [56].

By means of ATP-affinity chromatography it was shown that 5-LO from various sources can bind ATP, for review see [47]. Among LOs, only 5-LO may bind ATP and is activated by nucleotides. Database analysis revealed no consensus ATP- or nucleotide-binding sites in the amino-acid sequence, but two tryptophan residues (Trp-75 and Trp-201) have been identified that bound ATP-analogs specifically, and incorporation of these analogs inhibited ATP stimulation of 5-LO [59]. However, mutations of Trp-75 in the C2-like domain did not change apparent ATP affinity [60] but conferred decreased affinity for PC [19, 120]. Evidence for ATP-binding site in the catalytic domain is provided by experiments with truncation mutants of murine 5-LO (lacking the β-sandwich), where enzyme activity could still be upregulated by ATP [61].

5.3 Cell Membranes and PC

C2 domains often mediate Ca^{2+}-induced membrane association, and it is long known that the activity of purified 5-LO is stimulated microsomal membranes [62] or synthetic PC vesicles [57]. PC was required for Ca^{2+} stimulated 5-LO activity *in vitro* [52, 58] and 5-LO binding to synthetic PC liposomes was induced by Ca^{2+} [63]. The isolated C2-like β-sandwich had a higher affinity for zwitterionic PC vesicles than for anionic PS and phophatidylglycerol vesicles, and Trp-13, Trp-75 and Trp-102 were important for PC binding. It was suggested that the PC selectivity directs 5-LO to the nuclear envelope [19], which is in accordance with the requirement of the β-sandwich for translocation of GFP-5-LO constructs to the nuclear membrane in ionophore-stimulated HEK 293 cells [64].

Ca^{2+}-induced binding to PC stabilized the structures of both 5-LO protein and membrane, and it was found that 5-LO can bind also to cationic phospholipids [65]. This binding was stronger and occurred in absence of Ca^{2+}, though Ca^{2+} was required for 5-LO activity. Thus, it was suggested that 5-LO can bind to membranes in "productive/nonproductive modes", that is, membrane binding per se may not confer 5-LO activity. Increased membrane fluidity favoured 5-LO association, and it was argued that this should be the factor directing 5-LO to the AA enriched nuclear envelope [66]. Interestingly, addition of cholesterol to a membrane preparation reduced 5-LO activity *in vitro* by half [66].

5.4 Diacylglycerides

Besides phosphoplipids also various glycerides were found to activate 5-LO where OAG (1-oleoyl-2-acetyl-*sn*-glycerol) was most potent [67]. 5-LO stimulation occurred in absence of Ca^{2+}. In fact, Ca^{2+} prevented this stimulatory effect of OAG and also phospholipids or cellular membranes abolished OAG-induced stimulation of 5-LO. The mutant 5-LO-W13/75/102A was not stimulated by OAG suggesting that glycerides can bind to the C2-like domain of 5-LO. As Ca^{2+}, also OAG protected 5-LO against the inhibitory effect of GPx-1 [67]. 5-LO product formation was also stimulated by glycerides in intact leukocytes [68, 69]. Thus, inhibition of the diacylglyceride-forming pathway involving phospholipase D (PLD) and phosphatidic acid phosphatase (PA-P) decreased 5-LO product synthesis as well as 5-LO translocation, and exogenous supplementation of OAG reversed 5-LO suppression [69]. Interestingly, in monocytes from male human donors diacylglyceride formation was reduced due to inhibition of PLD by testosterone, resulting in lower LT biosynthesis compared to cells from female donors with high PLD activity [51].

5.5 Redox Regulation

The redox cycle of the active-site iron between ferrous and ferric forms is crucial for 5-LO catalysis. In the inactive state of 5-LO, the iron is in the ferrous form and requires conversion to the active ferric state that allows the enzyme to abstract a hydrogen from AA. This initial conversion of Fe^{2+} to Fe^{3+} is mediated by lipid hydroperoxides, addition of 5-HPETE, 12-HPETE, 15-HPETE or 13-HPODE results in the active ferric form, and the lag phase of the enzymatic reaction is shortened (for review, see [47]). However, under reducing conditions, i.e., presence of various GPX isoforms together with suitable thiols (GSH or DTT) 5-LO activity is prevented. In B-lymphocytes, RBL-2H3, and immature HL60 cells phospholipid hydroperoxide GPX-4 was shown to influence 5-LO activity, whereas in monocytic cells 5-LO product synthesis is regulated by GPX-1. As mentioned above, Ca^{2+} as well as DAG may counteract the inhibitory activity of GPX enzymes, seemingly by modulating the C2-like domain which changes the requirement of lipid hydroperoxides for 5-LO activation [55, 67].

5.6 5-LO Phosphorylation

5-LO is phosphorylated by various protein kinases, and phosphorylation of different residues may have divergent consequences for 5-LO in the cell (Fig. 2). *In vitro*, protein kinase (PK)A was found to phosphorylate 5-LO at Ser-523, p38 mitogen-activated protein kinase activated protein kinase (MAPKAPK, MK)-2/3, PKA, and CaMK-II at Ser-271, and ERK1/2 at Ser-663, see [4] for review. Recently, a regulatory role of 5-LO phosphorylation at Ser-271 with consequences on subcellular redistribution and activity was demonstrated *in vivo* [70].

Phosphorylations at Ser-271 and Ser-663 *in vitro* were strongly promoted by unsaturated fatty acids including AA, and these phosphorylations increased 5-LO product synthesis in neutrophils, monocytes and B lymphocytic cells [71–74]. It seems that dual phosphorylation at Ser-271 and Ser-663 work in conjunction to enhance 5-LO product synthesis, in particular when intracellular Ca^{2+} levels are low and thus, insufficient to fully activate 5-LO. Note that phosphorylation of 5-LO at Ser-271 and Ser-663 did not enhance the enzyme activity in cell-free assays.

Phosphorylation of 5-LO at Ser-523 by PKA, or substitution of Ser-523 with phosphomimetic glutamate, resulted in total loss of 5-LO activity in cell-free assays as well as in 5-LO transfected mouse NIH 3T3 fibroblasts [75]. This was connected with a cytoplasmic enrichment of the enzyme and prevention of nuclear import [76]. This most probably explains the suppression of 5-LO product synthesis by adenosine and cAMP-elevating agents in human neutrophils, which involves PKA and inhibition of 5-LO nuclear translocation in neutrophils [77].

Phosphorylation of 5-LO can modulate its subcellular localization (Fig. 2). However, this may depend on the cell type and on complex interplay with other cofactors that influence 5-LO activation (the interested reader is referred to [78]).

Quite recently, *in vitro* phosphorylation of 5-LO and subsequent detection of phosphorylated peptides revealed that 5-LO can also be a target for tyrosine kinases. Tyr-42, Tyr-53 and either Tyr-94 or Tyr-445 could be phosphorylated by the Src kinases Fgr, HCK and Yes, respectively [79].

5.7 Subcellular Localization

5-LO subcellular localization might be decisive for the capacity of cellular 5-LO to generate various products. In resting cells 5-LO is a soluble protein in the cytosol or inside the nucleoplasm, but in the activated cell 5-LO redistributes to the nuclear membrane and may need to access FLAP, thus co-localisation with $cPLA_2$, and presumably with LTC_4S and LTA_4H as well [80]. The subcellular locale for 5-LO in resting cells depends on the cell type and the cellular environment; for review see [47, 78].

Stimuli that evoke LT biosynthesis causes 5-LO subcellular redistribution to a membrane compartment within the perinuclear region irrespective of previous cytosolic or intranuclear localization. Apparently, cytoplasmic 5-LO associates with the ER and the outer nuclear membrane, whereas intranuclear 5-LO translocates to the inner membrane of the nuclear envelope [81]. Recent studies provide evidence that a multi-enzyme complex at the perinuclear region, involving $cPLA_2$, 5-LO, FLAP, and additional proteins, act to release AA and convert it to LTA_4 [29, 82]. For instance, FLAP that is distributed on both the inner and outer nuclear membrane may convert LTA_4 to LTC_4 by LTC_4S at the outer nuclear membrane, whereas at the inner membrane FLAP engages nuclear soluble LTA_4H to produce LTB_4 [29]. By FLIM analysis, formation of the 5-LO/FLAP complex in mouse neutrophils was reduced by inhibition of $cPLA_2$, indicating a role for AA in formation of this complex [82].

5.8 Gender-Related Differences in Leukotriene Formation

Gender-related differences caused by androgens in the regulation of 5-LO have been observed in primary human neutrophils and monocytes *in vitro* [50, 51] and during acute inflammation *in vivo* [83]. In neutrophils, androgens caused perinuclear localization of 5-LO in an ERK-dependent manner, which however conferred a reduced 5-LO product formation upon activation with ionophore or fMLP. Also, when whole blood from females was primed with LPS and stimulated with fMLP, formation of LTB_4 was three times higher compared to blood from males [50]. ERK activation by testosterone suppressed 5-LO product generation also in primary human monocytes, although 5-LO redistribution was hardly affected in these cells. Instead, the lower 5-LO product synthesis in male monocytes correlated to reduced PLD activity and could be restored by exogenous DAGs [51]. Such gender bias in LT synthesis may explain well-recognized sex differences in the incidence of LT-related disease such as asthma or allergic rhinitis which clearly dominate in females. Interestingly, female mice deficient in the BLT1 receptor were protected in a model of PAF-induced lethal shock, while male mice were sensitive [84]. Peculiarly, in a mouse model resembling human SLE (MRL-*lpr/lpr*) female mice had earlier mortality versus male mice, but the male advantage was abolished after 5-LO gene deletion [85]. Also, a gender-specific attenuation of atheroma formation was reported for dual 5-LO and 12/15-LO-KO mice in an experimental model of atherosclerosis [86].

6 Mechanisms in Regulation of 5-LO Gene Expression

The human gene encoding 5-LO (Fig. 3) is divided into 14 exons [87]. The promoter region contains eight GC-boxes, but lacks TATA and CAT boxes. Thus, it resembles promoters of typical house keeping genes. Nevertheless, 5-LO expression is mainly restricted to leukocytes. Granulocytes, monocytes/macrophages, mast cells, dendritic cells including Langerhans cells and B-lymphocytes express 5-LO.

6.1 DNA Methylation Determines Expression of 5-LO

DNA methylation leads to suppression of 5-LO expression [88]. Methylation-specific DNA sequencing showed methylation of the 5-LO core promoter in cell

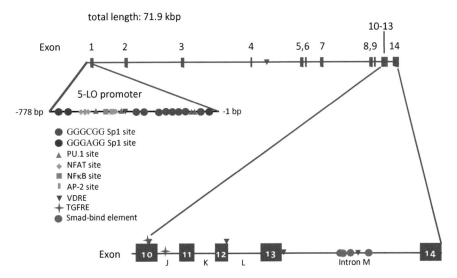

Fig. 3 The large human 5-LO gene (71.9 kb) is localized on chromosome 10, and it is divided in 14 exons. A common ancestral lipoxygenase gene is suggested by the identical intron/exon organization for human 5-, 12- and 15-LOs. Several transcription initiation sites have been described for the 5-LO gene, the major 65 nt upstream of ATG. Regulatory elements in the promoter, and in the 3' end of the gene, are indicated. The promoter is GC-rich, indicated by the multiple Sp1 binding sites

lines U-937 and HL-60TB (which do not express 5-LO protein), whereas it was unmethylated in HL-60 cells (express 5-LO protein when differentiated). *In vitro* methylation of the promoter inhibited reporter gene activity, and treatment of U-937 and HL-60TB cells with the demethylating agent 5-aza-2′deoxycytidine partially demethylated the 5-LO promoter and restored 5-LO expression. Using ChIP assay it was found that methyl-DNA binding proteins (MBD1, MBD2, MeCP2) bind to the methylated core promoter in U-937 cells [89]. It seems that DNA methylation is responsible for suppression of 5-LO expression in non-myeloid cell types and tissues, and that aberrant methylation could lead to upregulated 5-LO expression in tumor cells.

6.2 GC-Boxes in the 5-LO Gene, Effect of Trichostatin A

The proximal part of the human 5-LO gene promoter contains eight GC-boxes (GGGCGG). Five of these are arranged in tandem, and are recognized by transcription factors Sp1 and Egr-1. This proximal GC-rich part of the promoter was crucial for the expression of reporter genes [90–92]. Accordingly, expression of endogenous 5-LO (in Mono Mac 6 cells) was reduced by mithramycin, a drug blocking GC-boxes. The promoter also contains three Sp1 sequences (GGGAGG), one of these is located beside an Initiator-like sequence including the major TIS [93]. The histone deacetylase inhibitor trichostatin A increased 5-LO gene promoter activity [94]. This effect involved enhanced recruitment of Sp1/Sp3 to proximal GC boxes, while the GC-boxes in tandem seemed not to be involved [95]. By ChIP assays increased acetylation of histones H3 and H4 at the 5-LO core promoter was found to precede 5-LO mRNA induction, whereas 5-LO mRNA expression correlated with histone H3 lysine 4 trimethylation (H3K4me3, a marker for transcriptional activity of gene promoters) [96]. The mixed lineage leukemia protein (MLL) can catalyze trimethylation (H3K4me3), and the combined inhibition of HDAC1-3 induced 5-LO promoter activity in an MLL-dependent manner [96].

Naturally occurring mutations are found in the human 5-LO promoter. These consist either of the deletion of one or two, or the addition of one Sp1-binding site, to the normally five tandem GC boxes [91]. The effects of these mutations on 5-LO reporter gene expression were inconsistent, with different patterns in different cell lines. A possible function of these mutations *in vivo* emerged from a pharmacogenetic association between the mutant genotypes and responses to a 5-LO inhibitor (ABT-761) in an asthma clinical trial [97]. This finding implied that the mutations should lead to reduced 5-LO expression, and one study reported that eosinophils from asthmatics carrying a mutated non5/non5 genotype expressed less 5-LO mRNA and produced less LTC_4 [98]. But on the contrary, variations of the number of tandem repeats was associated with severity of airway hyperresponsiveness in AIA patients [99]. Also, children who were homozygous for variant alleles had higher urinary LTE_4 levels and trends towards worse asthma control [100]. In addition, one study reported no role for the tandem repeat polymorphisms in genetic susceptibility to asthma [101], while another stated that ALOX5 promoter polymorphisms had a clear influence in montelukast response in atopic moderate persistent asthma patients [102].

In relation to atherosclerosis mutations of the GC-boxes in tandem were connected with increased intima-media thickness and increased plasma level of C-reactive protein [103]. This would seem compatible with an increased production of pro-inflammatory LTs, due to upregulated 5-LO expression. However, in subsequent work three studies reported no support for association of the tandem repeat polymorphism and CAD/MI [104–106], while one did [107]. Furthermore, ALOX5 variants were associated with susceptibility to tuberculosis [108]. In summary, it is difficult to see clear patterns regarding the effect of mutations in this part of the human 5-LO gene promoter. Interestingly, in another species the GC-boxes in tandem are absent, the core part of the mouse 5-LO gene promoter contains only one Sp1/3 binding site [109].

6.3 Vitamin D Responsive Elements in the Promoter and Other Parts of the 5-LO Gene

Reporter gene assays using HeLa cells (do not express endogenous 5-LO) revealed two positive and two negative regulatory regions in the 5-LO gene promoter, and modestly enhanced transcription by phorbol ester [90]. Promoter analysis using Mono Mac 6 cells (which express 5-LO) revealed a positive regulatory region not detected in HeLa cells. This region containing several vitamin D response elements (VDREs) was located upstream of the TIS (-779 to -229). Binding of the vitamin D receptor was demonstrated by EMSA, DNA footprinting and chromatin immunoprecipitation assays [110]. More VDREs were found further upstream in the promoter, and in intron 4. Remarkably, the reporter gene response (in MCF-7 cells) suggested that the intron 4 VDRE (at +42,000) is one of the strongest of the human genome. The results suggested DNA looping of the basal promoter to very distant gene regions [111]. Subsequently, additional VDREs were characterised, in exons 10 and 12, and in intron M [112].

6.4 Upregulation of 5-LO Expression by TGFβ and 1,25(OH)₂D3

Although TGFß and $1,25(OH)_2D_3$ give about 100-fold upregulation of 5-LO protein in differentiating Mono Mac 6 cells [113] there was no induction of 5-LO promoter activity by TGFß and $1,25(OH)_2D_3$ in reporter gene assays. Apparently, the -6079 to $+53$ promoter region mediates basal 5-LO transcription [93, 94]. Also, no changes in 5-LO mRNA half life were observed. Instead, posttranscriptional events such as transcript elongation and maturation appeared to mediate the effect of TGFß and 1,25(OH)$_2$D$_3$ [114]. Insight into the mechanism was gained when 1,25 (OH)$_2$D$_3$ and TGFß were observed to upregulate the response for reporter gene plasmids containing the complete 5-LO coding sequence plus the introns J-M [115]. These effects of 1,25(OH)$_2$D$_3$ and TGFß were independent of the 5-LO promoter, and two functional response elements for the TGFß effectors Smad3 and 4 were identified in the distal part of the 5-LO gene (exons 10–14, Fig. 3) [115]. Later on, it was found that TGFß and 1,25(OH)$_2$D$_3$ induced chromatin modifications at exons 10 and 12, and at intron M, and this was related to the novel VDREs in these locations. ChIP assays revealed histone H4 K20 monomethylation and histone H3 K36 trimethylation, both markers for transcript elongation. Combined treatment with TGFß and 1,25(OH)$_2$D$_3$ also increased histone H4 acetylation, a marker for open chromatin, and the elongation form of RNA polymerase II was present at these sites [112]. It was concluded that chromatin opening and transcript elongation are important mechanisms for upregulation of 5-LO gene expression, by TGFß and 1,25(OH)$_2$D$_3$.

6.5 Possible Regulation by Alternative Splicing, and by miRNA

Splice variants of 5-LO have been described. In polymorphonuclear leukocytes and human myeloid cell lines, four 5-LO isoforms were found. When expressed in HEK293 cells (devoid of endogenous 5-LO) the isoforms were enzymatically inactive. When full-length 5-LO and an isoform were cotransfected, this reduced the 5-LO activity in intact HEK293 cells, but peculiarly not of the corresponding cell lysates [116]. In Mono Mac 6 cells, two truncated transcripts and four splice variants were found. The splice variant 5-LOΔ3, but not wt 5-LO, was regulated by nonsense-mediated mRNA decay [117].

Finally, 5-LO expression can also be regulated by miRNAs. In human PBMC derived macrophages transfected with mi-219-2, both 5-LO mRNA and protein were reduced [118]. In Mono Mac 6 cells, addition of antagomirs against miR-19a and miR-125b [during differentiation with TGFß and 1,25(OH)$_2$D$_3$] increased 5-LO protein about 1.8-fold. Also, when human T-lymphocytes (prepared from peripheral blood) were stimulated with PHA, 5-LO mRNA was downregulated in concert with induction of miR-19a-3p, while the antagomir upregulated 5-LO mRNA expression [119].

In summary, 5-LO is expressed primarily in various leukocytes, and this capacity is determined by DNA methylation which restricts transcription. When unmethylated, the core promoter region alone (6 kB) mediates basal transcription. Upregulation by the HDAC inhibitor trichostatin A underscored the importance of open chromatin for recruitment of Sp1/3 and RNA polymerase II to the proximal promoter. Differentiation of leukocytes is often connected with increased 5-LO expression, a well studied example is Mono Mac 6 cells. When these cells are differentiated with TGFß and $1,25(OH)_2D_3$, chromatin opening and transcript elongation are important mechanisms for the profound upregulation of 5-LO gene expression.

Acknowledgements Studies in our laboratories were supported by the Swedish Research Council (03X-217), and the DFG (SFB 1127 and FOR 1406).

References

1. Borgeat P, Hamberg M, Samuelsson B (1976) Transformation of arachidonic acid and homo-gamma-linolenic acid by rabbit polymorphonuclear leukocytes. Monohydroxy acids from novel lipoxygenases. J Biol Chem 251(24):7816–7820
2. Borgeat P, Samuelsson B (1979) Arachidonic acid metabolism in polymorphonuclear leukocytes: effects of ionophore A23187. Proc Natl Acad Sci USA 76(5):2148–2152
3. Murphy RC, Hammarstrom S, Samuelsson B (1979) Leukotriene C: a slow-reacting substance from murine mastocytoma cells. Proc Natl Acad Sci USA 76(9):4275–4279
4. Radmark O, Werz O, Steinhilber D, Samuelsson B (2007) 5-lipoxygenase: regulation of expression and enzyme activity. Trends Biochem Sci 32(7):332–341
5. Shimizu T, Rådmark O, Samuelsson B (1984) Enzyme with dual lipoxygenase activities catalyzes leukotriene A4 synthesis from arachidonic acid. Proc Natl Acad Sci USA 81(3):689–693
6. Abramovitz M, Wong E, Cox ME, Richardson CD, Li C, Vickers PJ (1993) 5-lipoxygenase-activating protein stimulates the utilization of arachidonic acid by 5-lipoxygenase. Eur J Biochem 215(1):105–111
7. Rakonjac M, Fischer L, Provost P, Werz O, Steinhilber D, Samuelsson B et al (2006) Coactosin-like protein supports 5-lipoxygenase enzyme activity and up-regulates leukotriene A4 production. Proc Natl Acad Sci USA 103(35):13150–13155
8. Powell WS, Rokach J (2005) Biochemistry, biology and chemistry of the 5-lipoxygenase product 5-oxo-ETE. Prog Lipid Res 44(2–3):154–183
9. Serhan CN, Chiang N (2002) Lipid-derived mediators in endogenous anti-inflammation and resolution: lipoxins and aspirin-triggered 15-epi-lipoxins. ScientificWorldJournal 2:169–204
10. Radmark O, Samuelsson B (2005) Regulation of 5-lipoxygenase enzyme activity. Biochem Biophys Res Commun 338(1):102–110
11. Haeggstrom JZ, Tholander F, Wetterholm A (2007) Structure and catalytic mechanisms of leukotriene A4 hydrolase. Prostaglandins Other Lipid Mediat 83(3):198–202
12. Lam BK, Austen KF (2002) Leukotriene C4 synthase: a pivotal enzyme in cellular biosynthesis of the cysteinyl leukotrienes. Prostaglandins Other Lipid Mediat 68–69:511–520
13. Gerstmeier J, Weinigel C, Barz D, Werz O, Garscha U (2014) An experimental cell-based model for studying the cell biology and molecular pharmacology of 5-lipoxygenase-activating protein in leukotriene biosynthesis. Biochim Biophys Acta 1840(9):2961–2969
14. Gillmor SA, Villasenor A, Fletterick R, Sigal E, Browner MF (1997) The structure of mammalian 15-lipoxygenase reveals similarity to the lipases and the determinants of substrate specificity. Nat Struct Biol 4(12):1003–1009
15. Gilbert NC, Bartlett SG, Waight MT, Neau DB, Boeglin WE, Brash AR et al (2011) The structure of human 5-lipoxygenase. Science 331(6014):217–219
16. Newcomer ME, Brash AR (2015) The structural basis for specificity in lipoxygenase catalysis. Protein Sci 24(3):298–309

17. Eek P, Jarving R, Jarving I, Gilbert NC, Newcomer ME, Samel N (2012) Structure of a calcium-dependent 11R-lipoxygenase suggests a mechanism for Ca2+ regulation. J Biol Chem 287(26):22377–22386
18. Rakonjac Ryge M, Tanabe M, Provost P, Persson B, Chen X, Funk CD et al (2014) A mutation interfering with 5-lipoxygenase domain interaction leads to increased enzyme activity. Arch Biochem Biophys 545:179–185
19. Kulkarni S, Das S, Funk CD, Murray D, Cho W (2002) A molecular basis of specific subcellular localization of the C2-like domain of 5-lipoxygenase. J Biol Chem 277 (15):13167–13174
20. Allard JB, Brock TG (2005) Structural organization of the regulatory domain of human 5-lipoxygenase. Curr Protein Pept Sci 6(2):125–131
21. Evans JF, Ferguson AD, Mosley RT, Hutchinson JH (2008) What's all the FLAP about?: 5-lipoxygenase-activating protein inhibitors for inflammatory diseases. Trends Pharmacol Sci 29(2):72–78
22. Jakobsson PJ, Morgenstern R, Mancini J, Ford-Hutchinson A, Persson B (2000) Membrane-associated proteins in eicosanoid and glutathione metabolism (MAPEG). A widespread protein superfamily. Am J Respir Crit Care Med 161(2 Pt 2):S20–24
23. Dixon RAF, Diehl RE, Opas E, Rands E, Vickers PJ, Evans JF et al (1990) Requirement of a 5-lipoxygenase-activating protein for leukotriene synthesis. Nature 343:282–284
24. Miller DK, Gillard JW, Vickers PJ, Sadowski S, Léveillé C, Mancini JA et al (1990) Identification and isolation of a membrane protein necessary for leukotriene production. Nature 343:278–281
25. Mancini JA, Waterman H, Riendeau D (1998) Cellular oxygenation of 12-hydroxyeicosatetraenoic acid and 15-hydroxyeicosatetraenoic acid by 5-lipoxygenase is stimulated by 5-lipoxygenase-activating protein. J Biol Chem 273(49):32842–32847
26. Plante H, Picard S, Mancini J, Borgeat P (2006) 5-lipoxygenase-activating protein homodimer in human neutrophils: evidence for a role in leukotriene biosynthesis. Biochem J 393(Pt 1):211–218
27. Mandal AK, Skoch J, Bacskai BJ, Hyman BT, Christmas P, Miller D et al (2004) The membrane organization of leukotriene synthesis. Proc Natl Acad Sci USA 101 (17):6587–6592
28. Ferguson AD, McKeever BM, Xu S, Wisniewski D, Miller DK, Yamin TT et al (2007) Crystal structure of inhibitor-bound human 5-lipoxygenase-activating protein. Science 317 (5837):510–512
29. Mandal AK, Jones PB, Bair AM, Christmas P, Miller D, Yamin TT et al (2008) The nuclear membrane organization of leukotriene synthesis. Proc Natl Acad Sci USA 105 (51):20434–20439
30. Strid T, Svartz J, Franck N, Hallin E, Ingelsson B, Soderstrom M et al (2009) Distinct parts of leukotriene C(4) synthase interact with 5-lipoxygenase and 5-lipoxygenase activating protein. Biochem Biophys Res Commun 381(4):518–522
31. Brock TG, Paine R, Petersgolden M (1994) Localization of 5-lipoxygenase to the nucleus of unstimulated rat basophilic leukemia cells. J Biol Chem 269(35):22059–22066
32. You HJ, Seo JM, Moon JY, Han SS, Ko YG, Kim JH (2007) Leukotriene synthesis in response to A23187 is inhibited by methyl-beta-cyclodextrin in RBL-2H3 cells. Mol Cells 23(1):57–63
33. Boyd RS, Jukes-Jones R, Walewska R, Brown D, Dyer MJ, Cain K (2009) Protein profiling of plasma membranes defines aberrant signaling pathways in mantle cell lymphoma. Mol Cell Proteomics 8(7):1501–1515
34. Jethwaney D, Islam MR, Leidal KG, de Bernabe DB, Campbell KP, Nauseef WM et al (2007) Proteomic analysis of plasma membrane and secretory vesicles from human neutrophils. Proteome Sci 5:12
35. Esser J, Gehrmann U, D'Alexandri FL, Hidalgo-Estevez AM, Wheelock CE, Scheynius A, et al (2010) Exosomes from human macrophages and dendritic cells contain enzymes for

leukotriene biosynthesis and promote granulocyte migration. J Allergy Clin Immunol 126 (5):1032–1040, 1040 e1031–1034
36. Hou X, Katahira T, Ohashi K, Mizuno K, Sugiyama S, Nakamura H (2013) Coactosin accelerates cell dynamism by promoting actin polymerization. Dev Biol 379(1):53–63
37. Provost P, Samuelsson B, Radmark O (1999) Interaction of 5-lipoxygenase with cellular proteins. Proc Natl Acad Sci USA 96(5):1881–1885
38. Provost P, Doucet J, Hammarberg T, Gerisch G, Samuelsson B, Radmark O (2001) 5-lipoxygenase interacts with coactosin-like protein. J Biol Chem 276(19):16520–16527
39. Feisst C, Pergola C, Rakonjac M, Rossi A, Koeberle A, Dodt G et al (2009) Hyperforin is a novel type of 5-lipoxygenase inhibitor with high efficacy in vivo. Cell Mol Life Sci 66 (16):2759–2771
40. Basavarajappa D, Wan M, Lukic A, Steinhilber D, Samuelsson B, Radmark O (2014) Roles of coactosin-like protein (CLP) and 5-lipoxygenase-activating protein (FLAP) in cellular leukotriene biosynthesis. Proc Natl Acad Sci USA 111(31):11371–11376
41. Perron MP, Provost P (2009) Protein components of the microRNA pathway and human diseases. Methods Mol Biol 487:369–385
42. Provost P, Dishart D, Doucet J, Frendewey D, Samuelsson B, Radmark O (2002) Ribonuclease activity and RNA binding of recombinant human Dicer. EMBO J 21(21):5864–5874
43. Dincbas-Renqvist V, Pepin G, Rakonjac M, Plante I, Ouellet DL, Hermansson A et al (2009) Human Dicer C-terminus functions as a 5-lipoxygenase binding domain. Biochim Biophys Acta 1789(2):99–108
44. Sala A, Folco G, Murphy RC (2010) Transcellular biosynthesis of eicosanoids. Pharmacol Rep 62(3):503–510
45. Werz O, Steinhilber D (2005) Development of 5-lipoxygenase inhibitors—lessons from cellular enzyme regulation. Biochem Pharmacol 70(3):327–333
46. Leslie CC (2004) Regulation of the specific release of arachidonic acid by cytosolic phospholipase A2. Prostaglandins Leukot Essent Fatty Acids 70(4):373–376
47. Werz O (2002) 5-lipoxygenase: cellular biology and molecular pharmacology. Curr Drug Targets Inflamm Allergy 1(1):23–44
48. Werz O, Burkert E, Samuelsson B, Radmark O, Steinhilber D (2002) Activation of 5-lipoxygenase by cell stress is calcium independent in human polymorphonuclear leukocytes. Blood 99(3):1044–1052
49. Krump E, Picard S, Mancini J, Borgeat P (1997) Suppression of leukotriene B-4 biosynthesis by endogenous adenosine in ligand-activated human neutrophils. J Exp Med 186 (8):1401–1406
50. Pergola C, Dodt G, Rossi A, Neunhoeffer E, Lawrenz B, Northoff H et al (2008) ERK-mediated regulation of leukotriene biosynthesis by androgens: a molecular basis for gender differences in inflammation and asthma. Proc Natl Acad Sci USA 105 (50):19881–19886
51. Pergola C, Rogge A, Dodt G, Northoff H, Weinigel C, Barz D et al (2011) Testosterone suppresses phospholipase D, causing sex differences in leukotriene biosynthesis in human monocytes. FASEB J 25(10):3377–3387
52. Reddy KV, Hammarberg T, Radmark O (2000) Mg2+ activates 5-lipoxygenase in vitro: dependency on concentrations of phosphatidylcholine and arachidonic acid. Biochemistry 39 (7):1840–1848
53. Hammarberg T, Radmark O (1999) 5-lipoxygenase binds calcium. Biochemistry 38 (14):4441–4447
54. Hammarberg T, Provost P, Persson B, Radmark O (2000) The N-terminal domain of 5-lipoxygenase binds calcium and mediates calcium stimulation of enzyme activity. J Biol Chem 275(49):38787–38793
55. Buerkert E, Arnold C, Hammarberg T, Radmark O, Steinhilber D, Werz O (2003) The C2-like {beta}-barrel domain mediates the Ca2+-dependent resistance of 5-lipoxygenase activity against inhibition by glutathione peroxidase-1. J Biol Chem 31:31

56. Noguchi M, Miyano M, Matsumoto T (1996) Physicochemical characterization of ATP binding to human 5-lipoxygenase. Lipids 31(4):367–371
57. Puustinen T, Scheffer MM, Samuelsson B (1988) Regulation of the human leukocyte 5-lipoxygenase: stimulation by micromolar calcium levels and phosphatidylcholine vesicles. Biochim Biophys Acta 960(3):261–267
58. Skorey KI, Gresser MJ (1998) Calcium is not required for 5-lipoxygenase activity at high phosphatidyl choline vesicle concentrations. Biochemistry 37(22):8027–8034
59. Zhang YY, Hammarberg T, Rådmark O, Samuelsson B, Ng CF, Funk CD et al (2000) Analysis of a nucleotide-binding site of 5-lipoxygenase by affinity labelling: binding characteristics and amino acid sequences. Biochem J 351(Pt 3):697–707
60. Okamoto H, Hammarberg T, Zhang YY, Persson B, Watanabe T, Samuelsson B et al (2005) Mutation analysis of the human 5-lipoxygenase C-terminus: support for a stabilizing C-terminal loop. Biochim Biophys Acta 1749(1):123–131
61. Walther M, Hofheinz K, Vogel R, Roffeis J, Kuhn H (2011) The N-terminal beta-barrel domain of mammalian lipoxygenases including mouse 5-lipoxygenase is not essential for catalytic activity and membrane binding but exhibits regulatory functions. Arch Biochem Biophys 516(1):1–9
62. Rouzer CA, Shimizu T, Samuelsson B (1985) On the nature of the 5-lipoxygenase reaction in human leukocytes: characterization of a membrane-associated stimulatory factor. Proc Natl Acad Sci USA 82:7505–7509
63. Noguchi M, Miyano M, Matsumoto T, Noma M (1994) Human 5-lipoxygenase associates with phosphatidylcholine liposomes and modulates LTA(4) synthetase activity. Biochim Biophys Acta 1215(3):300–306
64. Chen XS, Funk CD (2001) The N-terminal "beta-barrel" domain of 5-lipoxygenase is essential for nuclear membrane translocation. J Biol Chem 276(1):811–818
65. Pande AH, Moe D, Nemec KN, Qin S, Tan S, Tatulian SA (2004) Modulation of human 5-lipoxygenase activity by membrane lipids. Biochemistry 43(46):14653–14666
66. Pande AH, Qin S, Tatulian SA (2005) Membrane fluidity is a key modulator of membrane binding, insertion, and activity of 5-lipoxygenase. Biophys J 88(6):4084–4094
67. Hornig C, Albert D, Fischer L, Hornig M, Radmark O, Steinhilber D et al (2005) 1-Oleoyl-2-acetylglycerol stimulates 5-lipoxygenase activity via a putative (phospho)lipid binding site within the N-terminal C2-like domain. J Biol Chem 280(29):26913–26921
68. Albert D, Buerkert E, Steinhilber D, Werz O (2003) Induction of 5-lipoxygenase activation in polymorphonuclear leukocytes by 1-oleoyl-2-acetylglycerol. Biochim Biophys Acta 1631 (1):85–93
69. Albert D, Pergola C, Koeberle A, Dodt G, Steinhilber D, Werz O (2008) The role of diacylglyceride generation by phospholipase D and phosphatidic acid phosphatase in the activation of 5-lipoxygenase in polymorphonuclear leukocytes. J Leukoc Biol 83 (4):1019–1027
70. Fredman G, Ozcan L, Spolitu S, Hellmann J, Spite M, Backs J et al (2014) Resolvin D1 limits 5-lipoxygenase nuclear localization and leukotriene B4 synthesis by inhibiting a calcium-activated kinase pathway. Proc Natl Acad Sci USA 111(40):14530–14535
71. Werz O, Burkert E, Fischer L, Szellas D, Dishart D, Samuelsson B et al (2002) Extracellular signal-regulated kinases phosphorylate 5-lipoxygenase and stimulate 5-lipoxygenase product formation in leukocytes. FASEB J 16(11):1441–1443
72. Werz O, Klemm J, Radmark O, Samuelsson B (2001) p38 MAP kinase mediates stress-induced leukotriene synthesis in a human B-lymphocyte cell line. J Leukoc Biol 70 (5):830–838
73. Werz O, Klemm J, Samuelsson B, Rådmark O (2001) Phorbol ester up-regulates capacities for nuclear translocation and phosphorylation of 5-lipoxygenase in Mono Mac 6 cells and human polymorphonuclear leukocytes. Blood 97(8):2487–2495
74. Werz O, Klemm J, Samuelsson B, Radmark O (2000) 5-lipoxygenase is phosphorylated by p38 kinase dependent MAPKAP kinases. Proc Natl Acad Sci USA 97(10):5261–5266

75. Luo M, Jones SM, Phare SM, Coffey MJ, Peters-Golden M, Brock TG (2004) Protein kinase A inhibits leukotriene synthesis by phosphorylation of 5-lipoxygenase on serine 523. J Biol Chem 279(40):41512–41520
76. Luo M, Jones SM, Flamand N, Aronoff DM, Peters-Golden M, Brock TG (2005) Phosphorylation by protein kinase a inhibits nuclear import of 5-lipoxygenase. J Biol Chem 280 (49):40609–40616
77. Flamand N, Surette ME, Picard S, Bourgoin S, Borgeat P (2002) Cyclic AMP-mediated inhibition of 5-lipoxygenase translocation and leukotriene biosynthesis in human neutrophils. Mol Pharmacol 62(2):250–256
78. Radmark O, Werz O, Steinhilber D, Samuelsson B (2014) 5-lipoxygenase, a key enzyme for leukotriene biosynthesis in health and disease. Biochim Biophys Acta 1851(4):331–339
79. Markoutsa S, Surun D, Karas M, Hofmann B, Steinhilber D, Sorg BL (2014) Analysis of 5-lipoxygenase phosphorylation on molecular level by MALDI-MS. FEBS J 281 (8):1931–1947
80. Newcomer ME, Gilbert NC (2010) Location, location, location: compartmentalization of early events in leukotriene biosynthesis. J Biol Chem 285(33):25109–25114
81. Luo M, Jones SM, Peters-Golden M, Brock TG (2003) Nuclear localization of 5-lipoxygenase as a determinant of leukotriene B4 synthetic capacity. Proc Natl Acad Sci USA 100(21):12165–12170
82. Bair AM, Turman MV, Vaine CA, Panettieri RA Jr, Soberman RJ (2012) The nuclear membrane leukotriene synthetic complex is a signal integrator and transducer. Mol Biol Cell 23(22):4456–4464
83. Rossi A, Pergola C, Pace S, Radmark O, Werz O, Sautebin L (2014) In vivo sex differences in leukotriene biosynthesis in zymosan-induced peritonitis. Pharmacol Res 87:1–7
84. Haribabu B, Verghese MW, Steeber DA, Sellars DD, Bock CB, Snyderman R (2000) Targeted disruption of the leukotriene B(4) receptor in mice reveals its role in inflammation and platelet-activating factor-induced anaphylaxis. J Exp Med 192(3):433–438
85. Goulet JL, Griffiths RC, Ruiz P, Spurney RF, Pisetsky DS, Koller BH et al (1999) Deficiency of 5-lipoxygenase abolishes sex-related survival differences in MRL-lpr/lpr mice. J Immunol 163(1):359–366
86. Poeckel D, Zemski Berry KA, Murphy RC, Funk CD (2009) Dual 12/15- and 5-lipoxygenase deficiency in macrophages alters arachidonic acid metabolism and attenuates peritonitis and atherosclerosis in ApoE knock-out mice. J Biol Chem 284(31):21077–21089
87. Funk CD, Hoshiko S, Matsumoto T, Rdmark O, Samuelsson B (1989) Characterization of the human 5-lipoxygenase gene. Proc Natl Acad Sci USA 86(8):2587–2591
88. Uhl J, Klan N, Rose M, Entian KD, Werz O, Steinhilber D (2002) The 5-lipoxygenase promoter is regulated by DNA methylation. J Biol Chem 277(6):4374–4379
89. Katryniok C, Schnur N, Gillis A, von Knethen A, Sorg BL, Looijenga L et al (2010) Role of DNA methylation and methyl-DNA binding proteins in the repression of 5-lipoxygenase promoter activity. Biochim Biophys Acta 1801(1):49–57
90. Hoshiko S, Radmark O, Samuelsson B (1990) Characterization of the human 5-lipoxygenase gene promoter. Proc Natl Acad Sci USA 87(23):9073–9077
91. In KH, Silverman ES, Asano K, Beier D, Fischer AR, Keith TP et al (1999) Mutations in the human 5-lipoxygenase gene. Clin Rev Allergy Immunol 17(1-2):59–69
92. Silverman ES, Du J, De Sanctis GT, Radmark O, Samuelsson B, Drazen JM et al (1998) Egr-1 and Sp1 interact functionally with the 5-lipoxygenase promoter and its naturally occurring mutants. Am J Respir Cell Mol Biol 19(2):316–323
93. Dishart D, Schnur N, Klan N, Werz O, Steinhilber D, Samuelsson B et al (2005) GC-rich sequences in the 5-lipoxygenase gene promoter are required for expression in Mono Mac 6 cells, characterization of a novel Sp1 binding site. Biochim Biophys Acta 1738(1-3):37–47
94. Klan N, Seuter S, Schnur N, Jung M, Steinhilber D (2003) Trichostatin A and structurally related histone deacetylase inhibitors induce 5-lipoxygenase promoter activity. Biol Chem 384(5):777–785

95. Schnur N, Seuter S, Katryniok C, Radmark O, Steinhilber D (2007) The histone deacetylase inhibitor trichostatin A mediates upregulation of 5-lipoxygenase promoter activity by recruitment of Sp1 to distinct GC-boxes. Biochim Biophys Acta 1771(10):1271–1282
96. Ahmad K, Katryniok C, Scholz B, Merkens J, Loscher D, Marschalek R et al (2014) Inhibition of class I HDACs abrogates the dominant effect of MLL-AF4 by activation of wild-type MLL. Oncogenesis 3:e127
97. Drazen JM, Yandava CN, Dube L, Szczerback N, Hippensteel R, Pillari A et al (1999) Pharmacogenetic association between ALOX5 promoter genotype and the response to anti-asthma treatment. Nat Genet 22(2):168–170
98. Kalayci O, Birben E, Sackesen C, Keskin O, Tahan F, Wechsler ME et al (2006) ALOX5 promoter genotype, asthma severity and LTC production by eosinophils. Allergy 61 (1):97–103
99. Kim SH, Bae JS, Suh CH, Nahm DH, Holloway JW, Park HS (2005) Polymorphism of tandem repeat in promoter of 5-lipoxygenase in ASA-intolerant asthma: a positive association with airway hyperresponsiveness. Allergy 60(6):760–765
100. Mougey E, Lang JE, Allayee H, Teague WG, Dozor AJ, Wise RA et al (2013) ALOX5 polymorphism associates with increased leukotriene production and reduced lung function and asthma control in children with poorly controlled asthma. Clin Exp Allergy 43 (5):512–520
101. Sayers I, Barton S, Rorke S, Sawyer J, Peng Q, Beghe B et al (2003) Promoter polymorphism in the 5-lipoxygenase (ALOX5) and 5-lipoxygenase-activating protein (ALOX5AP) genes and asthma susceptibility in a Caucasian population. Clin Exp Allergy 33(8):1103–1110
102. Telleria JJ, Blanco-Quiros A, Varillas D, Armentia A, Fernandez-Carvajal I, Jesus Alonso M et al (2008) ALOX5 promoter genotype and response to montelukast in moderate persistent asthma. Respir Med 102(6):857–861
103. Dwyer JH, Allayee H, Dwyer KM, Fan J, Wu H, Mar R et al (2004) Arachidonate 5-lipoxygenase promoter genotype, dietary arachidonic acid, and atherosclerosis. N Engl J Med 350(1):29–37
104. Assimes TL, Knowles JW, Priest JR, Basu A, Volcik KA, Southwick A et al (2008) Common polymorphisms of ALOX5 and ALOX5AP and risk of coronary artery disease. Hum Genet 123(4):399–408
105. Gonzalez P, Reguero JR, Lozano I, Moris C, Coto E (2007) A functional Sp1/Egr1-tandem repeat polymorphism in the 5-lipoxygenase gene is not associated with myocardial infarction. Int J Immunogenet 34(2):127–130
106. Maznyczka A, Braund P, Mangino M, Samani NJ (2008) Arachidonate 5-lipoxygenase (5-LO) promoter genotype and risk of myocardial infarction: a case-control study. Atherosclerosis 199(2):328–332
107. Todur SP, Ashavaid TF (2012) Association of Sp1 tandem repeat polymorphism of ALOX5 with coronary artery disease in Indian subjects. Clin Transl Sci 5(5):408–411
108. Herb F, Thye T, Niemann S, Browne EN, Chinbuah MA, Gyapong J et al (2008) ALOX5 variants associated with susceptibility to human pulmonary tuberculosis. Hum Mol Genet 17 (7):1052–1060
109. Silverman ES, Le L, Baron RM, Hallock A, Hjoberg J, Shikanai T et al (2002) Cloning and functional analysis of the mouse 5-lipoxygenase promoter. Am J Respir Cell Mol Biol 26 (4):475–483
110. Sorg BL, Klan N, Seuter S, Dishart D, Radmark O, Habenicht A et al (2006) Analysis of the 5-lipoxygenase promoter and characterization of a vitamin D receptor binding site. Biochim Biophys Acta 1761(7):686–697
111. Seuter S, Vaisanen S, Radmark O, Carlberg C, Steinhilber D (2007) Functional characterization of vitamin D responding regions in the human 5-lipoxygenase gene. Biochim Biophys Acta 1771(7):864–872

112. Stoffers KL, Sorg BL, Seuter S, Rau O, Radmark O, Steinhilber D (2010) Calcitriol upregulates open chromatin and elongation markers at functional vitamin D response elements in the distal part of the 5-lipoxygenase gene. J Mol Biol 395(4):884–896
113. Steinhilber D (1999) 5-lipoxygenase: a target for antiinflammatory drugs revisited. Curr Med Chem 6(1):71–85
114. Harle D, Radmark O, Samuelsson B, Steinhilber D (1998) Calcitriol and transforming growth factor-beta upregulate 5-lipoxygenase mRNA expression by increasing gene transcription and mRNA maturation. Eur J Biochem 254(2):275–281
115. Seuter S, Sorg BL, Steinhilber D (2006) The coding sequence mediates induction of 5-lipoxygenase expression by Smads3/4. Biochem Biophys Res Commun 348(4):1403–1410
116. Boudreau LH, Bertin J, Robichaud PP, Laflamme M, Ouellette RJ, Flamand N et al (2011) Novel 5-lipoxygenase isoforms affect the biosynthesis of 5-lipoxygenase products. FASEB J 25(3):1097–1105
117. Ochs MJ, Sorg BL, Pufahl L, Grez M, Suess B, Steinhilber D (2012) Post-transcriptional regulation of 5-lipoxygenase mRNA expression via alternative splicing and nonsense-mediated mRNA decay. PLoS One 7(2):e31363
118. Fredman G, Li Y, Dalli J, Chiang N, Serhan CN (2012) Self-limited versus delayed resolution of acute inflammation: temporal regulation of pro-resolving mediators and microRNA. Sci Rep 2:639
119. Busch S, Auth E, Scholl F, Huenecke S, Koehl U, Suess B et al (2015) 5-lipoxygenase is a direct target of miR-19a-3p and miR-125b-5p. J Immunol 194(4):1646–1653
120. Esser J, Rakonjac M, Hofmann B, Fischer L, Provost P, Schneider G, Steinhilber D, Samuelsson B, Rådmark O (2009) Coactosin-like protein functions as a stabilizing chaperone for 5-lipoxygenase: role of tryptophan 102. Biochem J 425(1):265–274

Leukotriene A_4 Hydrolase and Leukotriene C_4 Synthase

Agnes Rinaldo-Matthis and Jesper Z. Haeggström

Abstract Leukotrienes are potent proinflammatory and immune modulating lipid mediators synthesized along the 5-lipoxygenase pathway of arachidonic acid metabolism. Leukotriene B_4 is one of the most potent chemotactic agents known while leukotriene C_4, D_4, and E_4 are a powerful smooth muscle contracting agents, particularly in the respiratory tract and microcirculation. The committed steps in the biosynthesis of leukotriene B_4 and C_4 are catalyzed by the key enzymes *leukotriene A_4 hydrolase* and *leukotriene C_4 synthase*, respectively. In this chapter we discuss the most recent advances in the understanding of these two enzymes at a structural, functional, and biological level.

Leukotrienes are lipid mediators generated from arachidonic acid through the 5-lipoxygenase pathway (Fig. 1). They are named after the cells ("leuko" from leukocytes) where they were originally discovered and the fact that they contain a conjugated triene in their chemical structure. Leukotrienes act as signaling molecules in inflammatory and allergic conditions both in the innate and adaptive immune response. The leukotrienes are formed by the sequential actions of several enzymes, first cytosolic phospholipase A_2 (cPLA$_2$), which releases arachidonic acid from membrane phospholipids, and second the 5-lipoxygenase (5-LO), supported by Five Lipoxygenase Activating Protein (FLAP), which dioxygenates arachidonic acid and dehydrates the resulting hydroperoxide to form the unstable epoxide intermediate, LTA_4. Here, the pathway branches such that LTA_4 is either converted by leukotriene (LT)A_4 hydrolase (LTA4H) or LTC_4 synthase (LTC4S) to form LTB_4 or LTC_4 respectively [1]. The GSH moiety of LTC_4 can be cleaved to generate LTD_4 and LTE_4 and together, these three molecules are referred to as cysteinyl-leukotrienes (cys-LTs).

LTB_4 is one of the most potent chemo-attractants for neutrophils known to date and cys-LTs are powerful smooth muscle contractile agents, particularly in the respiratory tract and microcirculation. The effects of leukotrienes are mediated through specific G-protein coupled receptors, two for LTB_4 (BLT1 and BLT2) and at least three for cys-LTs (CysLT1, CysLT2, and CysLT3) [1].

A. Rinaldo-Matthis • J.Z. Haeggström (✉)
Department of Medical Biochemistry and Biophysics, Karolinska Institutet, Scheelesväg 2, 17177 Stockholm, Sweden
e-mail: Jesper.Haeggstrom@ki.se

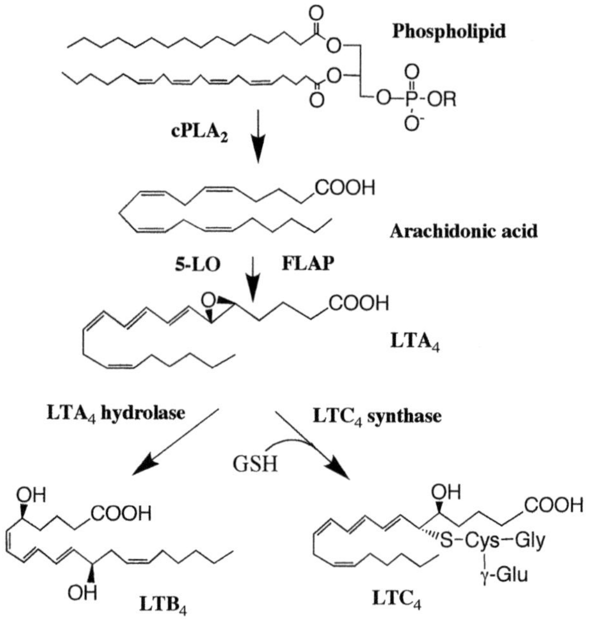

Fig. 1 cPLA$_2$ catalyzes the release of arachidonic acid from the membrane. Arachidonic acid is the substrate for 5-LO and FLAP that catalyze the formation of LTA$_4$, the substrate for both LTA4H and LTC4S

Due to the prominent bioactions of LTs, both LTA4H and LTC4S, are recognized as important drug targets in inflammation. Today there are two types of anti-leukotrienes on the market, both of which are used in the clinical management of asthma. One type is a biosynthesis inhibitor and targets 5-LO (Zileuton) and the other (exemplified by montelukast) blocks the actions of cys-LT signaling through antagonism at the CysLT1 receptor. However, yet no drugs have been developed targeting the enzymes LTC4S or LTA4H [2].

The X-ray structure of LTC4S [3, 4] as well as the newly discovered role of LTA4H in the resolution phase of inflammation now opens up new routes for drug development against these types of enzymes [5]. The purpose of this chapter is to review the recent advances in the structure and function of LTA4H and LTC4S as well as a to provide an update from recent drug design efforts.

1 Leukotriene A4 Hydrolase

Leukotriene A$_4$ hydrolase/aminopeptidase (LTA4H) (EC 3.3.2.6) is a 69 kDa monomeric zinc metalloenzyme with both an epoxide hydrolase and an aminopeptidase activity (Fig. 2). Both activities are dependent on a Zn^{2+} bound at the active site. The crystal structure of LTA4H has been determined in complex with both aminopeptidase substrates and inhibitors [6, 7]. It is a widely distributed enzyme that has been detected in almost all mammalian cells, albeit at significantly different levels [1].

LTB$_4$ is secreted by polymorphonuclear leukocytes (PMNs) that mediate inflammation by triggering chemotaxis and adherence of immune cells to the endothelium

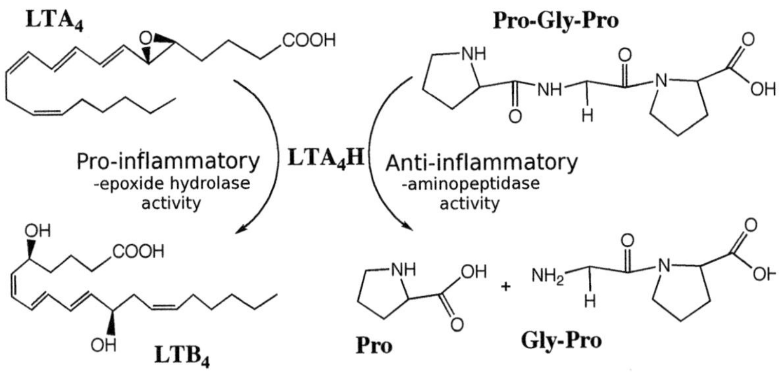

Fig. 2 LTA4H is a bifunctional enzyme with an epoxide hydrolase and an aminopeptidase activity

[8, 9]. LTB$_4$ also participates in host defense against infections and acts as a key mediator in Platelet-Activating Factor (PAF) induced lethal shock, vascular inflammation and arteriosclerosis [10].

In 2010 the physiological substrate for the aminopeptidase activity of LTA4H was identified. Snelgrove and co-workers discovered that extracellular LTA4H was responsible for the efficient hydrolysis and inactivation of the pro-inflammatory tripeptide Pro-Gly-Pro (PGP) during neutrophil-dependent inflammation in the respiratory tract [5]. PGP is a well-characterized chemotactic mediator for neutrophils generated from the enzymatic breakdown of extra cellular matrix (ECM) collagen. It is a biomarker for chronic obstructive pulmonary disease (COPD) and it has been implicated in neutrophil infiltration of the lung [11]. The two enzymatic activities of LTA4H, the epoxide hydrolase and the PGP aminopeptidase activity, have thus opposing roles during inflammatory reactions. During initiation of inflammation, LTB$_4$ is synthesized and governs neutrophil recruitment to the site of injury, whereas in the resolution phase of inflammation, LTA4H degrades the neutrophil chemo attractant PGP causing arrest in neutrophil recruitment [11].

1.1 Cellular Location

LTA4H is widely distributed in mammals and the enzyme have been found in practically all cells, organs and tissues in man, rat and guinea pig [12]. LTB$_4$ formation has also been observed in lower vertebrates including birds, fish and frogs. The frog LTA4H has been shown to form, in addition to LTB$_4$ an isomeric form of LTB$_4$ with unknown function [13]. LTB$_4$ has not been detected in non-vertebrates, although a yeast isoform has been described which are highly similar in sequence to the human LTA4H although it does not possess LTB$_4$ biosynthetic capacity [14].

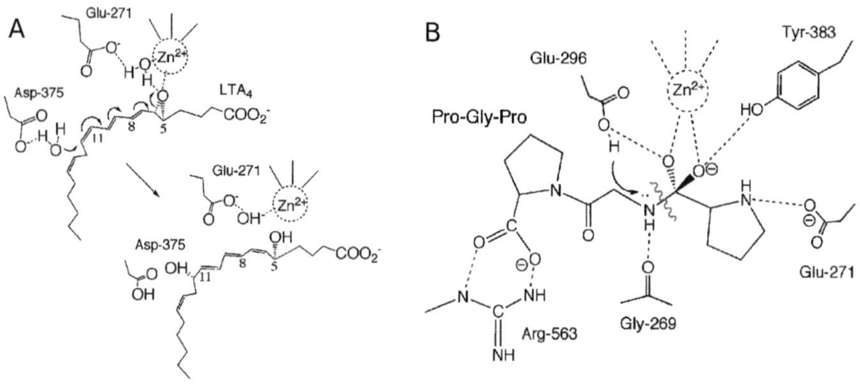

Fig. 3 The two enzymatic activities of LTA4H. In (**A**), the epoxide hydrolase activity is shown and in (**B**), the aminopeptidase Pro-Gly-Pro catalytic activity is shown

The enzyme LTA4H, as well as the product LTB_4, have also been detected in cells devoid of 5-LO; that is, cells unable to provide the substrate LTA_4 [1]. This broad distribution of LTA4H has been difficult to rationalize from a functional point of view, since leukotriene biosynthesis is generally regarded as a process restricted to white blood cells and bone marrow. An explanation for the uneven distribution of 5-LO and LTA4H has been the so-called trans cellular metabolism, a phenomenon that occurs *in vivo* and refers to the transfer of an intermediate or product, such as LTA_4, produced in one cell type, to another cell type for further enzymatic metabolism [15]. Another explanation considers the bi-functional property of LTA4H, where the peptide cleaving activity of LTA4H would account for the uneven distribution [5]. In human cells, LTA4H is present in the cytosol as well as in the extracellular space. The epoxide hydrolase activity is generally assumed to be cytosolic whereas the aminopeptidase activity is thought to play a major role extracellularly [12] (Fig. 3).

1.2 Structure

In 2001, the crystal structure of LTA4H in complex with the competitive inhibitor bestatin was determined to a resolution of 1.95 Ångström (Å) [6]. The structure revealed a protein with three domains, an N-terminal domain (residues 1–207), a Zn containing catalytic domain (208–450) and a α-helical C-terminal domain (461–610). The N-terminal domain is structurally similar to bacteriochlorophyll a, the catalytic domain is similar to that of thermolysin and the C-terminal domain has structural features resembling a so-called armadillo repeat or HEAT region, which implicates that the domain may take part in protein–protein interactions [6] (Fig. 4).

The active site is situated in a deep cleft in between the domains. The cleft consists of a hydrophilic part, where the Zn^{2+} is positioned, as well as a hydrophobic part located deeper in the protein. The Zn^{2+} is coordinated to His295, His299 and Glu318, constituting the Zn binding motif, $\underline{H}EXX\underline{H}$-(X)18-$\underline{E}$, that is conserved

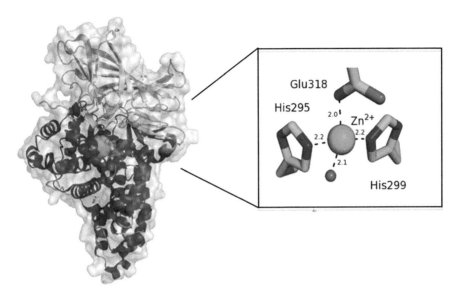

Fig. 4 The structure of LTA4H has three domains with the active site situated in a cleft between the domains close to where the Zn^{2+} (*green*) is bound. The *inset* illustrates the tetrahedral coordination of Zn^{2+} by His295, His299, Glu318 and a water molecule. The distances are measured in Ångström units

among members of the M1 family of metallopeptidases [16]. Besides LTA4H, other members in this family include aminopeptidase A, aminopeptidase B, and angiotensin-converting enzyme (ACE). Common to all members are the conserved stretch of amino acids containing the Zn binding ligands, a GXMEN peptide substrate binding motif as well as the catalytic mechanism by which the Zn^{2+} facilitates peptide hydrolysis.

Apart from the Zn ligands, the hydrophilic portion of the active site contains Glu271, Glu296 and Tyr383, residues important for the peptidase activity. The hydrophobic part of the active site is L-shaped and lined with aromatic and hydrophobic residues such as Leu and Phe. LTA4 manually modeled into the hydrophobic cavity of LTA4H has its carboxyl group coordinating Arg563, the epoxide coordinating the Zn^{2+} and C12, the site for hydroxyl group insertion, is close to Glu134, Asp375 and Tyr 267 [7]. The aminopeptidase and epoxide hydrolase activities are exerted at distinct but overlapping active sites. Certain resides such as Glu271, Arg563 and the Zn^{2+} are needed for both activities whereas Asp375 is critical only for the epoxide hydrolase reaction and Glu296 and Tyr383 are necessary for the aminopeptidase reaction (Fig. 5) [1]. In a recent paper the structure of frog LTA4H was described showing a similar structural arrangement as in the human enzyme [13]. The important active site residue Tyr 378 in the human enzyme is a Phe in the frog enzyme and has been described as being responsible for product selectivity in LTA4H [13].

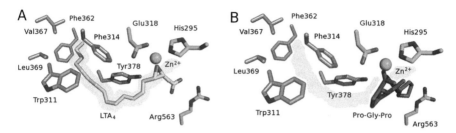

Fig. 5 (**A**) The active site of LTA4H, where the hydrophobic pocket is shown with *gray colored* residues, and the hydrophilic cleft, with *cyan colored* residues. The Zn^{2+} is *green* and a modeled LTA_4 is in *yellow*. (**B**) Shows the crystal structure complex of a Pro-Gly-Pro analogue, bound at the active site. The two catalytic activities occupy distinct but overlapping active sites

1.3 Catalytic Mechanism

LTA4H is unusually selective for its epoxide substrate LTA_4 and it only accepts a few other substrates such as the double bond isomers LTA_3 and LTA_5 although with low efficiency [17]. Hydrolysis of LTA_4 is believed to proceed according to a S_N1 mechanism [1]. The Zn^{2+} and Glu271 may participate in polarizing water promoting an acid induced opening of the epoxide, forming an unstable carbocation intermediate (Fig. 5a). The generated charge at the carbocation intermediate is delocalized over the conjugated triene system (C6–C12) leaving the planar sp^2 hybridized C12 open for nucleophilic attack from either side of the molecule. Subsequently, a water molecule is added at C12 in a stereo specific manner directed by Asp375 to generate the 12(R)-hydroxyl group in LTB_4. The epoxide hydrolase reaction of LTA4H is unique in the sense that the hydroxyl group is introduced in a stereospecific manner at a site distant (C12) from the epoxide moiety (C5 to C6).

The LTA4H epoxide hydrolase activity is suicide inactivated which leads to covalent attachment of the LTA_4 to the protein during catalysis [18]. Tyr378 was identified as the residue that binds LTA_4 during this process [18].

The aminopeptidase activity was characterized as an Arg specific tri-peptidase activity stimulated by monovalent anions, e.g., chloride ions [17]. The LTA4H aminopeptidase activity is less specific towards its substrate(s) as compared with the epoxide hydrolase activity although it prefers to hydrolyze positively charged tri-peptides such as Arg containing peptides [1]. However, these tripeptides only serve to map SAR and have no known endogenous counterparts.

The mechanism by which LTA4H hydrolyzes PGP [19] is similar as have been described previously for Arg-containing synthetic peptides [7]. The N-terminal part of the tri-peptide interacts with Glu271 and the C-terminus interacts with Arg563. The tri-peptidase activity follows a general base mechanism, similar to thermolysin, where Glu296 and Zn^{2+} activate a water molecule for nucleophilic attack on the carbonyl carbon of the scissile peptide bond (Fig. 5b).

The nucleophilic attack is facilitated by the induced partial positive charge at the carbonyl carbon. Zn^{2+} and Tyr383 stabilize the oxyanion intermediate. In the final step, Glu296 acts as an acid, protonating the N-terminus and facilitating the leaving group departure. The proposed mechanism of LTA4H is based on mutagenic and structural analysis of LTA4H as well as by similarities with thermolysin [7, 19].

2 LTA4H as a Drug Target

The product of LTA4H, LTB_4, is implicated in several inflammatory diseases such as inflammatory bowel disease (IBD), rheumatoid arthritis (RA), psoriasis, asthma and chronic obstructive pulmonary disease (COPD) [1]. An LTA4H deficient mouse was generated [20] and seen to develop normally supporting the notion that LTA4H is an attractive drug target. Studies using the LTA4H knock-out mice showed that they are resistant to lethal effects of systemic shock induced by PAF, thus identifying LTB_4 as a key mediator of this reaction [1]. Due to the importance of LTA4H in the generation of inflammatory mediators, several drug design programs have been initiated targeting LTA4H, both from academia as well as from industry [1]. Many efficient inhibitors have been identified, however, they block both the epoxide hydrolase and aminopeptidase activities. The relatively recent discovery that the aminopeptidase activity of LTA4H plays a crucial role in resolution of inflammation has changed the perspective in drug design against LTA4H [1] and now a selective blockade of the epoxide hydrolase activity while sparing the anti-inflammatory aminopeptidase activity has been considered as a novel route for inhibitor design. As of today it is currently unclear if PGP can be degraded by peptidases other than LTA4H.

ARM1 represents a novel type of LTA4H inhibitors that selectively blocks the epoxide hydrolase activity (LTB_4 synthesis) while sparing the aminopeptidase activity of LTA4H (PGP degradation) [19]. This type of inhibitor might have improved anti-inflammatory properties.

Receptor antagonists have also been developed blocking the LTB_4 signaling. Barchuk et al. showed that the BLT1 receptor antagonist, JNJ-40929837 decreased neutrophil recruitment suggesting that the BLT1 receptor antagonists might be beneficial in respiratory diseases characterized by neutrophilia, such as chronic obstructive pulmonary disease [21].

3 Leukotriene C_4 Synthase

Leukotriene (LT) C_4 synthase (LTC4S) (EC 4.4.1.20) is an 18 kDa integral membrane protein that catalyzes the formation of LTC_4 through a conjugation reaction between the epoxide LTA_4 and glutathione (GSH) (Fig. 6).

Once LTC_4 is formed by LTC4S it is exported by the multidrug resistance protein-1 to the extracellular space [22–24]. When LTC_4 is outside the cell, it is converted to LTD_4 and LTE_4 by γ-glutamyltranspeptidase and dipeptidases,

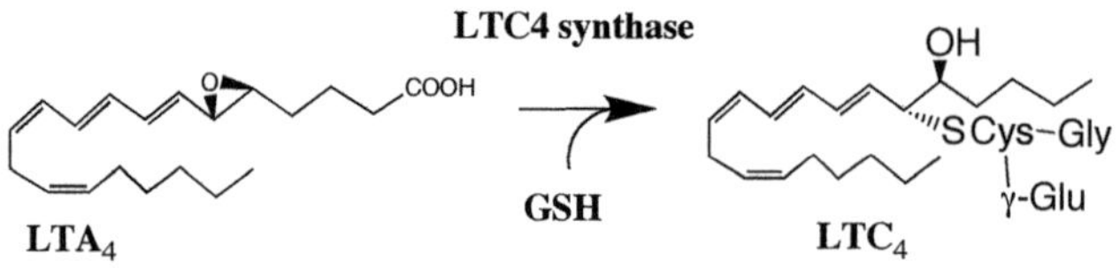

Fig. 6 LTC4S catalyzes the conjugation reaction between GSH and LTA_4 to form LTC_4

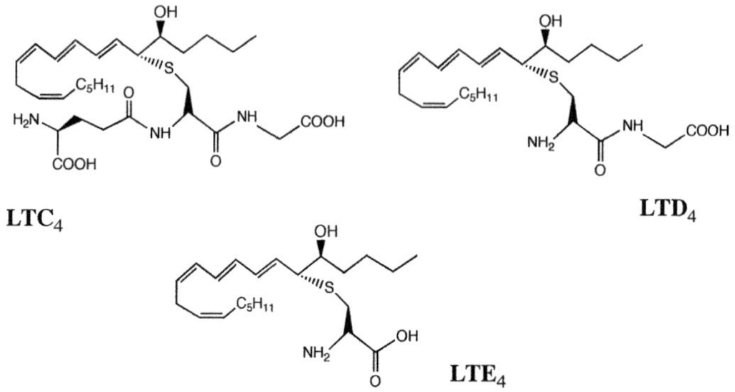

Fig. 7 A schematic representation of the cysteinyl leukotrienes, LTC_4, LTD_4 and LTE_4, is shown

respectively. LTE_4 is the most stable cysteinyl LT mediator, which easily can be detected in biological fluids such as in urine [25]. LTC_4, LTD_4 and LTE_4 are together called cysteinyl-leukotrienes, or cys-LTs (Fig. 7).

The bioactions of cysLTs were discovered already in 1938, long before these molecules and corresponding biosynthetic enzymes had been identified [26]. At that time cys-LTs were known as an entity called "slow reacting substance" (SRS) due to its effect as a slow but potent smooth muscle contracting agent, appearing in perfusates of guinea pig lungs treated with cobra venom [26]. In 1979 the chemical structure of LTC_4 was revealed, now known as the parent compound of cys-LTs [27].

LTC4S is found in a limited number of cell types such as eosinophils, mast cells, basophils and monocytes/macrophages [28]. Furthermore, LTC4S is also expressed in platelets although these cellular elements lack 5-lipoxygenase and cannot produce LTA_4 [29].

In the cell, LTC4S is localized to the peripheral endoplasmic reticulum (ER) and at the outer nuclear membrane but excluded from the inner nuclear membrane [30]. Studies, focused on the topology of LTC4S have shown that the active site is oriented towards the lumen of the nuclear envelope and endoplasmic reticulum [31].

LTC4S has been shown to interact with both the homologues FLAP and with the cytosolic 5-LO at the nuclear envelope [32].

The cys-LTs elicit their effect by binding to G protein coupled receptors, so called CysLT receptors [33]. At least three CysLT receptors have been identified.

CysLT1 has a high affinity for LTD_4 and is the target of "lukast" class of receptor antagonists (Montelukast, Zafirlukast, Pranlukast), used in asthma treatment [34]. The CysLT2 receptor is 38 % identical in sequence to CysLT1 and it binds LTC_4 and LTD_4 with equal affinity but weaker than CysLT1 [33]. CystLT2 is expressed in the same cells as CysLT1as well as in adrenal medulla cells, brain cells, cardiac Purkinje cells and endothelial cells [35]. Classical bioactions of cys-LTs are primarily mediated via CysLT1, whereas the role of CysLT2 is incompletely understood and seems to be related to the cardiovascular system [33]. Recently a new CysLT receptor was discovered when gpr99 was deorphanized as a receptor for LTE_4 with low nanomolar affinity. Its precise cellular distribution as well as its role in allergic inflammation needs to be determined [36].

3.1 Regulation of LTC4S

Regulation of LTC_4 biosynthesis has been an area of intense investigation due to the potent pro-inflammatory actions of cys-LTs. Studies have shown that the enzyme is regulated both at the transcriptional and post-translational level. At the transcriptional level, LTC4S expression can be induced by IL-3 and IL-5, in cells like human erythroleukemia and eosinophils [29, 37]. LTC4S expression is induced by TGFβ in monocytic THP-1 cells [38] and IL-4 has been seen to induce LTC4S expression in human mast cells [39].

Studies using eosinophils, neutrophils and monocytes have shown that the cellular LTC4S activity is down regulated by phorbol-12-myristate-13-acetate (PMA) treatments, suggesting that LTC4S is regulated by phosphorylation [40–42]. Furthermore, it was shown that only protein kinase C (PKC) was responsible for the reduced LTC4S activity during PMA treatments [1]. The presence of a conserved PKC consensus sequence in the N-terminal part of mouse and human LTC4S proteins suggests a direct role of PKC modulatory effect on LTC4S and evidences for direct phosphorylation of LTC4S protein by ribosomal protein S6 kinase (p70S6K) has been collected [43].

4 Other MAPEG Members

The protein sequence of LTC4S is most similar to microsomal GSH transferase 2 (MGST2) and they share 49 % sequence identity. MGST2 can produce LTC_4 although less efficient as compared with LTC4S [44].

LTC4S and MGST2, together with FLAP, microsomal prostaglandin E synthase 1 (mPGES-1), MGST1 and MGST3, constitute a protein family called Membrane-Associated Proteins in Eicosanoid and Glutathione Metabolism (MAPEG) all related in sequence by 20–50 % [45]. LTC4S, FLAP and mPGES-1 are involved in the generation of pro-inflammatory molecules derived from arachidonic acid whereas MGST1-3 are thought to mainly be involved in detoxification of xenobiotics. The MAPEG members are integral trimeric membrane proteins, localized to the nuclear and endoplasmic reticulum. The structures have been solved for four out of six MAPEG members. MGST1 and FLAP have afforded low-resolution structures, 3–4 Å [46, 47], whereas LTC4S and mPGES-1 have been solved to high resolution, between of 1.3 and 2 Å [3, 48]. With the exception of FLAP, which does not exhibit any known enzyme activity, the MAPEG members seem to share a similar catalytic mechanism, where the use of GSH as a cofactor or as a substrate is a common feature and a prerequisite for catalysis.

The structure of LTC4S was solved in 2007 to a resolution of 2.0 and 3.3 Å [3, 4]. The structure revealed a trimeric protein where each monomer consisted of five α-helices, four of which span the membrane (Fig. 3a). The active site is apparently situated at the interface between two monomers.

The GSH molecule is bound deep in the pocket (G-site) adjacent to and beneath a hydrophobic cleft. The GSH is coordinated by residues located on two adjacent monomers: Arg104, Tyr97, Tyr 93, Arg51, Asn55, Tyr59 and Glu58 from one monomer and Arg30 and Gln53 from the second monomer (Fig. 8b). The GSH adopts a unique horseshoe conformation also observed in the related protein mPGES-1 [48]. The sulfur of GSH was seen to be coordinated to Arg104 in the structure. Using mutagenesis studies as well as activity measurements, Arg104 was identified as essential for the activation of GSH [49].

Due to the unstable nature of the second substrate, LTA$_4$ ($t_{1/2}$ of 10 s), its exact binding site has not been identified. However, based on the binding of the detergent molecule Dodecylmaltoside (DDM) used for purification of the protein, a putative mode of LTA$_4$ binding was suggested (Fig. 8C). A detailed characterization of the exact position of DDM was obtained using a seleno-labeled DDM during structure determination [50]. The DDM is bound with its hydrocarbon tail in a hydrophobic cleft (H-site) formed at the interface between two monomers. In the deepest point of

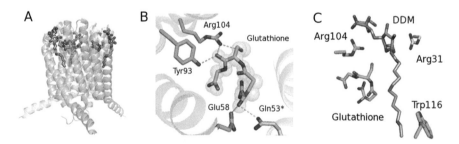

Fig. 8 In (**A**), the trimeric functional unit of LTC4S with the substrate GSH and a detergent bound in the active sites situated at interface between two monomers. In (**B**), important amino acids that coordinates the GSH is shown where *green color* represents amino acids located in one subunit and *yellow color* represents neighboring subunit. In (**C**), the detergent DDM that is believed to mimic the LTA$_4$ is bound close to GSH as well as Trp116

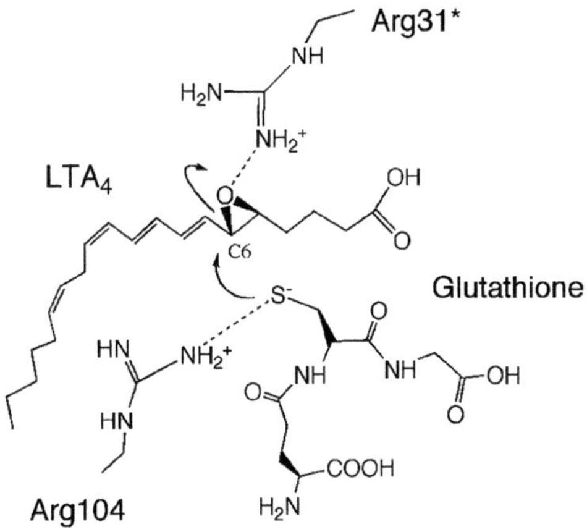

Fig. 9 A proposed catalytic mechanism of LTC4S where glutathione is activated/stabilized by Arg104 and the formed thiolate is conjugated with C6 of LTA$_4$. The epoxide is believed to be protonated by Arg31

the cleft the hydrocarbon tail interacts with Trp116 that presumably helps to hold the hydrophobic tail of LTA$_4$ within the active site [51].

The envisioned catalytic mechanism of LTC4S is similar to those of soluble GSH transferases where a thiolate anion is used for conjugation with the hydrophobic substrate LTA$_4$ to produce LTC$_4$ (Fig. 9). Details of the catalytic reaction have been obtained mainly based on mutational and structural studies. Several high-resolution crystal structures of LTC4S in complex with substrates, substrate analogs as well as different product analogs have been important for this work [3, 49, 51–53].

During catalysis, the bound substrate GSH is activated by Arg104 and a thiolate (GS$^-$) is formed that will be conjugated with the lipophilic substrate LTA$_4$ at position C6. To facilitate product formation, Arg31 has been suggested to protonate the epoxide of LTA$_4$ [53] and Trp116 was suggested to participate in the product release step [51].

The crystal structure of the mouse LTC4S, has been solved to resolution of 2.8 Å [54]. The sequence of the mouse and human enzymes share 88 % sequence identity. The structure as well as specific activity of the mouse LTC4S is similar to the human enzyme.

5 LTC4S as a Drug Target

The cys-LTs have been shown to play important roles in several diseases such as bronchial asthma and allergic rhinitis [55] and are implicated in the pathogenesis of acute respiratory distress syndrome [56, 57], glomerulonephritis [58], induced lung inflammation induced by hemorrhagic shock [59]. A recent study showed that respiratory syncytial virus infection up-regulates LTC4S mRNA expression and cys-LT synthesis in epithelial cells [60].

Furthermore, aspirin induced asthma (AIA) affecting 5–10 % of adults is characterized by an increased activity of LTC4S [61]. The AIA induced asthma has been correlated with polymorphism in the LTC4S promoter region but also with PKC dependent phosphorylation of the LTC4S protein.

LTC4S is a validated drug target both because it has a well-documented role in disease but also because targeted disruption of the LTC4S gene in mice did not affect growth or fertility [62]. LTC4S deficiency reduced or eliminated the capacity of different tissues to produce LTC_4, only the testis revealed normal LTC4S conjugating activity, probably reflecting a role of MGST2 as an LTC_4 producing enzyme [62]. As of today no drugs have yet been developed targeting LTC4S.

There are several challenges that need to be addressed during inhibitor development targeting LTC4S. One problem is the risk of cross-inhibition of related enzymes since the active site of LTC4S is to some extent conserved among MAPEG members. Another obstacle is related to the hydrophobic nature of LTC4S. The hydrophobic pocket of LTC4S and the energy required to bind the inhibitor relies on van derWaals and hydrophobic forces and not through electrostatic or hydrogen bond interactions. It is generally more difficult to obtain a high degree of specificity when hydrogen bonds or electrostatic forces cannot be used to obtain tight binding between the inhibitor and the protein. In spite of several difficulties encountered in exploiting LTC4S as a drug target, there has been some progress and a few inhibitors have been described [63]. Thus, helenalin (Fig. 10) has been seen to inhibit LTC4S activity in cell suspensions of platelets and human granulocytes [64]. Thymoquinone inhibits LTC4S activity in cell

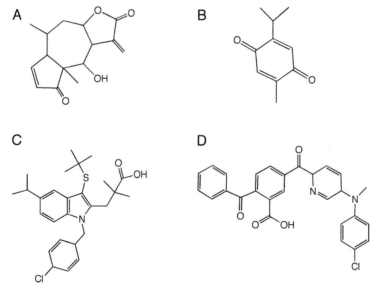

Fig. 10 Examples of LTC4S inhibitors where in (**A**), helenalin [69], in (**B**), thymoquinone [66], in (**C**), MK886 [67] and in (**D**), TK04 [54] are shown

suspensions of human platelets and granulocytes with an IC_{50} of 1.8 µM [65] and *in vivo* studies with this substance, using a mouse model of asthma, has shown an inhibitory effect on LTC4S activity [66]. MK886 is an inhibitor of LTC4S (IC_{50} of 3 µM) [67] as well as of other MAPEG members. However, *in vivo* studies have shown that even low concentrations of MK886 (1 µM) causes "permeability transition pore" (PTP) dependent mitochondrial dysfunction [68]. A series of patented bis-aromatic compounds have been seen to be efficient LTC4S inhibitors, some of them with IC_{50} values in the nM range. A related bis-aromatic compound, TK04, was seen to inhibit mouse and human LTC4S with an IC_{50} of about 400 nM [54].

References

1. Haeggstrom JZ, Funk CD (2011) Lipoxygenase and leukotriene pathways: biochemistry, biology, and roles in disease. Chem Rev 111:5866–5898
2. Haeggstrom JZ, Rinaldo-Matthis A, Wheelock CE, Wetterholm A (2010) Advances in eicosanoid research, novel therapeutic implications. Biochem Biophys Res Commun 396:135–139
3. Martinez Molina D, Wetterholm A, Kohl A, McCarthy AA, Niegowski D, Ohlson E, Hammarberg T, Eshaghi S, Haeggström JZ, Nordlund P (2007) Structural basis for synthesis of inflammatory mediators by human leukotriene C_4 synthase. Nature 448:613–616
4. Ago H, Kanaoka Y, Irikura D, Lam BK, Shimamura T, Austen KF, Miyano M (2007) Crystal structure of a human membrane protein involved in cysteinyl leukotriene biosynthesis. Nature 448:609–612
5. Snelgrove RJ, Jackson PL, Hardison MT, Noerager BD, Kinloch A, Gaggar A, Shastry S, Rowe SM, Shim YM, Hussell T, Blalock JE (2010) A critical role for LTA4H in limiting chronic pulmonary neutrophilic inflammation. Science 330:90–94
6. Thunnissen MM, Nordlund P, Haeggström JZ (2001) Crystal structure of human leukotriene A_4 hydrolase, a bifunctional enzyme in inflammation. Nat Struct Biol 8:131–135
7. Tholander F, Muroya A, Roques BP, Fournié-Zaluski MC, Thunnissen MM, Haeggström JZ (2008) Structure-based dissection of the active site chemistry of leukotriene A_4 hydrolase: implications for M1 aminopeptidases and inhibitor design. Chem Biol 15:920–929
8. Ford-Hutchinson AW (1990) Leukotriene B_4 in inflammation. Crit Rev Immunol 10:1–12
9. Samuelsson B (1983) Leukotrienes: mediators of immediate hypersensitivity reactions and inflammation. Science 220:568–575
10. Samuelsson B, Dahlen SE, Lindgren JA, Rouzer CA, Serhan CN (1987) Leukotrienes and lipoxins: structures, biosynthesis, and biological effects. Science 237:1171–1176
11. Snelgrove RJ (2011) Leukotriene A_4 hydrolase: an anti-inflammatory role for a proinflammatory enzyme. Thorax 66:550–551
12. Haeggström JZ (1998) Leukotriene A_4 hydrolase. In: Holgate S, Dahlen S-E (eds) 5-lipoxygenase products in asthma. Marcel Dekker, New York, NY, pp 51–76
13. Stsiapanava A, Tholander F, Kumar RB, Qureshi AA, Niegowski D, Hasan M, Thunnissen M, Haeggstrom JZ, Rinaldo-Matthis A (2014) Product formation controlled by substrate dynamics in leukotriene A_4 hydrolase. Biochim Biophys Acta 1844:439–446
14. Helgstrand C, Hasan M, Uysal H, Haeggstrom JZ, Thunnissen MM (2011) A leukotriene A_4 hydrolase-related aminopeptidase from yeast undergoes induced fit upon inhibitor binding. J Mol Biol 406:120–134
15. Sala A, Folco G, Murphy RC (2010) Transcellular biosynthesis of eicosanoids. Pharmacol Rep 62:503–510

16. Barret AJ, Rawlings ND, Woessner JF (eds) (1998) Handbook of proteolytic enzymes. Academic, London, San Diego
17. Haeggstrom JZ (2000) Structure, function, and regulation of leukotriene A_4 hydrolase. Am J Respir Crit Care Med 161:S25–31
18. Orning L, Gierse J, Duffin K, Bild G, Krivi G, Fitzpatrick FA (1992) Mechanism-based inactivation of leukotriene A_4 hydrolase/aminopeptidase by leukotriene A_4. Mass spectrometric and kinetic characterization. J Biol Chem 267:22733–22739
19. Stsiapanava A, Olsson U, Wan M, Kleinschmidt T, Rutishauser D, Zubarev RA, Samuelsson B, Rinaldo-Matthis A, Haeggstrom JZ (2014) Binding of Pro-Gly-Pro at the active site of leukotriene A_4 hydrolase/aminopeptidase and development of an epoxide hydrolase selective inhibitor. Proc Natl Acad Sci USA 111:4227–4232
20. Byrum RS, Goulet JL, Snouwaert JN, Griffiths RJ, Koller BH (1999) Determination of the contribution of cysteinyl leukotrienes and leukotriene B_4 in acute inflammatory responses using 5-lipoxygenase- and leukotriene A_4 hydrolase-deficient mice. J Immunol 163:6810–6819
21. Barchuk W, Lambert J, Fuhr R, Jiang JZ, Bertelsen K, Fourie A, Liu X, Silkoff PE, Barnathan ES, Thurmond R (2014) Effects of JNJ-40929837, a leukotriene A_4 hydrolase inhibitor, in a bronchial allergen challenge model of asthma. Pulm Pharmacol Ther 29:15–23
22. Lam BK, Owen WF Jr, Austen KF, Soberman RJ (1989) The identification of a distinct export step following the biosynthesis of leukotriene C_4 by human eosinophils. J Biol Chem 264:12885–12889
23. Leier I, Jedlitschky G, Buchholz U, Cole SP, Deeley RG, Keppler D (1994) The MRP gene encodes an ATP-dependent export pump for leukotriene C_4 and structurally related conjugates. J Biol Chem 269:27807–27810
24. Wijnholds J, Evers R, van Leusden MR, Mol CA, Zaman GJ, Mayer U, Beijnen JH, van der Valk M, Krimpenfort P, Borst P (1997) Increased sensitivity to anticancer drugs and decreased inflammatory response in mice lacking the multidrug resistance-associated protein. Nat Med 3:1275–1279
25. Murphy RC, Sala A, Voelkel N, Maclouf J (1991) Appearance of urinary metabolites of LTE_4 in human subjects. Ann N Y Acad Sci 629:105–111
26. Feldberg W, Kellaway CH (1938) Liberation of histamine and formation of lysocithin-like substances by cobra venom. J Physiol 94:187–226
27. Murphy RC, Hammarstrom S, Samuelsson B (1979) Leukotriene C: a slow-reacting substance from murine mastocytoma cells. Proc Natl Acad Sci USA 76:4275–4279
28. Lam BK, Austen KF (2002) Leukotriene C_4 synthase: a pivotal enzyme in cellular biosynthesis of the cysteinyl leukotrienes. Prostaglandins Other Lipid Mediat 68–69:511–520
29. Soderstrom M, Mannervik B, Garkov V, Hammarstrom S (1992) On the nature of leukotriene C_4 synthase in human platelets. Arch Biochem Biophys 294:70–74
30. Mandal AK, Jones PB, Bair AM, Christmas P, Miller D, Yamin TT, Wisniewski D, Menke J, Evans JF, Hyman BT, Bacskai B, Chen M, Lee DM, Nikolic B, Soberman RJ (2008) The nuclear membrane organization of leukotriene synthesis. Proc Natl Acad Sci USA 105:20434–20439
31. Christmas P, Weber BM, McKee M, Brown D, Soberman RJ (2002) Membrane localization and topology of leukotriene C_4 synthase. J Biol Chem 277:28902–28908
32. Strid T, Svartz J, Franck N, Hallin E, Ingelsson B, Soderstrom M, Hammarstrom S (2009) Distinct parts of leukotriene C_4 synthase interact with 5-lipoxygenase and 5-lipoxygenase activating protein. Biochem Biophys Res Commun 381:518–522
33. Nakamura M, Shimizu T (2011) Leukotriene receptors. Chem Rev 111:6231–6298
34. Lynch KR, O'Neill GP, Liu Q, Im DS, Sawyer N, Metters KM, Coulombe N, Abramovitz M, Figueroa DJ, Zeng Z, Connolly BM, Bai C, Austin CP, Chateauneuf A, Stocco R, Greig GM, Kargman S, Hooks SB, Hosfield E, Williams DL Jr, Ford-Hutchinson AW, Caskey CT, Evans JF (1999) Characterization of the human cysteinyl leukotriene CysLT1 receptor. Nature 399:789–793

35. Heise CE, O'Dowd BF, Figueroa DJ, Sawyer N, Nguyen T, Im DS, Stocco R, Bellefeuille JN, Abramovitz M, Cheng R, Williams DL Jr, Zeng Z, Liu Q, Ma L, Clements MK, Coulombe N, Liu Y, Austin CP, George SR, O'Neill GP, Metters KM, Lynch KR, Evans JF (2000) Characterization of the human cysteinyl leukotriene 2 receptor. J Biol Chem 275:30531–30536
36. Kanaoka Y, Maekawa A, Austen KF (2013) Identification of GPR99 protein as a potential third cysteinyl leukotriene receptor with a preference for leukotriene E_4 ligand. J Biol Chem 288:10967–10972
37. Boyce JA, Lam BK, Penrose JF, Friend DS, Parsons S, Owen WF, Austen KF (1996) Expression of LTC_4 synthase during the development of eosinophils in vitro from cord blood progenitors. Blood 88:4338–4347
38. Schroder O, Sjostrom M, Qiu H, Jakobsson PJ, Haeggstrom JZ (2005) Microsomal glutathione S-transferases: selective up-regulation of leukotriene C_4 synthase during lipopolysaccharide-induced pyresis. Cell Mol Life Sci 62:87–94
39. Hsieh FH, Lam BK, Penrose JF, Austen KF, Boyce JA (2001) T helper cell type 2 cytokines coordinately regulate immunoglobulin E-dependent cysteinyl leukotriene production by human cord blood-derived mast cells: profound induction of leukotriene C_4 synthase expression by interleukin 4. J Exp Med 193:123–133
40. Ali A, Ford-Hutchinson AW, Nicholson DW (1994) Activation of protein kinase C downregulates leukotriene C_4 synthase activity and attenuates cysteinyl leukotriene production in an eosinophilic substrain of HL-60 cells. J Immunol 153:776–788
41. Sjolinder M, Tornhamre S, Werga P, Edenius C, Lindgren JA (1995) Phorbol ester-induced suppression of leukotriene C_4 synthase activity in human granulocytes. FEBS Lett 377:87–91
42. Tornhamre S, Edenius C, Lindgren JA (1995) Receptor-mediated regulation of leukotriene C_4 synthase activity in human platelets. Eur J Biochem/FEBS 234:513–520
43. Esser J, Gehrmann U, Salvado MD, Wetterholm A, Haeggstrom JZ, Samuelsson B, Gabrielsson S, Scheynius A, Radmark O (2011) Zymosan suppresses leukotriene C_4 synthase activity in differentiating monocytes: antagonism by aspirin and protein kinase inhibitors. FASEB J 25:1417–1427
44. Ahmad S, Niegowski D, Wetterholm A, Haeggstrom JZ, Morgenstern R, Rinaldo-Matthis A (2013) Catalytic characterization of human microsomal glutathione S-transferase 2: identification of rate-limiting steps. Biochemistry 52:1755–1764
45. Bresell A, Weinander R, Lundqvist G, Raza H, Shimoji M, Sun TH, Balk L, Wiklund R, Eriksson J, Jansson C, Persson B, Jakobsson PJ, Morgenstern R (2005) Bioinformatic and enzymatic characterization of the MAPEG superfamily. FEBS J 272:1688–1703
46. Holm P, Morgenstern R, Fujiyoshi Y, Hebert H (2001) The 3-D structure of microsomal glutathione transferase 1 at 6 angstrom resolution as determined by electron crystallography of p22(1)2(1) crystals. Chem-Biol Interact 133:68–70
47. Ferguson AD, McKeever BM, Xu S, Wisniewski D, Miller DK, Yamin TT, Spencer RH, Chu L, Ujjainwalla F, Cunningham BR, Evans JF, Becker JW (2007) Crystal structure of inhibitor-bound human 5-lipoxygenase-activating protein. Science 317:510–512
48. Sjogren T, Nord J, Ek M, Johansson P, Liu G, Geschwindner S (2013) Crystal structure of microsomal prostaglandin E2 synthase provides insight into diversity in the MAPEG superfamily. Proc Natl Acad Sci USA 110:3806–3811
49. Rinaldo-Matthis A, Wetterholm A, Molina DM, Holm J, Niegowski D, Ohlson E, Nordlund P, Morgenstern R, Haeggstrom JZ (2010) Arginine 104 is a key catalytic residue in leukotriene C_4 synthase. J Biol Chem 285:40771–40776
50. Saino H, Ago H, Ukita Y, Miyano M (2011) Seleno-detergent MAD phasing of leukotriene C_4 synthase in complex with dodecyl-beta-D-selenomaltoside. Acta Crystallogr Sect F Struct Biol Cryst Commun 67:1666–1673
51. Niegowski D, Kleinschmidt T, Olsson U, Ahmad S, Rinaldo-Matthis A, Haeggstrom JZ (2014) Crystal structures of leukotriene C_4 synthase in complex with product analogs, implications for the enzyme mechanism. J Biol Chem 289:5199–5207

52. Rinaldo-Matthis A, Ahmad S, Wetterholm A, Lachmann P, Morgenstern R, Haeggstrom JZ (2012) Pre-steady-state kinetic characterization of thiolate anion formation in human leukotriene C_4 synthase. Biochemistry 51:848–856
53. Saino H, Ukita Y, Ago H, Irikura D, Nisawa A, Ueno G, Yamamoto M, Kanaoka Y, Lam BK, Austen KF, Miyano M (2011) The catalytic architecture of leukotriene C_4 synthase with two arginine residues. J Biol Chem 286:16392–16401
54. Niegowski D, Kleinschmidt T, Ahmad S, Qureshi AA, Marback M, Rinaldo-Matthis A, Haeggstrom JZ (2014) Structure and inhibition of mouse leukotriene C_4 synthase. PLoS One 9:e96763
55. Hay DW, Torphy TJ, Undem BJ (1995) Cysteinyl leukotrienes in asthma: old mediators up to new tricks. Trends Pharmacol Sci 16:304–309
56. Amat M, Barcons M, Mancebo J, Mateo J, Oliver A, Mayoral JF, Fontcuberta J, Vila L (2000) Evolution of leukotriene B_4, peptide leukotrienes, and interleukin-8 plasma concentrations in patients at risk of acute respiratory distress syndrome and with acute respiratory distress syndrome: mortality prognostic study. Crit Care Med 28:57–62
57. Matthay MA, Eschenbacher WL, Goetzl EJ (1984) Elevated concentrations of leukotriene D_4 in pulmonary edema fluid of patients with the adult respiratory distress syndrome. J Clin Immunol 4:479–483
58. Petric R, Ford-Hutchinson A (1995) Inhibition of leukotriene biosynthesis improves renal function in experimental glomerulonephritis. J Lipid Mediat Cell Signal 11:231–240
59. Al-Amran FG, Hadi NR, Hashim AM (2011) Leukotriene biosynthesis inhibition ameliorates acute lung injury following hemorrhagic shock in rats. J Cardiothorac Surg 6:81
60. Sun LH, Chen AH, Yang ZF, Chen JJ, Guan WD, Wu JL, Qin S, Zhong NS (2013) Respiratory syncytial virus induces leukotriene C_4 synthase expression in bronchial epithelial cells. Respirology 18 Suppl 3:40–46
61. Szczeklik A, Stevenson DD (1999) Aspirin-induced asthma: advances in pathogenesis and management. J Allergy Clin Immunol 104:5–13
62. Lam BK, Austen KF (2002) Leukotriene C_4 synthase: a pivotal enzyme in cellular biosynthesis of the cysteinyl leukotrienes. Prostaglandins Other Lipid Mediat 68–69:511–520
63. Devi NS, Doble M (2012) Leukotriene C_4 synthase: upcoming drug target for inflammation. Curr Drug Targets 13:1107–1118
64. Tornhamre S, Schmidt TJ, Nasman-Glaser B, Ericsson I, Lindgren JA (2001) Inhibitory effects of helenalin and related compounds on 5-lipoxygenase and leukotriene C_4 synthase in human blood cells. Biochem Pharmacol 62:903–911
65. Mansour M, Tornhamre S (2004) Inhibition of 5-lipoxygenase and leukotriene C_4 synthase in human blood cells by thymoquinone. J Enzyme Inhib Med Chem 19:431–436
66. El Gazzar M, El Mezayen R, Nicolls MR, Marecki JC, Dreskin SC (2006) Downregulation of leukotriene biosynthesis by thymoquinone attenuates airway inflammation in a mouse model of allergic asthma. Biochim Biophys Acta 1760:1088–1095
67. Lam BK, Penrose JF, Freeman GJ, Austen KF (1994) Expression cloning of a cDNA for human leukotriene C_4 synthase, an integral membrane protein conjugating reduced glutathione to leukotriene A_4. Proc Natl Acad Sci USA 91:7663–7667
68. Gugliucci A, Ranzato L, Scorrano L, Colonna R, Petronilli V, Cusan C, Prato M, Mancini M, Pagano F, Bernardi P (2002) Mitochondria are direct targets of the lipoxygenase inhibitor MK886. A strategy for cell killing by combined treatment with MK886 and cyclooxygenase inhibitors. J Biol Chem 277:31789–31795
69. Lyss G, Knorre A, Schmidt TJ, Pahl HL, Merfort I (1998) The anti-inflammatory sesquiterpene lactone helenalin inhibits the transcription factor NF-kappaB by directly targeting p65. J Biol Chem 273:33508–33516

Catalytic Multiplicity of 15-Lipoxygenase-1 Orthologs (ALOX15) of Different Species

Hartmut Kühn, Felix Karst, and Dagmar Heydeck

Abstract Lipoxygenases (LOX) form a family of lipid peroxidizing enzymes, which have been implicated in a number of physiological processes and in the pathogenesis of inflammatory, hyperproliferative and neurodegenerative diseases. They occur in bacteria and eucarya and the human genome involves six functional LOX genes, which encode for six different LOX isoforms. One of these isoforms is ALOX15, which has first been described in rabbits in 1974 as an enzyme capable of oxidizing membrane phospholipids during the maturational breakdown of mitochondria in immature red blood cells. During the following decades ALOX15 orthologs have extensively been characterized and their biological functions have been studied in a number of cellular in vitro systems as well as in various whole animal disease models. This review is aimed at summarizing the current knowledge on the protein-chemical, molecular biological and enzymatic properties of ALOX15 orthologs of various mammalian species (rabbit, pig, human, nonhuman primates, mouse, rat). Because of space limitations the biological roles of ALOX15 orthologs have not been addressed since this topic has extensively been covered in a previous review (Kuhn et al., Biochim Biophys Acta 1851:308–330, 2015).

List of Non-standard Abbreviations

LOX	Lipoxygenase
AA	Arachidonic acid
13S-H(p)ODE	(13S,9Z,11E)-13-hydro(pero)xyoctadeca-9,11-dienoic acid
15S-H(p)ETE	(15S,5Z,8Z,11Z,13E)-15-hydro(pero)xyeicosa-5,8,11,13-tetraenoic acid
12S-H(p)ETE	(12S,5Z,8Z,10E,14Z)-12-hydro(pero)xyeicosa-5,8,10,14-tetraenoic acid
SAXS	Small angle X-ray scattering

H. Kühn (✉) • F. Karst • D. Heydeck
Institute of Biochemistry, Charité—University Medicine Berlin, Charitéplatz 1, CCO-Building, Virchowweg 6, 10117 Berlin, Germany
e-mail: hartmut.kuehn@charite.de

1 Introduction

Lipoxygenases (LOXs) are non-heme iron-containing fatty acid dioxygenases [1, 2] that catalyze the dioxygenation of polyunsaturated fatty acids to the corresponding hydroperoxy derivatives. They are widely distributed in higher plants [3] and mammals [4] but except for rare cases their biological activities are not well understood. Although there are reports on the expression of true LOXs in lower multi- and single cellular organisms, a systematic search of the publically available genomic databases for LOX-like sequences indicated that these enzymes rarely occur in primitive living beings [5]. A rough estimate of possible LOXs in bacterial species suggested that LOX-like sequences might only occur in <0.5 % of all bacterial species sequenced so far. The first true LOX was discovered in dried soybean seeds more than 60 years ago and pioneer work has been carried out on this enzyme to explore the molecular details of the catalytic properties of LOXs in general. The first animal LOX [6] was described in human blood platelets in 1974 and this enzyme was characterized as arachidonic acid 12-lipoxygenase (ALOX12). Some months later a different LOX-isoenzyme was reported in the lysate of immature rabbit red blood cells [7]. Since these early days the number of well-characterized LOXs has constantly been growing, which may be related to the fact that more and more LOX-like sequences are being identified in the different genome projects. The corresponding enzymes can be expressed as recombinant proteins in heterologous overexpression systems, which allows detailed characterization of their protein-chemical and enzymatic properties. The major substrates of all LOX-isoforms characterized so far are free polyenoic fatty acids. In mammalian cells, linoleic acid (C18:$\Delta^{9,12}$,n − 6) and arachidonic acid (C20:$\Delta^{5,8,11,14}$,n − 6) are the dominant LOX-substrates. In the absence of specific alimentary supplementation (fish oil diet) eicosapentaenoic acid (C20:$\Delta^{5,8,11,14,17}$,n − 3) and docosahexaenoic acid (C22:$\Delta^{4,7,10,13,16,19}$,n − 3) only occur in lower abundance. In many plants, alpha- (C18:$\Delta^{9,12,15}$,n − 3) and gamma- (C18:$\Delta^{6,9,12}$,n − 6) linolenic acid are the most abundant polyenoic fatty acids.

Among the six different human LOX-isoforms ALOX5 [8] and ALOX15 [2] are probably the best characterized isoenzymes. The biological relevance of mammalian LOX-isoforms has recently been reviewed [4] and thus, there is no need for a detailed discussion of the biological activities of ALOX15 orthologs. This paper is aimed at summarizing the enzymatic properties of ALOX15 orthologs of different mammalian species with particular emphasis on their catalytic multiplicity. This aspect is sometimes neglected in the current LOX literature. Among the ALOX15 orthologs of different mammalian species rabbit ALOX15 (rabALOX15) has been characterized most comprehensively. However, because of the high degree of structural and functional homology between rabALOX15 and the orthologs from other mammals, including humans, most of our knowledge on rabALOX 15 can be transferred to the human enzyme (humALOX) or to the orthologs of other mammalian species.

Writing a review about well-characterized enzymes, which have been discovered many years ago, is always selective and strongly depends on the personal perspective of the authors. Although we did our best to balance the selection of citations we might have overlooked important contributions. We want to apologize to those distinguished colleagues who significantly contributed to the field but whose work could not be referenced because of space limitations.

2 Lipoxygenase Multiplicity and ALOX15 Orthologs

Traditionally, animal LOXs have been classified according to their reaction specificity using arachidonic acid as model substrate. When oxygen is introduced at carbon atom 5 of the fatty acid backbone, the corresponding enzyme was called 5-LOX. If the substrate is oxygenated at carbon 15, a 15-LOX was predicted as catalyst. In the early days of LOX-research this simple classification system was extremely helpful, but after more and more LOX sequences became available this nomenclature became confusing since it does not consider the evolutionary and functional relatedness of different LOX-isoforms. Moreover, it leads to confusions since evolutionary closely related LOX orthologs are subgrouped in different classes [2]. In recent years newly discovered LOX-isoforms are frequently classified according to their sequence similarity with any of the human LOX-isoforms. This classification concept works well for most mammalian LOXs but because of the low degree of sequence conservation in LOX-isoforms of lower plants and animals it is sometimes problematic to classify LOX-isoforms of evolutionary more distant living beings. For instance, in zebrafish a number of LOX genes have been identified, but unequivocal assignment of these genes to the human ALOX-isoforms is not always possible because of the rather low degree of amino acid conservation (see section "Zebrafish LOX Genes and LOX-Isoforms").

2.1 Human LOX Genes and LOX-Isoforms

The human genome involves six functional LOX genes (ALOX15, ALOX15B, ALOX12, ALOX12B, ALOXE3, ALOX5) and their expression leads to six functionally distinct LOX-isoforms. This structural multiplicity does not mirror a high degree of functional redundancy since all human isoforms fulfill different biological functions. In other words, the functional alterations induced by structural disruption of a certain LOX gene cannot be compensated for by upregulation of another LOX isoform. For instance, human ALOX5 has been implicated in the biosynthesis of leukotrienes of the 5,6-series, which play an important role as inflammatory and allergic mediators [1, 9, 10]. If this biosynthetic pathway is corrupted, 5,6-leukotriene biosynthesis cannot be restored by upregulation of ALOX15, ALOX12B or ALOX12. The genes for ALOX15, ALOX15B,

Table 1 Bioactivities of ALOX15 orthologs

LOX-isoform	Knockout mice (KO) Overexpression (TG)	Bioactivity
ALOX15	KO, TG	Hematopoiesis, osteogenesis, inflammation, stroke, atherogenesis, tumorgenesis, adipogenesis
ALOX15b	–	Skin development, hair growth, tumorgenesis
ALOX12	KO	Skin development, platelet aggregation
ALOX12b	KO	Adipogenesis, skin development, ichthyosis
ALOXE3	KO	Skin development, ichthyosis
ALOX5	KO	Immune regulation, inflammation, bronchial asthma, rhinitis

A detailed review of the bioactivity of different ALOX15 orthologs has recently been published and the references describing the bioactivities can be obtained from this paper (Kuhn et al., Biochim. Biophys. Acta 1851, 308–330). Availability of knockout mice (KO) and overexpressing transgenics (TG) is indicated.

ALOX12, ALOX12B and ALOXE3 are tandemly arranged in a joint LOX-gene cluster on the short arm of chromosome 17. In contrast, the ALOX5 gene is localized on the long arm of chromosome 10. The biological functions of the different LOX-isoforms are very diverse and have recently been reviewed in detail [4]. In Table 1 a brief summary of the most important biological roles of the different human LOX-isoforms is given.

ALOX15 is one of the six functional human LOX-isoforms. It oxygenates arachidonic acid with dual reaction specificity to a mixture of 15S-HpETE (90 %) and 12S-HpETE (10 %). Because of this dual reaction specificity ALOX15 was formerly named 12/15-LOX. The oxygenation of arachidonic acid involves hydrogen abstraction from two different bisallylic methylenes (C13 and C10 of arachidonic acid) and as for other LOX-isoforms there is an antarafacial relation between hydrogen abstraction and oxygen insertion.

2.2 Mouse LOX Genes and LOX-Isoforms

In contrast to the human genome, the mouse genome contains seven functional LOX genes. An orthologous gene exists in mice for each of the human LOX genes (Alox15, Alox15b, Alox12, Alox12b, Aloxe3, Alox5) but in addition, a functional Aloxe12 gene was found. This gene, which is expressed mainly in the epidermis (the e in the name indicates the major expression site), is present in humans as corrupted pseudogen, which does not encode for a functional enzyme. As in humans, except for the Alox5 gene all other LOX genes including the Aloxe12 gene are localized in a joint LOX cluster in a syntenic region on chromosome 11. The genes encoding for Alox15 [11], Alox12 [12], Alox5 [13], Alox12b [14] and Aloxe3 [15] have been targeted by functional inactivation but Alox15, Alox12 and

Alox5 knockout mice do not show major phenotypic alterations unless challenged otherwise. In contrast, Alox12b and Aloxe3 knockout mice show a dramatic epidermal phenotype. Aloxe3 knockout mice have a normal embryogenesis but later on develop typical symptoms of a human skin disease called congenital ichthyosis [16]. Alox12b knockout mice survive embryogenesis without major defects but die after birth because of excessive dehydration [14].

Mouse Alox15 shares a high degree of amino acid sequence similarity with the human ortholog and does also exhibit dual reaction specificity with arachidonic acid. However, for the murine enzyme the dominant arachidonic acid oxygenation product is 12S-HpETE (85 %) and 15S-HETE was only formed in smaller amounts (15 %). The molecular basis for the different reaction specificities of the two enzymes has been explored in detail and a single amino acid exchange (Leu353Phe) switched the dominant 12-lipoxygenase activity of the enzyme to 15-lipoxygenation [17].

2.3 Zebrafish LOX Genes and LOX-Isoforms

The zebrafish is frequently employed as model organism for vertebrate development. Because of its extracorporal embryogenesis and its transperancy embryo development can optically be followed. Moreover, targeted gene expression silencing can easily be induced applying the morpholino oligonucleotide method. Completion of the zebrafish genome indicated that for more than 75 % of the genes found in the zebrafish there is a human ortholog [18]. The zebrafish genome involves various LOX-like genes, which are localized on different chromosomes (Table 2). Amino acid sequence alignments indicated that only the zbfLOX2 gene exhibits a high degree of sequence similarity (>75 %) with human ALOX5. All other zbfALOX genes only show low degrees of sequence similarity (<35 %) with any of the human LOX-isoforms. Thus, on the basis of amino acid sequence comparisons it is impossible to conclude orthology relations between human and zebrafish ALOX genes. This is particularly the case for human ALOX15 (humALOX15). Even if one includes other genomic parameters such as exon–intron organization, promoter structures and chromosomal localization, it is impossible to assign any of the zbfALOX gene to the human ALOX15 gene. Thus, for the time being it remains unclear whether there is an ALOX15 ortholog in zebrafish. This is clearly not the case for humALOX5 since the high degree of amino acid sequence similarity between humALOX5 and zbfALOX2 strongly suggest an orthology relation between the two enzymes.

Unfortunately, most of the zebrafish ALOX isoforms have not been characterized. The zbfALOX1 was expressed as recombinant protein and some of its protein-chemical and enzymatic properties have been determined [20]. This enzyme oxygenates arachidonic acid almost exclusively to 12S-HpETE and does not follow the triade concept, which explains the reaction specificity of ALOX15 orthologs. In fact, mutations of the triade determinants, which induce alterations in the reaction

Table 2 ALOX genes in the zebrafish genome

Gene	Accession number	Chr.	Homology with mammalian ALOX orthologs
zbfLOX1	**NP_955912.1**	**7**	**Highest homology with hum- (17.9 %) und mouALOX5 (17.3 %); [19, 20]**
zbfLOX2	**NP_001290191.1**	**13**	**Highest homology with humALOX5 (74.5 %) and mouALOX5 (73.9 %); [21]**
zbfALOX3	XP_009289657.1	15	Highest homology with humALOX5 (19.3 %) und mouALOX5 (19.1 %)
zbfALOX4	NP_001038796.1	15	Highest homology with humALOX5 (20.2 %) and mouALOX5 (20.2 %)
zbfALOX5	NP_001018414.1	15	Highest homology with hum- und mouALOX5 (20 %)
zbfLOX6	XP_009293707.1	21	Highest homology with mouAloxe12 (8.6 %), humALOX15 (8 %), humALOX15B (7.5 %), mouAlox15b (7.5 %)
zbfALOX7	NP_001070814	21	Highest homology with humALOX12 (26.5 %), humALOX15 (23 %), mouAloxe12 (21.8 %)
zbfLOX8	XP_009293408.1	21	Highest homology with humALOX12 (6.6 %), humALOX15B (6.5 %)
zbfALOX9	XP_001920640.3	8	Highest homology with humALOX15 (7.3 %)

Characterized enzymes are indicated in bold

specificity of ALOX12, ALOX15 and ALOX5 orthologs of various species, did not alter the reaction specificity of this enzyme [20]. Moreover, this enzyme carries a Gly at a critical position of its primary structure and thus, should function as 12R-LOX. However, detailed product analysis indicated dominant formation of 12S-HpETE [19, 20]. Expression silencing of zbfALOX1 leads to a defective phenotype (abnormal development of brain, eyes, and tail; pericardial and yolk sac edema) and the authors concluded that this enzyme might be considered the ortholog of mouse Alox12b [19]. However, this conclusion is not well supported by structural (low degree of amino acid conservation between the two enzymes) and functional data (arachidonic acid 12S-lipoxygenation by zbfALOX1, but 12R-lipoxygenation by mouAlox12b). Moreover, in mice expression silencing of Alox12b induces a defective epidermal phenotype with little impact on brain and eye development.

On the basis of amino acid sequence comparison zbfALOX2 is likely to be the zebrafish ortholog of human ALOX5. The corresponding enzyme was recently expressed and characterized in our lab [21]. The enzyme oxygenated arachidonic acid almost exclusively to 5S-HpETE. As for humALOX5 and mouAlox5 multiple mutagenesis of the triade determinants of zbfALOX2 inverted the reaction specificity from 5S- to 15S-lipoxygenation. For humALOX5 phosphorylation mimicking mutants have been suggested to convert the reaction specificity of arachidonic acid oxygenation from 5S- to 15S-lipoxygenation [22]. However, other groups could not confirm these data [23]. Applying a similar strategy for zbfALOX2 we found that phosphorylation-mimicking mutants at two different positions do not alter the reaction specificity of this enzyme [23].

2.4 Bacterial LOX Genes and LOX-Isoforms

LOX-isoforms occur in bacteria [24] but they are not widely distributed [5]. A rough estimate of the occurrence frequency of LOX genes in bacteria suggested that among the bacterial genomes sequenced to date, <0.5 % carry potential LOX genes. This number may even be an overestimate since the majority of hits were labeled LOX-like sequences in the databases. However, a closer look for the occurrence of functionally essential amino acid residues suggested a partial lack of iron ligands. From these data one may conclude that the majority of the bacterial LOX-like sequences may not encode for true LOX enzymes. Bacterial species containing LOX-like sequences include firmicutes, different types of proteobacteria, cyanobacteria, actinobacteria and representatives of the CFB group [5]. Although the functionality of bacterial LOXs has not been characterized for any bacterial species, the observation that <0.5 % of the bacterial genomes contain potential LOX genes, suggest that these enzymes only sporadically occur in bacteria. In fact, most human pathogenic bacteria including *Escherichia coli* (bacterial model organism) do not carry LOX genes [5].

The most comprehensively characterized bacterial LOX-isoform is that from *Pseudomonas aeruginosa* [25–27]. This enzyme (psaALOX1) oxygenates arachidonic acid to 15S-HpETE and thus must be classified as arachidonic acid 15-LOX. However, since the degree of amino acid conservation with mammalian ALOX15 orthologs is rather low, psaALOX1 may not be considered the *P. aeruginosa* ortholog of humALOX15. *P. aeruginosa*, a facultative human pathogen, expresses psaALOX1 as secreted protein (the native protein carries a secretion signal peptide) but its biological role has not been studied in detail. The enzyme can be expressed as recombinant non-secreted protein at high levels (more than 5 mg purified protein per 50 ml liquid culture) in *E. coli* and its protein-chemical and enzymatic properties have been characterized [26, 28]. It has been crystallized and the X-ray structure was solved to a resolution of 1.7–2.1 Å [29]. Its structure differs from that of mammalian LOX-isoforms, since the single polypeptide chain does not fold into the classical two-domain structure. Moreover, the substrate-binding pocket is much bigger than that of all mammalian LOX-isoforms and in the X-ray structure a phospholipid molecule is bound at the active site in close proximity to the non-heme iron. These structural data suggested that this enzyme might be capable of oxygenating phospholipids and we have recently confirmed this catalytic activity by analyzing the specific oxygenation products. However, when compared with free fatty acid the phospholipid oxygenase activity of psaLOX1 was <1 %. Thus, although the enzyme binds phospholipids at the active site and although it oxygenates different phospholipid classes (phosphatidylcholine, phosphatidylethanolamine, phosphatidylserine, phosphatidylinositol) to specific oxygenation products, the phospholipid oxygenase activity of psaLOX1 is rather inefficient. However, if the enzyme is expressed and secreted in vivo in large amounts, it may well oxygenate biomembranes and thus, may contribute to the pathogenicity of this bacterium. In fact, long-termm (24 h) in vitro incubation of the purified enzyme with human red blood cells leads to hemolysis [30].

3 Enzymology of ALOX15 Orthologs of Different Mammalian Species

In 1975 the first mammalian ALOX15 ortholog was discovered in rabbit reticulocytes [7]. Later on a similar enzyme was described in porcine polymorphonuclear leukocytes and according to the specificity-based nomenclature this enzyme was called porcine leukocyte-type 12-LOX [31]. Because of the different reaction specificity of the two enzymes an orthologous relation between them has not been suggested. Meanwhile ALOX15 orthologs from a number of other mammalian species including mouse, rat, cattle, different apes and men have been characterized and key properties of these enzymes are briefly summarized below.

3.1 Rabbit ALOX15

When lysates of immature rabbit red blood cells were incubated in vitro with isolated rat liver mitochondria, the structure of these organelles was disrupted [32]. These structural changes were accompanied by inactivation of the respiratory chain and by a loss of matrix enzymes (malate dehydrogenase). Unfortunately, the mechanistic basis for these structural and functional alterations could not be identified at that time, but proteases and phospholipases have been suggested as mitochondria lysis factor(s) in the reticulocyte cytosol. In the early 1970s LOXs have been believed to be restricted to plants and thus, these enzymes were not an option to explain the observations. In late 1974 a true LOX was described in human blood platelets [6, 33] and this discovery prompted Schewe and Rapoport to test whether a LOX-isoform may be expressed in rabbit reticulocytes. To do so they partially purified the mitochondrial lysis factor by ammonium sulfate precipitation and incubated an aliquot of the precipitate with intact rat liver mitochondria. After different time intervals, malonyldialdehyde (MDA) was measured as readout parameter to quantify lipid peroxidation. In the presence of native enzyme preparations large amounts of MDA were detected but minimal MDA amounts were found when a heat-inactivated enzyme preparation was used. These data suggested that rabbit reticulocytes express a LOX capable of oxidizing phospholipids of mitochondrial membranes [7]. The enzyme was next purified to electrophoretic homogeneity and characterized with respect to its protein-chemical and enzymatic properties [34]. The enzyme possesses a molecular weight of 75 kDa and a native isoelectric point of 5.5. It oxidizes different free polyenoic fatty acids [34, 35], phospholipids [7], biomembranes [36] and lipoproteins [37].

When the reaction kinetics of rabALOX15 with polyenoic fatty acids were explored, the enzyme exhibited two peculiarities. (1) During the early phases of fatty acid oxygenation (first 2 min) the enzyme shows auto-catalytic behavior as indicated by the kinetic lag-period [38]. During this lag-period the reaction rate was increased with time, suggesting that the products of fatty acid oxygenation may activate the enzyme. Addition of hydroperoxy fatty acids abolished the kinetic lag-phase. When isolated from native and/or recombinant sources, ALOX15 is present as catalytically silent ferrous enzyme. To initiate fatty acid oxygenation, the enzyme must first be oxidized to a ferric form, capable of initiating hydrogen abstraction. This enzyme oxidation occurs during the kinetic lag-phase. (2) At later stages of fatty acid oxygenation, rabALOX15 undergoes suicidal inactivation [39]. Unfortunately, the molecular basis for this enzyme inactivation still remains unclear. Initially, it has been suggested that hydroperoxy fatty acids may oxidize catalytically relevant amino acids at the active site. In fact, treatment of pure rabbit ALOX15 with 13S-HpODE induced selective oxidation of a methionine residue [40]. However, site-directed mutagenesis of this methionine to an oxidation resistant alanine did not reduce the degree of suicidal inactivation [41]. Covalent modification of rabALOX15 was reported when the enzyme was incubated with 15S-HpETE [42] and separation of proteolytic cleavage peptides by two-dimensional-gel electrophoresis confirmed this hypothesis [43]. Despite these descriptive experimental data the molecular basis for suicidal enzyme inactivation of ALOX-isoforms remains unclear.

Oxygenation of free fatty acids by rabALOX15 is strongly pH-dependent and in the absence of detergents a pH-optimum between pH 7.0 and 7.4 was determined. The temperature optimum for linoleic acid oxygenation ranged between 20 and 25 °C and an activation energy of about 12 kJ/mol was determined. Thermal stability assays [44] indicated that in the absence of substrate, rabbit ALOX15 is stable over long time intervals only at temperatures <10 °C. At temperatures >20 °C it undergoes structural fluctuations and loses catalytic activity [44]. In refolding experiments the structural alterations induced by short-time exposure to 30 °C were completely reversible, but further temperature elevations caused irreversible changes [44].

In the presence of mM calcium concentrations rabALOX15 binds to biomembranes [45, 46] and membrane binding strongly augments the specific fatty acid oxygenase activity without impacting the reaction specificity [45, 47]. The enzyme does not contain a high affinity calcium binding site but site directed mutagenesis indicated that surface exposed hydrophobic amino acids of the N-terminal ß-barrel (Tyr15, Phe70, Leu71) and the catalytic domain (Trp181, Leu195) are involved in membrane binding [46].

As all ALOX15 orthologs characterized so far, rabALOX15 exhibits dual positional specificity for arachidonic acid oxygenation. It converts arachidonic acid to a mixture of 12S- and 15S-HpETE in a ratio of about 1:10 [48]. When the dual specificity was first described, there was a debate about whether a single enzyme might exhibit such dual reaction specificity or whether the 12-HpETE formation may be related to contaminations of the enzyme preparation with blood platelet ALOX12. Co-purification of 12- and 15-lipoxygenase activity during the entire purification procedure of the enzyme, which involved ammonium sulfate precipitation, anion exchange chromatography and isoelectric focusing, strongly suggested that the two catalytic activities reside on a single enzyme molecule [49].

3.2 Porcine ALOX15

In 1986 a LOX was purified to near homogeneity from the cytosolic fraction of porcine leukocytes by consecutive ammonium sulfate fractionation, ion exchange chromatography and immunoaffinity chromatography [31]. As rabALOX15 this enzyme showed a kinetic lag-phase, which was abolished by the addition of 12-HpETE. Linoleic acid and gamma-linolenic acid were also oxygenated. When 15-HpETE was used as substrate, a complex mixture of arachidonic acid double oxygenation products involving 8S,15S-DiH(p)ETE and 14R,15S-DiH(p)ETE were produced. Moreover, the formation of 14,15-leukotriene A4 was inferred from the characteristic pattern of its hydrolysis products. Thus, the enzyme exhibited both, 12-lipoxygenase and 14,15-leukotriene A4 synthase activity. As rabALOX15 the porcine enzyme contained 1 mol iron per mol enzyme and iron chelators suppressed the catalytic activity [50]. The enzyme is expressed in high amounts in porcine mixed peripheral leukocytes and in the parenchymal cells of the anterior pituitary of pigs [51]. More detailed investigations on the cellular distribution of this enzyme indicated its presence in neutrophils and monocytes but it was not detected in other peripheral blood cells such as lymphocytes, platelets and erythrocytes. In several other organs such as the GI-tract (ileum, jejunum), lymphatic tissues (spleen, lymph node, thymus), ovary, lung and liver the enzyme was detected in resident mast cells and in infiltrating monocytes but there was no evidence for major expression of the enzyme in the parenchymal cells of these organs [52]. The porcine ALOX15 (pigALOX15) cDNA was cloned in 1990 [53] and the amino acid sequence revealed a high degree (86 %) of sequence conservation with rabALOX15. Two years later the gene encoding for pigALOX15 was cloned [54]. It spans approximately 8 kb and consists of 14 exons and 13 introns. The putative promoter contains nine GC-boxes as potential Sp1-binding sites and two AP-2 binding sequences. As for rabALOX15, there were neither typical TATA nor CCAAT boxes and Southern blot analysis suggested a single copy gene in the haploid genome. Overall sequence comparisons revealed striking similarities in genomic organization and promoter sequences between the pigALOX15 and rabALOX15 and these data suggested for the first time that the two enzymes might be evolutionarily related despite their different reaction specificity. More detailed comparison of catalytic activities of the two enzymes provided further evidence for such functional relatedness. (1) Dual reaction specificity: As rabALOX15 pigALOX15 exhibits a dual reaction specificity since it converts arachidonic acid to 12S- and 15S-HpETE in a ratio of about 10:1 [55, 56]. For rabALOX15 an inverse product pattern has been identified [48]. When Val418 and Val419 of pigALOX15 were mutated to Ile and Met (amino acids present at these positions in rabALOX15) 15-H(p)ETE was identified as major oxygenation product. An inverse alteration of the reaction specificity (from 15S- to 12S-lipoxygenation) was observed when Ile418 of rabALOX15 was mutated to smaller amino acid residues [57]. These data indicate that although the reaction specificities of rabALOX15 and pigALOX15 are different, both enzymes follow the triade concept, which explains the reaction specificity of ALOX15 orthologs. (2) Suicidal inactivation: As rabALOX15, but in contrast to humALOX12, pigALOX15 undergoes suicidal inactivation during fatty acid oxygenation. Although the detailed mechanism of suicidal inactivation still remains unclear it has been suggested that reactive intermediates might covalently bind to the protein inducing catalytic silence [58]. (3) Lipoxin synthase activity: As rabALOX15 the porcine ortholog (pigALOX15) exhibits lipoxin synthase activity [59]. When incubated with 5,15-DiHETE 5S,14R,15S-TriHETE (lipoxin B) was identified as major oxygenation product and ^{18}O experiments indicated that the additional hydroxy group at C14 originated from atmospheric oxygen. A similar specific product pattern is formed when pure rabALOX15 was incubated with 5S,15S-DiHETE and 15S-HETE methyl ester [60]. (4) Membrane oxygenase activity: As rabALOX15 pigALOX15 is capable of oxygenating complex substrates such as phospholipids and biomembranes [61]. When pigALOX15 was incubated with 1-palmitoyl-2-arachidonoyl-sn-glycero-3-phosphocholine, the 12-hydroperoxy derivative was specifically formed. Oxygenation of rat liver mitochondrial and endoplasmic membranes led to the formation of esterified 12S-H(p)ETE and 13S-H(p)ODE. Interestingly, human ALOX12 was almost inactive with esterified polyenoic fatty acids [61]. (5) Lipoprotein oxygenase activity: As rabALOX15 pigALOX15 is capable of specifically oxidizing phospholipids and cholesterol esters present in low density lipoproteins [62]. In comparative experiments humALOX12 and humALOX5 were much less effective. As major oxygenation products esterified 13S-H(p)ODE and 15S-H(p)ETE were identified. In conclusion, these data suggest that pigALOX15 and rabALOX15 are functional equivalents despite their different reaction specificity of arachidonic acid oxygenation.

3.3 Human ALOX15

The human ortholog of rabALOX15 (humALOX15) was first described in 1988. The enzyme was purified to homogeneity from human eosinophil-enriched leukocytes using a complex purification strategy involving ammonium sulfate precipitation, hydrophobic interaction chromatography and high pressure liquid chromatography on hydroxyapatite and cation-exchange columns [63]. Comparison of the N-terminal amino acid sequence revealed a 71 % sequence identity with rabALOX15. To explore the primary structure of this enzyme its full-length cDNA was cloned from a human reticulocyte cDNA library and the deduced amino acid sequence indicated a high degree (>85 %) of sequence similarity with rabALOX15 [64]. When compared with human ALOX5, the overall degree of sequence similarity was <40 % but there were two distinct regions of higher sequence identity, which included four of the five potential proteinogenic iron ligands. Later on the humALOX15 gene including the promoter region was cloned [65, 66] and alternative transcripts were identified [66]. The gene was mapped to a joint LOX gene cluster on chromosome 17p13.3, which also contains the genes of other human

LOX-isoforms except ALOX5. This gene cluster is located in close proximity to the tumor-suppressor gene p53 and gene expression regulation studies suggested that there might be a functional link between p53 and ALOX15 expression [67]. Because of this connection humALOX15 has been implicated in the development of prostate adenocarcinoma [68].

Human ALOX15 is constitutively expressed at high levels in immature red blood cells, in eosinophils and in airway epithelial cells [69]. Lower expression levels have been reported for polymorphonuclear leukocytes [70, 71], alveolar macrophages [72], vascular cells [73], uterus [74], various parts of the brain [75, 76] and for atherosclerotic lesions [77]. Human peripheral blood monocytes do not express ALOX15. However, the Th2-cytokines interleukin-4 and interleukin-13 (IL4, IL13) [78, 79] strongly upregulate humALOX15 expression in these cells and microarray experiments indicated that the *ALOX15* gene is the most strongly upregulated gene product of the IL4 response in human peripheral monocytes [80]. Although the mechanism of IL4 induced upregulation of *ALOX15* expression is not completely understood, several constituents of the intracellular signal transduction cascade have been identified. Competition assays with an IL4 receptor antagonist suggested involvement of the IL-4/13 cell surface receptor [81]. Moreover, phosphorylation and acetylation of the transcription factor STAT6 by histone acetyltransferase CREB-binding protein/p300 has been implicated [82]. The *ALOX15* promoter involves putative STAT6 binding sites [83] and serial promoter deletion studies suggested their functionality. Additional regulatory events include phosphorylation of Jak2 and Tyk2, p38 MAPK induced phosphorylation of STAT1 and STAT3 and activation of PKC [84–86]. More recently, ERK1/2 protein kinase as well as the transcription factors Elk1, Egr-1 and CREB have been implicated in the IL13 induced signaling cascade [87]. IL4 does not induce *ALOX15* expression in all peripheral monocytes since 10–40 % of cells remain ALOX15 negative [88]. The reasons for this heterogeneity are unclear, but may be related to the maturation stage of the cells and/or their metabolic states [89]. IL4 does also induce upregulation of *ALOX15* expression in A549 airway epithelial cells [81] and in orbital fibroblasts [90]. In A549 cells, the Ku antigen, which is induced in response to IL4/13 stimulation, binds to the *ALOX15* promoter and induces expression of *ALOX15* [91]. However, this is clearly not the only mechanism for IL4/13 induced transcriptional upregulation of *ALOX15*. In a recent study [92] a role of histone H3 methylation was suggested. Following IL4 stimulation, demethylation of H3 was observed and this reaction was catalyzed by the H3K27me2/3-specific demethylase UTX. In fact, siRNA induced expression silencing of UTX attenuated IL4-induced *ALOX15* expression. These data indicate that epigenetic processes are involved in IL4-induced expression regulation of the *ALOX15* gene.

Human ALOX15 was overexpressed in various pro- and eukaryotic systems [93, 94] and the recombinant enzyme was characterized with respect to its catalytic properties. As the native enzyme, recombinant humALOX exhibits a dual positional specificity with arachidonic acid as substrate. For humALOX15 Ile418 and Met419 were identified as major specificity determinants and mutagenesis data indicated that introduction of amino acid residues with less bulky side chains at these positions alter the reaction specificity in favor of 12-lipoxygenation [95, 96]. Linoleic acid was also accepted as substrate and 13S-HpODE was identified as major linoleic acid oxygenation product [93]. In the baculovirus/insect cell system humALOX15 was expressed at very high levels (approximately 20 % of cellular protein) and the recombinant enzyme was purified to apparent electrophoretic homogeneity at a yield of 25–50 mg of pure enzyme per L of liquid culture [93]. The specific activity (molecular turnover rate at substrate saturation) of the final enzyme preparation varied between 8 and 25 s^{-1} (variations between 5 and 50 s^{-1} for rabALOX15) depending on the quality of enzyme preparation and a native isoelectric point of 5.9 was determined for the human enzyme. Here again, purified humALOX15 exhibited a dual positional specificity with arachidonic acid and double oxygenation products, such as 14R,15S-DiH(p)ETE and various 8S,15S-DiH(p)ETE isomers were also formed. With linoleic acid as substrate, a pH-optimum of 7.0 and a K_M of 3 µM were determined. The enzyme undergoes suicidal inactivation during fatty acid oxygenation, is sensitive to standard lipoxygenase inhibitors, and oxygenates phospholipids, cholesterol esters, biomembranes and human low-density lipoprotein [93].

Comparison of the nuclear genomes of extinct human subspecies [97, 98] with that of *H. sapiens* suggested that both, *H. neanderthalensis* and *H. denisovans* expressed functional ALOX15 isoforms. When we aligned on the amino acid level the humALOX15 sequences with those of *H. neanderthalensis* and *H denisovan* we only observed subtle structural differences [21, 99]. Neither of the detected amino acid exchanges was at positions impacting iron coordination or reaction specificity. Some of the observed mutations were localized at positions, for which genetic variability (SNPs or rare mutations) has been observed among different *H. sapiens* individuals [100]. Since only a few *H. neanderthalensis* fossils and a single *H. denisovan* individual have been sequenced, it remains unclear whether the observed amino acid exchanges are characteristic for all individuals of these extinct human subspecies or whether they are unique for the sequenced individuals.

3.4 ALOX15 Orthologs of Nonhuman Primates

ALOX15 genes have been identified in a number of non-human primates and the corresponding enzymes (macaca, baboon, orangutan, chimpanzee) have been expressed as recombinant proteins [101, 102]. Analyzing the sequence data of these ALOX15 orthologs it was predicted on the basis of the triade concept that ALOX15 orthologs of most mammals including lower non-human primates (macaca, baboon) catalyze dominant 12-lipoxygenation. In contrast, the orthologs of more developed non-human primates (orangutan, chimpanzee) and humans convert arachidonic acid mainly to 15-H(p)ETE [5]. When the ALOX15 orthologs of *M. mulatta* (rhesus monkey), *P. anubis* (baboon), *P. abelii* (orangutan) and *P. troglodytes* (chimpanzee) were expressed as recombinant enzymes, this prediction was confirmed [101, 102]. Gibbons, which are flanked in evolution by macacas (12-lipoxygenating ALOX15) and orangutans (15-lipoxygenating ALOX15), express an ALOX15 ortholog with pronounced dual specificity [21]. In fact, according to own unpublished data the recombinant gibbon ALOX15 converted arachidonic acid to almost equal amounts of 12S- and 15S-HpETE and thus may be considered a transition ALOX15 ortholog interconnecting major 12- and 15-lipoxygenating enzymes. These unpublished data suggest that there might be a targeted alteration of the positional specificity of ALOX15 orthologs during late primate evolution. The only exception from this rule is the rabbit, but in this mammalian species 12- and 15-lipoxygenating ALOX15 isoforms are expressed in a tissue specific manner [103]. If there is an evolutionary concept for targeted alteration of ALOX15 reaction specificity during late primate development it remains to be worked out, which might be the evolutionary driving forces for this concept. Is there anything the 15-lipoxygenating ALOX15 orthologs of higher primates can do better than the 12-lipoxygenating counterparts of lower primates? Work is in progress in our lab to answer this question.

3.5 Murine ALOX15 Orthologs (mouAlox15, ratAlox15)

The mouse genome (*Mus musculus*) involves seven functional LOX genes: Alox5, Alox15, Alox15b, Alox12, Alox12b, Aloxe3, Aloxe12 [104]. The gene and the cDNA for mouAlox15, which was formerly called leukocyte-type 12-LOX, was cloned from different tissues [105–107]. The mouAlox15 gene, which spans 7.5 kbp is divided into 14 exons and 13 introns. It was mapped to a central region of chromosome 17, which also involves all other ALOX genes except Alox5. Expression of the recombinant mouAlox15 in HEK [105] or COS [106] cells indicated dual positional specificity of arachidonic acid oxygenation with 12-HpETE as major oxygenation product. The catalytic activity of mouAlox15 with complex substrates (phospholipids, biomembranes, lipoproteins) has not been explored in detail. It may be predicted that these substrates are well accepted but currently no quantitative data are available to prove this prediction.

The tissue specific expression of mouAlox15 has not been well characterized. The major cellular sources of this enzyme are residential peritoneal macrophages. Interestingly, thioglycollate elicitation in vivo decreased the share of Alox15 positive cells to about 10 % [88]. Murine peripheral monocytes, alveolar macrophages and bone marrow derived macrophages express Alox15 only at low levels [88]. On the other hand, human peritoneal macrophages (prepared from human ascitis puncture fluid) do not express ALOX15 at high levels and these data suggest species-specific differences in the tissue-specific expression patterns of human and mouse ALOX15 orthologs.

The question, which of the mouse LOX genes might constitute the functional equivalent of humALOX15, has been a matter of discussion for several years.

Because of the functional differences between the two enzymes (mouAlox15 is a dominantly 12-lipoxygenating isoform whereas humALOX15 is dominantly 15-lipoxygenating) it has been suggested that mouse Alox15 may functionally be more closely related to human ALOX12. However, a functional relatedness of these two enzymes can only be suggested if ALOX15 orthologs exhibit their bioactivity via the formation of arachidonic acid oxygenation products [12S-H(p)ETE or secondary products derived from this metabolite]. In contrast, when ALOX15 orthologs exhibit their biological function by oxygenating complex lipid-protein assemblies (lipoproteins, biomembranes) there is hardly any functional similarity between mouse Alox15 and human ALOX12. Genomic sequence alignments, chromosomal localization and comparison of the enzyme properties strongly suggest that the mouse leukocyte-type 12-LOX (old nomenclature) and the human reticulocyte-type 12/15-LOX (old nomenclature) are orthologous enzymes. Usually, enzyme orthologs fulfill similar functions in different organisms and thus mouse Alox15 may constitute the functional equivalent of human ALOX15 despite their different reaction specificity of arachidonic acid oxygenation.

The Alox15 ortholog of rat has been identified as arachidonic acid 12-lipoxygenating enzyme [108, 109]. Its cDNA was cloned in 1993 [108]. The open reading frame indicated a protein of 662 amino acids with a theoretical molecular mass of 75 kDa. On the amino acid level, the enzyme displayed a high degree of sequence conservation with humALOX15 (75 % identity) and pigALOX15 (71 % identity). The recombinant enzyme expressed in *E. coli* exhibited major 12-lipoxygenase activity with 15-HpETE being a minor oxygenation product. When the Sloane determinants were mutated in ratALOX15 (Lys416Gln, Ala417Ile, Met418Val) there was no alteration in the reaction specificity of the enzyme. However, later mutagenesis studies indicated that as for mouAlox15 the Sloane-determinants do not play a major role for ratAlox15 specificity. For these two enzymes the amino acid present at position 353 appears to be much more important and corresponding mutagenesis studies confirmed this conclusion for mouse [17] and rat Alox15 [109]. Recently, rat Alox15 was high level expressed as recombinant his-tag fusion protein in *E. coli* and tested for its capability of oxygenating complex lipid substrates such as phospholipids, biomembranes and lipoproteins [109]. Multiple site directed mutagenesis studies suggested the applicability of the triad concept with particular importance of Leu353 and Ile593 as specificity determinants. Wildtype rat Alox15 and its 15-lipoxygenating Leu353Phe mutant were capable of oxygenating ester lipids of biomembranes and high-density lipoproteins. For the wildtype enzyme 13S-H(p)ODE and 12S-H(p)ETE were identified as major oxygenation products but for the Leu353Phe mutant 13S-H(p)ODE and 15S-H(p)ETE prevailed [109]. When normalized to similar arachidonic acid oxygenase activity, the 12-lipoxygenating wildtype enzyme exhibited a lower membrane oxygenase activity when compared with the 15-lipoxygenating Leu353Phe mutant.

Table 3 Degree of homology of mammalian ALOX15 orthologs (comparison with humALOX15)

Species	Amino acid identity (%)	Amino acid differences (number)
Homo sapiens (man)	100	0
Pan troglodytes (chimpanzee)	99	2
Gorilla gorilla (gorilla)	99	7
Pongo abelii (orangutan)	99	9
Nomascus leucogenys (gibbon)	97	21
Papio anubis (baboon)	95	35
Macaca mulatta (rhesus monkey)	95	36
Bos taurus (cattle)	87	88
Sus scrofa (pig)	86	90
Rattus norvegicus (rat)	75	168
Mus musculus (mouse)	74	174
Monodelphis domestica (opossum)	72	186

4 Structural Biology of ALOX15 Orthologs

The degree of amino acid sequence conservation between the ALOX15 orthologs of various mammalian species is rather high (>70 %). Mammalian species, which are ranked high in evolution share a higher degree of amino acid similarity than the orthologs ranked low and some examples are given in Table 3. Since the amino acid sequence is the most decisive parameter impacting the 3D-structure of a protein, the structures of all ALOX15 orthologs should be similar. So far, two crystal structures have been solved for mammalian ALOX15 orthologs (rabALOX15, pigALOX15).

4.1 Structural Biology of rabALOX15

4.1.1 Crystal Structure

Rabbit ALOX15 has already been crystallized in 1990 [110], but the crystals did not well defract in X-ray analysis. In 1997 the crystal structure of an enzyme inhibitor complex was solved [111]. Although the electron density maps satisfactorily mirrored the overall structure, data were not highly refined (2.4 Å). Some structural elements could not be specified and thus, were modeled in. More recent re-evaluation of the original X-ray coordinates suggested a mixture of two structurally distinct conformers: (1) A ligand-free conformer (conformer A), in which the central cavity of the enzyme harboring the non-heme iron was empty. (2) A ligand-bound conformer (conformer B), in which the central cavity of the enzyme was occupied by the exogenous inhibitor [112]. Monomeric rabbit ALOX15 has a cylindrical shape (height of 10 nm) with an elliptic ground square (longer diameter 6.1 nm, shorter one of 4.5 nm). The single polypeptide chain folds into a two-domain structure, which involves a small N-terminal β-barrel domain and a larger mostly helical catalytic subunit. The two domains are covalently linked by a flexible oligopeptide. The small N-terminal domain comprises 110 amino acids and is composed of 8 β-sheets. Gene technical deletion of the β-barrel domain impaired the membrane binding capacity of the recombinant enzyme [113] and more detailed mutagenesis studies suggested that surface-exposed hydrophobic amino acids in both domains are involved in membrane binding [46, 114]. N-terminal truncation did also reduce the catalytic efficiency and led to more rapid suicidal inactivation. Thus, the N-terminal β-barrel domain may also play a role for regulation of the catalytic activity [114, 115].

The C-terminal catalytic domain consists of 21 helices, which are interrupted by a small β-sheet sub-domain [111]. The center of the C-terminal domain involves two long helices, which carry four of the five protein iron ligands. The putative substrate-binding pocket is a boot-shaped cavity, which is accessible from the protein surface. The walls of this cavity are lined by 23 predominantly hydrophobic amino acids from six different helices ($\alpha 2$, $\alpha 7$, $\alpha 9$, $\alpha 10$, $\alpha 16$ and $\alpha 18$) and by the loop interconnecting helices $\alpha 9$ and $\alpha 10$. The side chains of Phe353, Ile418/Met419 and Ile593 (triade determinants) define the bottom of the cavity. According to the X-ray data [112] ligand binding at the active site induces two major structural alterations: (1) The surface helix $\alpha 2$ is dislocated. (2) Helix $\alpha 18$ retreats from the cavity enlarging the volume of the substrate-binding pocket.

4.1.2 Solution Structure

In aqueous solutions macromolecules are more flexible and this allows structural rearrangement in response to alterations of the external milieu (temperature, pH, protein concentration). A number of spectroscopic studies, such as small angle X-ray scattering [116, 117], dynamic fluorescence studies and fluorescence resonance energy transfer measurements [44, 118], as well as various computational methods [117, 119] have been applied to explore the motional flexibility of rabALOX15 in aqueous solutions. Summarizing these results one may conclude that three elements contribute to structural flexibility: (1) Ligand binding induces conformational alterations: Binding of an inhibitor (R7) at the active site induces conformational alterations and helices $\alpha 2$- and $\alpha 18$ are particularly affected [112]. In the inhibitor-bound form the $\alpha 18$ helix is displaced from the substrate binding pocket providing more space for ligand binding [120]. It should be stressed at this point that this "induced-fit mechanism" was not confirmed by the X-ray data of pigALOX15 [121]. (2) Interdomain movement: The two domains of rabbit ALOX15 are covalently interconnected by a flexible linker peptide, which does not fold into a stable secondary structure. Initial small angle X-ray scattering (SAXS) measurements on aqueous solutions of wildtype rabALOX15 suggested a high degree of interdomain movement [116]. Since SAXS data can be interpreted in different ways [122] and since no interdomain movement was observed for soybeanALOX1 [123] the initial SAXS measurements on rabALOX15 were repeated under variable experimental conditions [117]. At pH 6.8 and in the absence of salt, interdomain movement was largely suppressed and the low resolution SAXS structure did match the crystal structure of the ligand-free conformer. In contrast, at pH 8.0 and in the presence of salt (200 mM NaCl), the N-terminal domain appears to swing away from the catalytic domain resulting in significant expansion of the molecule. More recent molecular dynamics simulation [124] and site directed mutagenesis studies at the interdomain interface [125] were consistent with the principal possibility of interdomain motion. (3) Reversible dimerization: For a long time LOX-isoforms were believed to function as catalytically active monomers. However, recent SAXS experiments suggested a monomer–dimer equilibration of rabALOX15 in aqueous solutions. At low protein concentrations (<1 mg/ml), at low ionic strength and at low pH (6.8) ALOX15 is mainly present as hydrated monomer. At higher protein concentrations (>1 mg/ml), in the presence of salt (200 mM NaCl) and at higher pH (8.0) the monomer–dimer equilibrium was shifted toward ALOX15 dimers [117]. Moreover, ligand binding at the active site appears to impact the monomer–dimer equilibrium. Ligand-free rabALOX15 is present predominantly (85 %) as protein monomer but addition of an active site ligand (13-HODE) strongly shifted the equilibrium towards dimers (95 %).

4.2 Structural Biology of pigALOX15

A truncated version of pigALOX15 (lack of the N-terminal ß-barrel domain) was crystallized as enzyme-inhibitor [4-(2-oxapentadeca-4-yne)phenylpropanoic acid] complex [121]. The crystals diffract at a resolution of 1.9 Å and the electron density map indicated that the inhibitor occupied a U-shaped channel, which is open at one end to the surface of the protein and extends past the redox-active iron. In models, the channel accommodated arachidonic acid, defining the potential substrate binding site. There was a void volume adjacent to the inhibitor binding site connecting to the surface of the enzyme. This cavity was discussed as putative access channel for oxygen. The structure of pigALOX15 was different from the structures of rabALOX15 in an important aspect. For rabALOX15 it was suggested that inhibitor binding at the active site induced significant conformational alteration (see section "Solution Structure") resembling an "induced fit mechanism" [112]. The structure of the pigALOX15-inhibitor complex indicated that no significant conformational change was required in order to bind the inhibitor at the active site. In fact, according to the X-ray data there is enough space available in the substrate binding pocket to accommodate the inhibitor without substantial conformational alterations.

5 Catalytic Multiplicity of ALOX15 Orthologs (Moonlighting Character)

Various LOX-isoforms including rabALOX15 [126] exhibit multiple catalytic activities. They oxygenate polyenoic fatty acids to hydroperoxy derivatives but also exhibit a lipohydroperoxidase activity (sometimes also called hydroperoxide isomerase activity) [39, 126, 127] which converts lipid hydroperoxides to secondary lipid peroxidation products. The oxygenase activity involves hydrogen abstraction from a bisallylic methylene of the fatty acid substrate as rate limiting step (Fig. 1). In contrast, the lipohydroperoxidase activity is initiated by a homolytic cleavage of the hydroperoxy bond, which formally leads to the formation of alkoxy and hydroxy radicals (Fig. 2). In addition, rabALOX15 [127] and pigALOX15 [128] exhibit leukotriene synthase activity (Fig. 3), converting hydroperoxy fatty acids containing a conjugated diene system to epoxy eicosanoids carrying conjugated trienes.

5.1 Lipoxygenase Activity

5.1.1 Monooxygenation of Free Fatty Acids

As other LOX-isoforms, ALOX15 orthologs of different species prefer free polyenoic fatty acids as substrates. However, these enzymes also oxygenate phospholipids, cholesterol esters and even complex lipid protein assemblies, such as biomembranes and lipoproteins. With all substrates rabALOX15 shows non-linear reaction kinetics [38]. A typical kinetic progress curve of fatty acid oxygenation

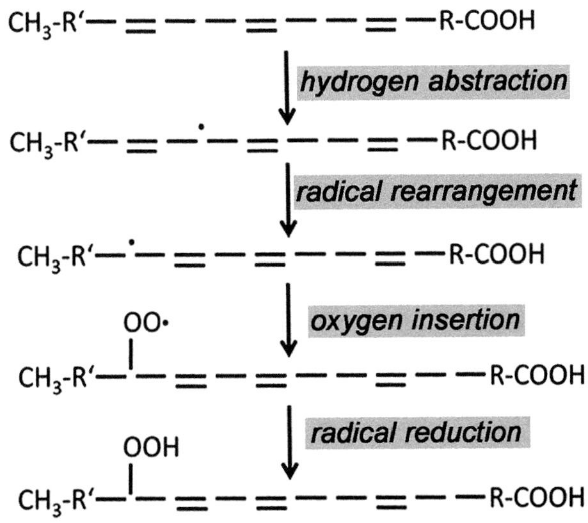

Fig. 1 Mechanistic scheme of the lipoxygenase reaction. Lipoxygenation consists of four elementary reactions (hydrogen abstraction, radical rearrangement, dioxygen insertion, peroxy radical reduction) and involves a valency shuttling between ferrous and ferric LOX (not shown). To initiate the reaction, the ferrous LOX is first activated by peroxide-dependent oxidation to a ferric form

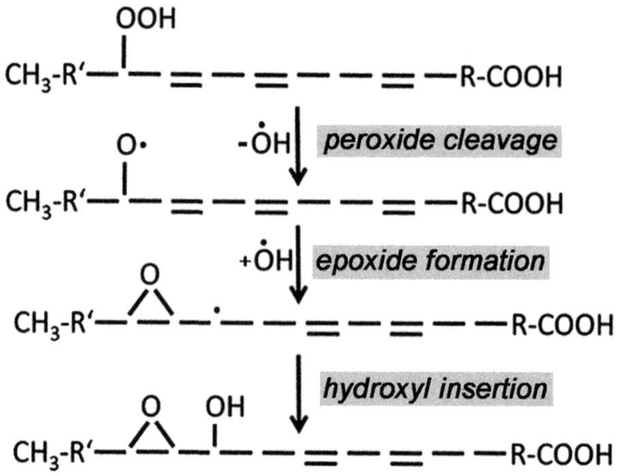

Fig. 2 Mechanistic scheme of the lipohydroperoxidase reaction. The lipohydroperoxidase reaction, which leads to the formation of epoxy hydroxy compounds, formally consists of three elementary reactions (peroxide cleavage, epoxide formation, hydroxyl insertion). In the presence of hydroperoxy fatty acids and non-oxygenated fatty acids, this reaction competes with the lipoxygenase reaction. For soybean ALOX1 it has been detected under anaerobic conditions but for rabALOX15 it does also occur in the presence of oxygen. However, under anaerobic conditions it becomes dominating also for rabALOX15. The reaction leading to the formation of keto dienes, fatty acid dimers, short chain aldehydes and alkanes follows different mechanisms after initial cleavage of the hydroperoxide bond

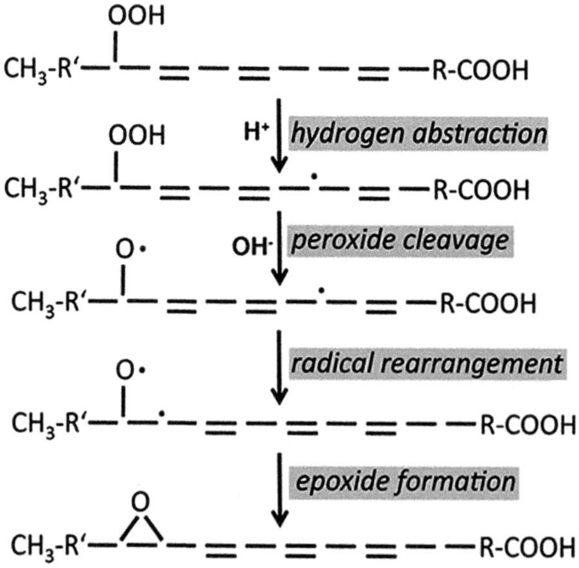

Fig. 3 Mechanistic scheme of the leukotriene synthase reaction. Here the four catalytic elementary reactions are hydrogen abstraction, peroxide bond cleavage, radical rearrangement, epoxide formation

exhibits sigmoid shape. It starts with a kinetic lag-phase, which is followed by a linear progression phase. At later stages the activity is decreasing owing to suicidal inactivation. The kinetic lag-phase can be abolished by the addition of 13-HpODE or 15-HpETE. Although the molecular mechanism for hydroperoxide dependent activation of ALOX15 orthologs is still unclear, an oxidation of the catalytically silent ferrous ALOX15 to an active ferric form has been suggested. However, the chemistry of peroxide dependent LOX-activation appears to be more complex than simple enzyme oxidation, since the duration of the kinetic lag-period depends on the presence of molecular dioxygen. The kinetic progress curve for rabALOX15 catalyzed oxygenation of (19R/S,5Z,8Z,11Z,14Z)-19-hydroxyeicosa-5,8,11,14-tetraenoic acid under normoxic conditions was characterized by an extensive (more than 30 min) kinetic lag phase [129]. However, under hyperoxic conditions a much shorter lag-phase was observed. Thus, molecular dioxygen serves not only as a lipoxygenase substrate, but also impacts peroxide-dependent enzyme activation [129]. A similar oxygen dependence of the LOX activation was recently reported for ALOXE3 [130].

Rabbit ALOX15 undergoes suicidal inactivation during the oxygenation of polyenoic fatty acids [39]. Initially, it has been suggested that the hydroperoxy fatty acids formed may oxidize catalytically relevant amino acids at the active site. Indeed, treatment of pure rabALOX15 with 13S-HpODE induced selective oxidation of one methionine residue to the corresponding sulfoxide [40]. However, site-directed mutagenesis of this Met to an oxidation resistant Ala did not reduce the degree of suicidal inactivation [41]. As alternative explanation of suicidal inactivation, covalent modification of the enzyme by reactive intermediates was suggested [42]. However, despite these descriptive experimental data the molecular basis for suicidal enzyme inactivation remains unclear and several mechanisms may contribute.

Among the naturally occurring polyenoic fatty acids linoleic acid, alpha- and gamma-linolenic acid, arachidonic acid, eicosapentaenoic acid and docosahexaenoic acid are well accepted as substrates. Monoenoic fatty acids (oleic acid) and saturated fatty acids of comparable chain length (stearic acid, arachidic acid) are not oxygenated but function as weak competitive inhibitors. The molar turnover numbers of purified rabbit and human ALOX15 vary between 5 and 50 s^{-1} depending on the quality of enzyme preparations. When a 1:1 mixture of linoleic acid and arachidonic acid was used as substrate for recombinant rabALOX15 both fatty acids are oxygenated with comparable efficiency (Fig. 4). Here a 15-H(p)ETE/13H(p)ODE ratio of 42:58 is formed during a 5 min incubation period. In contrast, arachidonic acid is preferred by the 12-lipoxygenating Ile410Ala mutant (15-H(p)ETE/13H(p)ODE ratio of 79:21). According to MD simulations [120, 131] linoleic acid and arachidonic acid share a common overall orientation at the active site. However, arachidonic acid is bound closer to the active site helix α18 and has a limited degree of motional freedom. The tail of linoleic acid fluctuates more freely and adopts a number of energetically similar conformations at the active site [120, 131].

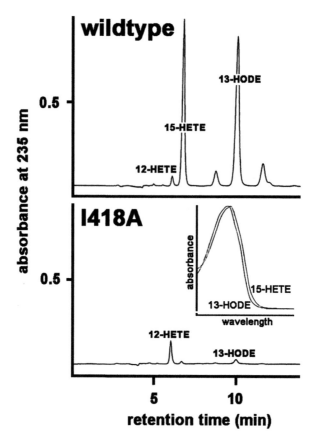

Fig. 4 Fatty acid oxygenation by wildtype rabALOX15 and its 12-lipoxygenating Ile418Ala mutant. Recombinant rabALOX15 and its 12-lipoxygenating Ile418Ala mutant were incubated for 2 min at room temperature with a 1:1 molar mixture of linoleic acid and arachidonic acid (160 μM each). The reaction was terminated by the addition of sodium borohydride to reduce the hydroperoxy fatty acids formed to the corresponding alcohols, an equal volume of methanol was added (1 ml) and protein precipitate was spun down. Conjugated dienes were purified by RP-HPLC and further analyzed by SP-HPLC with a solvent system consisting of hexane:2-propanol:acetic acid (100:2:0.1, by vol.) at a flow rate of 1 ml/min. The retention times of authentic standards are given above the chromatographic traces. *Inset*: UV-spectra recorded for 15-HETE and 13-HODE formed by the wildtype enzyme

5.1.2 Double and Triple Oxygenation of Fatty Acids (Lipoxin Synthase Activity)

When the primary products of LOX-catalyzed fatty acid oxygenation contain bisallylic methylenes, they constitute substrates for double and triple oxygenation. For instance, 15S-HETE and its methyl ester are converted by rabALOX15 to a complex mixture of double and triple oxygenation products involving a specific lipoxin B4 isomer [60, 132]. Polyunsaturated fatty acids appear to penetrate the active site of rabALOX15 with their methyl end ahead, but oxygenated fatty acids carrying a hydrophilic hydroxy group close to its methyl terminus may show an inverse head-to-tail substrate orientation [133, 134]. There may be a binding equilibrium between "normal" (methyl terminus ahead) and inverse substrate alignment (carboxylate ahead) and the relative shares of normally and inversely bound substrate depends on the functional groups on either end of the fatty acid backbone [132].

Lipoxins are anti-inflammatory mediators, which have been implicated in inflammatory resolution [135]. In principle there are at least two different biosynthetic mechanisms for lipoxin biosynthesis: (1) The triple oxygenation pathway converts arachidonic acid via three consecutive lipoxygenase reactions catalyzed by different ALOX-isoforms to different lipoxin isomers (5,6,15-TriHETE or 5,14,15-TriHETE). For instance, arachidonic acid can be double oxygenated to 14,15-DiH(p)ETE by ALOX15 and this substrate is subsequently oxygenated by ALOX5 to 5,14,15-TriH(p)ETE (lipoxin A4 isomer). Alternatively, ALOX5 first produces 5-H(p)ETE, which is subsequently double oxygenated by ALOX15 to 5,14,15-TriHETE. When pure rabALOX15 is incubated with 15S-HETE or 5S,15S-DiHETE methyl ester, large amounts of 5S,14R,15S-TriHETE (a specific lipoxin B4 isomer) are formed and experiments with atmospheric oxygen isotops indicated multiple oxygenation of the substrate as biosynthetic mechanism [60]. (2) The epoxy leukotriene pathway of lipoxin biosynthesis implicated the involvement of epoxy leukotrienes as reaction intermediates. According to this mechanistic scenario, the primary products of the ALOX5 or ALOX15 reaction (5S-HpETE, 15S-HpETE) are converted to 5,6- or 14,15-epoxy leukotrienes and hydrolysis of these epoxy compounds leads to the formation of vicinal diols (5,6-DiHETE or 14,15-DiHETE). 5,6-DiHETE can then be oxygenated by ALOX15 to 5,6,15-TriHETE (lipoxin A4 isomer) and this has been shown in our lab for recombinant human and rabbit ALOX15. 14,15-DiHETE can in principle be oxidized by ALOX5 to 5,14,15-TriHETE (lipoxin B4 isomer).

Pure pigALOX15 also exhibits lipoxin synthase activity. It converts 5,15-DiHpETE to a mixture of various conjugated tetraenes [59, 136]. The major tetraene was identified as 5S,14R,15S-trihydroperoxy-6E,8Z,10E,12E-eicosatetraenoic acid, which was reduced to 5S,14R,15S-TriHETE (lipoxin B4 isomer). Experiments with oxygen isotops indicated that the additional hydroperoxy group introduced during this reaction originated from atmospheric oxygen. There were several other tetraene products, which might have been synthesized via the epoxy leukotriene pathway involving hydrolysis of an unstable epoxy intermediate [59].

5.1.3 Oxygenation of Phospholipids, Cholesterol Ester, Biomembranes and Lipoproteins

Purified rabALOX15 oxygenates phospholipids [7] and cholesterol esters [137] containing polyunsaturated fatty acids. Although the reaction rates of phospholipid oxidation are lower than the rate of free polyenoic fatty acid oxygenation, specific reaction products have been detected. These data indicate, that as for free fatty acids, ALOX15 tightly controls the oxygenation reaction. On the other hand, oxygenation of phospholipids by ALOX15 is somewhat surprising, since *in silico* docking studies suggest that binding of a phospholipid molecule in the substrate binding pocket is hardly possible without significant rearrangement of the active site structure. This is apparently not the case for the secretable LOX of *Pseudomonas aeruginosa* (psaLOX1). The active site of this enzyme is uniquely shaped so that phospholipid binding is possible [29]. On the other hand, when we recently compared in side-by-side experiments the phospholipid oxygenase activity of psaLOX1 and rabALOX15 we found that when normalized to similar linoleic acid oxygenase activities, the phospholipid oxygenase activity of the rabbit enzyme is significantly higher than that of psaLOX1. However, rabALOX15 does not induce hemolysis of red blood cells when the purified enzyme was incubated with the cells in vitro. In contrast, psaLOX1 caused almost complete hemolysis under identical conditions [30]

The ALOX15 orthologs of rabbits [36], humans [93] and pigs [61] are capable of directly oxygenating complex lipid protein assemblies such as biomembranes and lipoproteins. Specific ALOX15 products have been detected in membranes of rabbit reticulocytes [138] indicating that membrane oxygenation is not an in vitro artifact but actually also proceeds in vivo during red cell maturation. Addition of rabALOX15 to purified rat liver mitochondrial membranes in vitro induced disruption of the organelle, inactivation of the respiratory chain and the release of matrix enzymes [32]. In other in vitro models of ALOX15-membrane interaction the enzyme integrated into the membranes of various organelles, allowing release of proteins from the organelle lumen and access of proteases to both, lumenal and integral membrane proteins [139]. As discussed for the phospholipid oxygenase activity of ALOX15 the structural basis of membrane oxygenation by rabALOX15 remains unclear. The enzyme is capable of binding to biomembranes [46] but the mechanisms by which the membrane phospholipids are oxygenated to specific products remain elusive. Intact membrane phospholipids do not fit into the substrate-binding pocket of rabALOX15 without major rearrangement of its 3D-structure and hydrolysis and reesterification processes are quite unlikely in the reconstituted in vitro systems.

Pure rabALOX15 is also capable of oxidizing the ester lipids of low-density lipoproteins to specific ALOX15 products [37]. The hydroperoxy lipids subsequently induce free-radical mediated secondary reactions, which render the product pattern more unspecific at longer incubation periods [140, 141]. The possible involvement of ALOX15 in the formation of oxidized low-density lipoprotein was the basis for the pro-atherogenic character of the enzyme and expression silencing studies in mice confirmed this activity in various mouse atherosclerosis models [142–144].

5.2 Lipohydroperoxidase Activity

Under certain conditions (anaerobiosis, hypoxia, limited fatty acid supply) LOXs are capable of decomposing hydroperoxylipids to an array of secondary lipid peroxidation products. This catalytic activity, which has extensively been studied for soyALOX1 [145], was called lipohydroperoxidase activity and the product mixture involved ketodienes, epoxy hydroxy compounds, short chain aldehydes, volatile hydrocarbons and mixed oxygenated and non-oxygenated fatty acid dimers [146, 147]. Mechanistic experiments indicated that this reaction does also proceed under aerobic conditions in the presence of guaiacol serving as electron donor [148]. As the lipoxygenase reaction, the reaction sequence of the lipohydroperoxidase activity involves a valency change of the non-heme iron. Although the catalytic efficiency of the lipohydroperoxidase of soyALOX1 is somewhat lower than the efficiency of the lipoxygenase activity, it constitutes a true catalytic activity, which was also described for other LOX-isoforms. For instance, lipoxygenase activity of human ALOXE3 [130, 149] is largely suppressed under normoxic conditions but the enzyme exhibits a strong aerobic lipohydroperoxidase (hydroperoxide isomerase) activity. In vitro, the recombinant enzyme is capable of decomposing lipohydroperoxides to secondary products such as epoxy hydroxy compounds (hepoxilins). This catalytic activity is of in vivo relevance since the hepoxilin synthase activity of ALOXE3 has been implicated in the formation of the water barrier of the skin [150].

A lipohydroperoxidase activity has also been described for rabALOX15 [39, 151]. As for humALOXE3 this catalytic activity was not restricted to anaerobic conditions, but also occurs when oxygen is present [152, 153]. Thus, lipoxygenase and lipohydroperoxidase activities are present in parallel under certain conditions. From these data it might be concluded that the ratio between lipoxygenase and lipohydroperoxidase activity of a given LOX-isoform depends on the enzyme properties, the oxygen pressure and the reaction conditions. Unfortunately, little is known about the lipohydroperoxidase activity of other ALOX15 orthologs and it remains unclear whether there are differences between the different ALOX15 orthologs. When we analyzed the product pattern of arachidonic acid oxygenation (long term incubations) by the recombinant ALOX15 orthologs of chimpanzee, orangutan and macaca under aerobic conditions, we detected the formation of small amounts conjugated keto dienes, which are typical lipohydroperoxidase products. Although the biosynthetic mechanism of these keto dienes has not been explored in detail they might originate from an aerobic hydroperoxidase activity of the ALOX15 orthologs.

5.3 Leukotriene Synthase Activity

ALOX5 [10], Alox12 [154] and ALOX15 orthologs [127, 128] are capable of converting hydroperoxy fatty acids such as 5- or 15-HpETE to epoxy leukotrienes. The mechanism of this catalytic activity involves both, stereoselective hydrogen abstraction from a bisallylic methylene (elementary reaction of the lipoxygenase activity) and homolytic cleavage of the hydroperoxy group (elementary reaction of the lipohydroperoxidase activity). Thus, leukotriene synthase activity may be considered a combination of the two catalytic activities of LOXs, which formally leads to the formation of an oxygenated fatty acid biradical. This biradical may then stabilize via epoxide formation (Fig. 3).

Pure rabALOX15 oxygenates arachidonic acid to a 10:1 mixture of 15S- and 12S-HpETE [48]. At higher temperatures additional products such as 13-hydroxy-14,15-epoxy-5,8,11-eicosatrienoic acid (lipohydroperoxidase product), 8,15-diHpETE isomers, 5,15-HpETE and diastereoisomers of 8,15-DiHETE were detected. The diHpETEs isomers are formed as lipoxygenation products since each incorporated 2 molecules of atmospheric $^{18}O_2$. The 8,15-diHETE isomers originated from the hydrolysis of the 14,15-epoxy leukotriene A_4. The conversion of 15-HpETE to 14,15-leukotriene A_4 was inhibited by lipoxygenase inhibitors indicating direct involvement of the enzyme in 14,15-leukotriene A_4 biosynthesis [127].

A similar leukotriene synthase activity was reported for the purified pigALOX15 [128, 155]. Here 15S-HpETE was converted to a mixture of 8S,15S-DiHpETE, 14R,15S-DiHpETE and a series of 8,15- and 14,15-diols. $^{18}O_2$ experiments indicated the biosynthetic mechanism for the formation of these compounds. Formation of 14,15-DiHpETE and 14,15-LTA$_4$ was associated with stereoselective abstraction of the pro-L hydrogen indicating the enzymatic origin [128].

6 Perspectives

ALOX15 was discovered in rabbit reticulocytes because of its capability of oxygenating membrane bound phospholipids [7] and later experiments indicated that rabALOX15- and pigALOX15 catalyzed membrane oxygenation leads to the formation of specific esterified oxygenation products, such as 15S-H(p)ETE, 12-H(p)ETE and 13S-H(p)ODE [36, 61, 137]. These data suggested a direct lipoxygenation of membrane bound phospholipids. However, recent molecular docking studies suggested that binding of a phospholipid molecule at the active site of rabALOX15 is sterically impossible without major rearrangements of the active site architecture. Although rabALOX15 exhibits a high degree of motional flexibility [117] it is hard to believe that such substancial structural rearrangement may take place to allow proper binding of phospholipids at the active site. A similar problem exists for the 15-lipoxygenating soyALOX1, which appears to be structurally more stable than the rabbit enzyme [123]. This enzyme is also capable of specifically oxidizing phospholipids in the presence of detergents [156]. Here again docking studies indicate steric clushes for phospholipid binding. In this system it remains unclear how the detergent might alter the 3D-structure of this enzyme so that phospholipid oxygenation becomes possible. Direct X-ray data on an ALOX15-phospholipid complex would shed light on the structural rearrangements that are required for ALOX15-catalyzed specific phospholipid oxygenation.

Among mammalian ALOX15 orthologs, membrane oxygenase activity has only been demonstrated for the 12-lipoxygenating ALOX15 orthologs of pigs and rats and for the 15-lipoxygenating orthologs of rabbits and men. For the other orthologs corresponding data are lacking, but it is highly probable that these orthologs may also exhibit a membrane oxygenase activity. However, an interesting question, which should be addressed in the future, is whether there is a quantitative difference between the membrane oxygenase activities of 12- and 15-lipoxygenating ALOX15 orthologs. We recently found that the 15-lipoxygenating rabALOX15 did not show any preference between linoleic acid and arachidonic acid when the two substrates are simultaneously provided at equal concentrations (Fig. 4). In contrast, for the 12-lipoxygenating Ile418Ala mutant of this enzyme a clear preference of arachidonic acid was observed. Since membranes of mammalian cells usually contain both, linoleic acid and arachidonic acid in variable concentrations the extent of membrane oxygenase activity may depend on the reaction specificity of the enzyme and the fatty acid composition of the membranes. For the time being, it is impossible to say whether 12- or 15-lipoxygenating ALOX15 orthologs have a higher membrane oxygenase activity since comparative side-by-side experiments using different ALOX15 orthologs and the same membrane preparation have not been carried out.

LOXs are oxygen-metabolizing enzymes but little is known about the fate of oxygen during the LOX-reaction. There is no evidence for direct oxygen binding at the ligand-free enzyme and thus, it was hypothesized that atmospheric oxygen is accepted by the fatty acid radical formed during initial hydrogen abstraction. However, it remains unclear how atmospheric oxygen reaches the fatty acid radical at the active site since the radical is shielded by the protein. In silico studies on rabALOX15 [157], experiments on the soyALOX1 [158, 159] and structural data on pigALOX15 [120] suggested the existance of oxygen diffusion channels within the enzymes but direct experimental confirmation of these data is still pending. To obtain such data, X-ray crystallographic studies under high oxygen or xenon concentrations (pumpimg studies)should be carried out [160, 161] to identify gas diffusion channels and areas with high oxygen affinity.

References

1. Haeggstrom JZ, Funk CD (2011) Lipoxygenase and leukotriene pathways: biochemistry, biology, and roles in disease. Chem Rev 111:5866–5898
2. Ivanov I, Heydeck D, Hofheinz K, Roffeis J, O'Donnell VB, Kuhn H, Walther M (2010) Molecular enzymology of lipoxygenases. Arch Biochem Biophys 503:161–174
3. Andreou A, Brodhun F, Feussner I (2009) Biosynthesis of oxylipins in non-mammals. Prog Lipid Res 48:148–170
4. Kuhn H, Banthiya S, van Leyen K (2015) Mammalian lipoxygenases and their biological relevance. Biochim Biophys Acta 1851:308–330
5. Horn T, Adel S, Schumann R, Sur S, Kakularam KR, Polamarasetty A, Redanna P, Kuhn H, Heydeck D (2014) Evolutionary aspects of lipoxygenases and genetic diversity of human leukotriene signaling. Prog Lipid Res 57C:13–39
6. Hamberg M, Samuelsson B (1974) Prostaglandin endoperoxides. Novel transformations of arachidonic acid in human platelets. Proc Natl Acad Sci USA 71:3400–3404
7. Schewe T, Halangk W, Hiebsch C, Rapoport SM (1975) A lipoxygenase in rabbit reticulocytes which attacks phospholipids and intact mitochondria. FEBS Lett 60:149–152
8. Radmark O, Werz O, Steinhilber D, Samuelsson B (2015) 5-lipoxygenase, a key enzyme for leukotriene biosynthesis in health and disease. Biochim Biophys Acta 1851:331–339

9. Kanaoka Y, Boyce JA (2014) Cysteinyl leukotrienes and their receptors; emerging concepts. Allergy Asthma Immunol Res 6:288–295
10. Samuelsson B, Dahlen SE, Lindgren JA, Rouzer CA, Serhan CN (1987) Leukotrienes and lipoxins: structures, biosynthesis, and biological effects. Science 237:1171–1176
11. Sun D, Funk CD (1996) Disruption of 12/15-lipoxygenase expression in peritoneal macrophages. Enhanced utilization of the 5-lipoxygenase pathway and diminished oxidation of low density lipoprotein. J Biol Chem 271:24055–24062
12. Johnson EN, Brass LF, Funk CD (1998) Increased platelet sensitivity to ADP in mice lacking platelet-type 12-lipoxygenase. Proc Natl Acad Sci USA 95:3100–3105
13. Chen XS, Sheller JR, Johnson EN, Funk CD (1994) Role of leukotrienes revealed by targeted disruption of the 5-lipoxygenase gene. Nature 372:179–182
14. Epp N, Fürstenberger G, Müller K, de Juanes S, Leitges M, Hausser I, Thieme F, Liebisch G, Schmitz G, Krieg P (2007) 12R-lipoxygenase deficiency disrupts epidermal barrier function. J Cell Biol 177:173–182
15. Krieg P, Rosenberger S, de Juanes S, Latzko S, Hou J, Dick A, Kloz U, van der Hoeven F, Hausser I, Esposito I et al (2013) Aloxe3 knockout mice reveal a function of epidermal lipoxygenase-3 as hepoxilin synthase and its pivotal role in barrier formation. J Invest Dermatol 133:172–180
16. Krieg P, Furstenberger G (2014) The role of lipoxygenases in epidermis. Biochim Biophys Acta 1841:390–400
17. Borngraber S, Kuban RJ, Anton M, Kuhn H (1996) Phenylalanine 353 is a primary determinant for the positional specificity of mammalian 15-lipoxygenases. J Mol Biol 264:1145–1153
18. Howe K, Clark MD, Torroja CF, Torrance J, Berthelot C, Muffato M, Collins JE, Humphray S, McLaren K, Matthews L et al (2013) The zebrafish reference genome sequence and its relationship to the human genome. Nature 496:498–503
19. Haas U, Raschperger E, Hamberg M, Samuelsson B, Tryggvason K, Haeggstrom JZ (2011) Targeted knock-down of a structurally atypical zebrafish 12S-lipoxygenase leads to severe impairment of embryonic development. Proc Natl Acad Sci USA 108:20479–20484
20. Jansen C, Hofheinz K, Vogel R, Roffeis J, Anton M, Reddanna P, Kuhn H, Walther M (2011) Stereocontrol of arachidonic acid oxygenation by vertebrate lipoxygenases: newly cloned zebrafish lipoxygenase 1 does not follow the Ala-versus-Gly concept. J Biol Chem 286:37804–37812
21. Adel S, Kakularam KR, Horn T, Reddanna P, Kuhn H, Heydeck D (2015) Leukotriene signaling in the extinct human subspecies Homo denisovan and Homo neanderthalensis. Structural and functional comparison with Homo sapiens. Arch Biochem Biophys 565:17–24
22. Gilbert NC, Rui Z, Neau DB, Waight MT, Bartlett SG, Boeglin WE, Brash AR, Newcomer ME (2012) Conversion of human 5-lipoxygenase to a 15-lipoxygenase by a point mutation to mimic phosphorylation at Serine-663. FASEB J 26:3222–3229
23. Adel S, Hofheinz K, Heydeck D, Kuhn H, Häfner AK (2014) Phosphorylation mimicking mutations of ALOX5 orthologs of different vertebrates do not alter reaction specificities of the enzymes. Biochim Biophys Acta 1841:1460–1466
24. Hansen J, Garreta A, Benincasa M, Fuste MC, Busquets M, Manresa A (2013) Bacterial lipoxygenases, a new subfamily of enzymes? A phylogenetic approach. Appl Microbiol Biotechnol 97:4737–4747
25. Busquets M, Carpena X, Fita I, Fusté C, Garreta A, Manresa Á (2011) Crystallization of the lipoxygenase of Pseudomonas aeruginosa 42A2, evolution and phylogenetic study of the subfamilies of the lipoxygenases. Transworld Research Network, Trivandrum, India
26. Lu X, Zhang J, Liu S, Zhang D, Xu Z, Wu J, Li J, Du G, Chen J (2013) Overproduction, purification, and characterization of extracellular lipoxygenase of Pseudomonas aeruginosa in Escherichia coli. Appl Microbiol Biotechnol 97:5793–5800

27. Vance RE, Hong S, Gronert K, Serhan CN, Mekalanos JJ (2004) The opportunistic pathogen Pseudomonas aeruginosa carries a secretable arachidonate 15-lipoxygenase. Proc Natl Acad Sci USA 101:2135–2139
28. Xu Z, Liu S, Lu X, Rao S, Kang Z, Li J, Wang M, Chen J (2014) Thermal inactivation of a recombinant lipoxygenase from Pseudomonas aeruginosa BBE in the absence and presence of additives. J Sci Food Agric 94:1753–1757
29. Garreta A, Val-Moraes SP, Garcia-Fernandez Q, Busquets M, Juan C, Oliver A, Ortiz A, Gaffney BJ, Fita I, Manresa A et al (2013) Structure and interaction with phospholipids of a prokaryotic lipoxygenase from Pseudomonas aeruginosa. FASEB J 27:4811–4821
30. Banthiya S, Pekarova M, Kuhn H, Heydeck D (2015) Secreted lipoxygenase from *Pseudomonas aeruginosa* exhibits biomembrane oxygenase activity and induces hemolysis in human red blood cells. Arch Biochem Biophys 584:116–124
31. Yokoyama C, Shinjo F, Yoshimoto T, Yamamoto S, Oates JA, Brash AR (1986) Arachidonate 12-lipoxygenase purified from porcine leukocytes by immunoaffinity chromatography and its reactivity with hydroperoxyeicosatetraenoic acids. J Biol Chem 261:16714–16721
32. Rapoport SM, Schewe T (1986) The maturational breakdown of mitochondria in reticulocytes. Biochim Biophys Acta 864:471–495
33. Nugteren DH (1975) Arachidonate lipoxygenase in blood platelets. Biochim Biophys Acta 380:299–307
34. Rapoport SM, Schewe T, Wiesner R, Halangk W, Ludwig P, Janicke-Hohne M, Tannert C, Hiebsch C, Klatt D (1979) The lipoxygenase of reticulocytes. Purification, characterization and biological dynamics of the lipoxygenase; its identity with the respiratory inhibitors of the reticulocyte. Eur J Biochem 96:545–561
35. Kuhn H, Sprecher H, Brash AR (1990) On singular or dual positional specificity of lipoxygenases. The number of chiral products varies with alignment of methylene groups at the active site of the enzyme. J Biol Chem 265:16300–16305
36. Kuhn H, Belkner J, Wiesner R, Brash AR (1990) Oxygenation of biological membranes by the pure reticulocyte lipoxygenase. J Biol Chem 265:18351–18361
37. Belkner J, Wiesner R, Rathman J, Barnett J, Sigal E, Kuhn H (1993) Oxygenation of lipoproteins by mammalian lipoxygenases. Eur J Biochem 213:251–261
38. Ludwig P, Holzhutter HG, Colosimo A, Silvestrini MC, Schewe T, Rapoport SM (1987) A kinetic model for lipoxygenases based on experimental data with the lipoxygenase of reticulocytes. Eur J Biochem 168:325–337
39. Hartel B, Ludwig P, Schewe T, Rapoport SM (1982) Self-inactivation by 13-hydroperoxylinoleic acid and lipohydroperoxidase activity of the reticulocyte lipoxygenase. Eur J Biochem 126:353–357
40. Rapoport S, Hartel B, Hausdorf G (1984) Methionine sulfoxide formation: the cause of self-inactivation of reticulocyte lipoxygenase. Eur J Biochem 139:573–576
41. Gan QF, Witkop GL, Sloane DL, Straub KM, Sigal E (1995) Identification of a specific methionine in mammalian 15-lipoxygenase which is oxygenated by the enzyme product 13-HPODE: dissociation of sulfoxide formation from self-inactivation. Biochemistry 34:7069–7079
42. Wiesner R, Suzuki H, Walther M, Yamamoto S, Kuhn H (2003) Suicidal inactivation of the rabbit 15-lipoxygenase by 15S-HpETE is paralleled by covalent modification of active site peptides. Free Radic Biol Med 34:304–315
43. Kuhn H, Saam J, Eibach S, Holzhutter HG, Ivanov I, Walther M (2005) Structural biology of mammalian lipoxygenases: enzymatic consequences of targeted alterations of the protein structure. Biochem Biophys Res Commun 338:93–101
44. Mei G, Di Venere A, Nicolai E, Angelucci CB, Ivanov I, Sabatucci A, Dainese E, Kuhn H, Maccarrone M (2008) Structural properties of plant and mammalian lipoxygenases. Temperature-dependent conformational alterations and membrane binding ability. Biochemistry 47:9234–9242

45. Brinckmann R, Schnurr K, Heydeck D, Rosenbach T, Kolde G, Kuhn H (1998) Membrane translocation of 15-lipoxygenase in hematopoietic cells is calcium-dependent and activates the oxygenase activity of the enzyme. Blood 91:64–74
46. Walther M, Wiesner R, Kuhn H (2004) Investigations into calcium-dependent membrane association of 15-lipoxygenase-1. Mechanistic roles of surface-exposed hydrophobic amino acids and calcium. J Biol Chem 279:3717–3725
47. Lankin VZ, Kuhn H, Hiebsch C, Schewe T, Rapoport SM, Tikhaze AK, Gordeeva NT (1985) On the nature of the stimulation of the lipoxygenase from rabbit reticulocytes by biological membranes. Biomed Biochim Acta 44:655–664
48. Bryant RW, Bailey JM, Schewe T, Rapoport SM (1982) Positional specificity of a reticulocyte lipoxygenase. Conversion of arachidonic acid to 15-S-hydroperoxy-eicosatetraenoic acid. J Biol Chem 257:6050–6055
49. Kuhn H, Wiesner R, Schewe T, Rapoport SM (1983) Reticulocyte lipoxygenase exhibits both n-6 and n-9 activities. FEBS Lett 153:353–356
50. Matsuda S, Suzuki H, Yoshimoto T, Yamamoto S, Miyatake A (1991) Analysis of non-heme iron in arachidonate 12-lipoxygenase of porcine leukocytes. Biochim Biophys Acta 1084:202–204
51. Ueda N, Hiroshima A, Natsui K, Shinjo F, Yoshimoto T, Yamamoto S, Ii K, Gerozissis K, Dray F (1990) Localization of arachidonate 12-lipoxygenase in parenchymal cells of porcine anterior pituitary. J Biol Chem 265:2311–2316
52. Maruyama T, Ueda N, Yoshimoto T, Yamamoto S, Komatsu N, Watanabe K (1989) Immunohistochemical study of arachidonate 12-lipoxygenase in porcine tissues. J Histochem Cytochem 37:1125–1131
53. Yoshimoto T, Suzuki H, Yamamoto S, Takai T, Yokoyama C, Tanabe T (1990) Cloning and sequence analysis of the cDNA for arachidonate 12-lipoxygenase of porcine leukocytes. Proc Natl Acad Sci USA 87:2142–2146
54. Arakawa T, Oshima T, Kishimoto K, Yoshimoto T, Yamamoto S (1992) Molecular structure and function of the porcine arachidonate 12-lipoxygenase gene. J Biol Chem 267:12188–12191
55. Reddy RG, Yoshimoto T, Yamamoto S, Marnett LJ (1994) Expression, purification, and characterization of porcine leukocyte 12-lipoxygenase produced in the methylotrophic yeast, Pichia pastoris. Biochem Biophys Res Commun 205:381–388
56. Suzuki H, Kishimoto K, Yoshimoto T, Yamamoto S, Kanai F, Ebina Y, Miyatake A, Tanabe T (1994) Site-directed mutagenesis studies on the iron-binding domain and the determinant for the substrate oxygenation site of porcine leukocyte arachidonate 12-lipoxygenase. Biochim Biophys Acta 1210:308–316
57. Borngraber S, Browner M, Gillmor S, Gerth C, Anton M, Fletterick R, Kuhn H (1999) Shape and specificity in mammalian 15-lipoxygenase active site. The functional interplay of sequence determinants for the reaction specificity. J Biol Chem 274:37345–37350
58. Kishimoto K, Nakamura M, Suzuki H, Yoshimoto T, Yamamoto S, Takao T, Shimonishi Y, Tanabe T (1996) Suicide inactivation of porcine leukocyte 12-lipoxygenase associated with its incorporation of 15-hydroperoxy-5,8,11,13-eicosatetraenoic acid derivative. Biochim Biophys Acta 1300:56–62
59. Ueda N, Yamamoto S, Fitzsimmons BJ, Rokach J (1987) Lipoxin synthesis by arachidonate 5-lipoxygenase purified from porcine leukocytes. Biochem Biophys Res Commun 144:996–1002
60. Kuhn H, Wiesner R, Alder L, Fitzsimmons BJ, Rokach J, Brash AR (1987) Formation of lipoxin B by the pure reticulocyte lipoxygenase via sequential oxygenation of the substrate. Eur J Biochem 169:593–601
61. Takahashi Y, Glasgow WC, Suzuki H, Taketani Y, Yamamoto S, Anton M, Kuhn H, Brash AR (1993) Investigation of the oxygenation of phospholipids by the porcine leukocyte and human platelet arachidonate 12-lipoxygenases. Eur J Biochem 218:165–171

62. Kuhn H, Belkner J, Suzuki H, Yamamoto S (1994) Oxidative modification of human lipoproteins by lipoxygenases of different positional specificities. J Lipid Res 35:1749–1759
63. Sigal E, Grunberger D, Craik CS, Caughey GH, Nadel JA (1988) Arachidonate 15-lipoxygenase (omega-6 lipoxygenase) from human leukocytes. Purification and structural homology to other mammalian lipoxygenases. J Biol Chem 263:5328–5332
64. Sigal E, Craik CS, Highland E, Grunberger D, Costello LL, Dixon RA, Nadel JA (1988) Molecular cloning and primary structure of human 15-lipoxygenase. Biochem Biophys Res Commun 157:457–464
65. Kelavkar U, Wang S, Montero A, Murtagh J, Shah K, Badr K (1998) Human 15-lipoxygenase gene promoter: analysis and identification of DNA binding sites for IL-13-induced regulatory factors in monocytes. Mol Biol Rep 25:173–182
66. Kritzik MR, Ziober AF, Dicharry S, Conrad DJ, Sigal E (1997) Characterization and sequence of an additional 15-lipoxygenase transcript and of the human gene. Biochim Biophys Acta 1352:267–281
67. Kelavkar UP, Badr KF (1999) Effects of mutant p53 expression on human 15-lipoxygenase-promoter activity and murine 12/15-lipoxygenase gene expression: evidence that 15-lipoxygenase is a mutator gene. Proc Natl Acad Sci USA 96:4378–4383
68. Kelavkar U, Cohen C, Eling T, Badr K (2002) 15-lipoxygenase-1 overexpression in prostate adenocarcinoma. Adv Exp Med Biol 507:133–145
69. Nadel JA, Conrad DJ, Ueki IF, Schuster A, Sigal E (1991) Immunocytochemical localization of arachidonate 15-lipoxygenase in erythrocytes, leukocytes, and airway cells. J Clin Invest 87:1139–1145
70. Narumiya S, Salmon JA, Flower RJ, Vane JR (1982) Purification and properties of arachidonate-15-lipoxygenase from rabbit peritoneal polymorphonuclear leukocytes. Adv Prostaglandin Thromboxane Leukot Res 9:77–82
71. Vanderhoek JY, Bailey JM (1984) Activation of a 15-lipoxygenase/leukotriene pathway in human polymorphonuclear leukocytes by the anti-inflammatory agent ibuprofen. J Biol Chem 259:6752–6756
72. Levy BD, Romano M, Chapman HA, Reilly JJ, Drazen J, Serhan CN (1993) Human alveolar macrophages have 15-lipoxygenase and generate 15(S)-hydroxy-5,8,11-cis-13-trans-eicosatetraenoic acid and lipoxins. J Clin Invest 92:1572–1579
73. Takayama H, Gimbrone MA Jr, Schafer AI (1987) Vascular lipoxygenase activity: synthesis of 15-hydroxyeicosatetraenoic acid from arachidonic acid by blood vessels and cultured vascular endothelial cells. Thromb Res 45:803–816
74. Lei ZM, Rao CV (1992) The expression of 15-lipoxygenase gene and the presence of functional enzyme in cytoplasm and nuclei of pregnancy human myometria. Endocrinology 130:861–870
75. Giannopoulos PF, Joshi YB, Chu J, Pratico D (2013) The 12-15-lipoxygenase is a modulator of Alzheimer's-related tau pathology in vivo. Aging Cell 12:1082–1090
76. Haynes RL, van Leyen K (2013) 12/15-lipoxygenase expression is increased in oligodendrocytes and microglia of periventricular leukomalacia. Dev Neurosci 35:140–154
77. Yla-Herttuala S, Rosenfeld ME, Parthasarathy S, Glass CK, Sigal E, Witztum JL, Steinberg D (1990) Colocalization of 15-lipoxygenase mRNA and protein with epitopes of oxidized low density lipoprotein in macrophage-rich areas of atherosclerotic lesions. Proc Natl Acad Sci USA 87:6959–6963
78. Conrad DJ, Kuhn H, Mulkins M, Highland E, Sigal E (1992) Specific inflammatory cytokines regulate the expression of human monocyte 15-lipoxygenase. Proc Natl Acad Sci USA 89:217–221
79. Nassar GM, Morrow JD, Roberts LJ 2nd, Lakkis FG, Badr KF (1994) Induction of 15-lipoxygenase by interleukin-13 in human blood monocytes. J Biol Chem 269:27631–27634

80. Chaitidis P, O'Donnell V, Kuban RJ, Bermudez-Fajardo A, Ungethuem U, Kuhn H (2005) Gene expression alterations of human peripheral blood monocytes induced by medium-term treatment with the TH2-cytokines interleukin-4 and -13. Cytokine 30:366–377
81. Brinckmann R, Topp MS, Zalan I, Heydeck D, Ludwig P, Kuhn H, Berdel WE, Habenicht JR (1996) Regulation of 15-lipoxygenase expression in lung epithelial cells by interleukin-4. Biochem J 318(Pt 1):305–312
82. Shankaranarayanan P, Chaitidis P, Kuhn H, Nigam S (2001) Acetylation by histone acetyltransferase CREB-binding protein/p300 of STAT6 is required for transcriptional activation of the 15-lipoxygenase-1 gene. J Biol Chem 276:42753–42760
83. Liu C, Schain F, Han H, Xu D, Andersson-Sand H, Forsell P, Claesson HE, Bjorkholm M, Sjoberg J (2012) Epigenetic and transcriptional control of the 15-lipoxygenase-1 gene in a Hodgkin lymphoma cell line. Exp Cell Res 318:169–176
84. Roy B, Cathcart MK (1998) Induction of 15-lipoxygenase expression by IL-13 requires tyrosine phosphorylation of Jak2 and Tyk2 in human monocytes. J Biol Chem 273:32023–32029
85. Xu B, Bhattacharjee A, Roy B, Feldman GM, Cathcart MK (2004) Role of protein kinase C isoforms in the regulation of interleukin-13-induced 15-lipoxygenase gene expression in human monocytes. J Biol Chem 279:15954–15960
86. Xu B, Bhattacharjee A, Roy B, Xu HM, Anthony D, Frank DA, Feldman GM, Cathcart MK (2003) Interleukin-13 induction of 15-lipoxygenase gene expression requires p38 mitogen-activated protein kinase-mediated serine 727 phosphorylation of Stat1 and Stat3. Mol Cell Biol 23:3918–3928
87. Bhattacharjee A, Shukla M, Yakubenko VP, Mulya A, Kundu S, Cathcart MK (2013) IL-4 and IL-13 employ discrete signaling pathways for target gene expression in alternatively activated monocytes/macrophages. Free Radic Biol Med 54:1–16
88. Kuhn H, O'Donnell VB (2006) Inflammation and immune regulation by 12/15-lipoxygenases. Prog Lipid Res 45:334–356
89. Tsao CH, Shiau MY, Chuang PH, Chang YH, Hwang J (2014) Interleukin-4 regulates lipid metabolism by inhibiting adipogenesis and promoting lipolysis. J Lipid Res 55:385–397
90. Chen B, Tsui S, Boeglin WE, Douglas RS, Brash AR, Smith TJ (2006) Interleukin-4 induces 15-lipoxygenase-1 expression in human orbital fibroblasts from patients with Graves disease. Evidence for anatomic site-selective actions of Th2 cytokines. J Biol Chem 281:18296–18306
91. Kelavkar UP, Wang S, Badr KF (2000) Ku autoantigen (DNA helicase) is required for interleukins-13/-4-induction of 15-lipoxygenase-1 gene expression in human epithelial cells. Genes Immun 1:237–250
92. Han H, Xu D, Liu C, Claesson HE, Bjorkholm M, Sjoberg J (2014) Interleukin-4-mediated 15-lipoxygenase-1 trans-activation requires UTX recruitment and H3K27me3 demethylation at the promoter in A549 cells. PLoS One 9:e85085
93. Kühn H, Barnett J, Grunberger D, Baecker P, Chow J, Nguyen B, Bursztyn-Pettegrew H, Chan H, Sigal E (1993) Overexpression, purification and characterization of human recombinant 15-lipoxygenase. Biochim Biophys Acta 1169:80–89
94. Sloane DL, Dixon RA, Craik CS, Sigal E (1991) Expression of cloned human 15-lipoxygenase in eukaryotic and prokaryotic systems. Adv Prostaglandin Thromboxane Leukot Res 21A:25–28
95. Sloane DL, Leung R, Barnett J, Craik CS, Sigal E (1995) Conversion of human 15-lipoxygenase to an efficient 12-lipoxygenase: the side-chain geometry of amino acids 417 and 418 determine positional specificity. Protein Eng 8:275–282
96. Sloane DL, Leung R, Craik CS, Sigal E (1991) A primary determinant for lipoxygenase positional specificity. Nature 354:149–152
97. Green RE, Krause J, Briggs AW, Maricic T, Stenzel U, Kircher M, Patterson N, Li H, Zhai W, Fritz MH et al (2010) A draft sequence of the Neandertal genome. Science 328:710–722

98. Prüfer K, Racimo F, Patterson N, Jay F, Sankararaman S, Sawyer S, Heinze A, Renaud G, Sudmant PH, de Filippo C et al (2014) The complete genome sequence of a Neanderthal from the Altai Mountains. Nature 505:43–49
99. Chaitidis P, Adel S, Anton M, Heydeck D, Kuhn H, Horn T (2013) Lipoxygenase pathways in Homo neanderthalensis: functional comparison with Homo sapiens isoforms. J Lipid Res 54:1397–1409
100. Horn T, Reddy Kakularam K, Anton M, Richter C, Reddanna P, Kuhn H (2013) Functional characterization of genetic enzyme variations in human lipoxygenases. Redox Biol 1:566–577
101. Johannesson M, Backman L, Claesson HE, Forsell PK (2010) Cloning, purification and characterization of non-human primate 12/15-lipoxygenases. Prostaglandins Leukot Essent Fatty Acids 82:121–129
102. Vogel R, Jansen C, Roffeis J, Reddanna P, Forsell P, Claesson HE, Kuhn H, Walther M (2010) Applicability of the triad concept for the positional specificity of mammalian lipoxygenases. J Biol Chem 285:5369–5376
103. Berger M, Schwarz K, Thiele H, Reimann I, Huth A, Borngraber S, Kuhn H, Thiele BJ (1998) Simultaneous expression of leukocyte-type 12-lipoxygenase and reticulocyte-type 15-lipoxygenase in rabbits. J Mol Biol 278:935–948
104. Funk CD, Chen XS, Johnson EN, Zhao L (2002) Lipoxygenase genes and their targeted disruption. Prostaglandins Other Lipid Mediat 68–69:303–312
105. Chen XS, Kurre U, Jenkins NA, Copeland NG, Funk CD (1994) cDNA cloning, expression, mutagenesis of C-terminal isoleucine, genomic structure, and chromosomal localizations of murine 12-lipoxygenases. J Biol Chem 269:13979–13987
106. Freire-Moar J, Alavi-Nassab A, Ng M, Mulkins M, Sigal E (1995) Cloning and characterization of a murine macrophage lipoxygenase. Biochim Biophys Acta 1254:112–116
107. Kinzig A, Fürstenberger G, Bürger F, Vogel S, Müller-Decker K, Mincheva A, Lichter P, Marks F, Krieg P (1997) Murine epidermal lipoxygenase (Aloxe) encodes a 12-lipoxygenase isoform. FEBS Lett 402:162–166
108. Watanabe T, Medina JF, Haeggstrom JZ, Radmark O, Samuelsson B (1993) Molecular cloning of a 12-lipoxygenase cDNA from rat brain. Eur J Biochem 212:605–612
109. Pekárová M, Kuhn H, Bezáková L, Ufer C, Heydeck D (2015) Mutagenesis of triad determinants of rat Alox15 alters the specificity of fatty acid and phospholipid oxygenation. Arch Biochem Biophys 571:50–507
110. Sloane DL, Browner MF, Dauter Z, Wilson K, Fletterick RJ, Sigal E (1990) Purification and crystallization of 15-lipoxygenase from rabbit reticulocytes. Biochem Biophys Res Commun 173:507–513
111. Gillmor SA, Villasenor A, Fletterick R, Sigal E, Browner MF (1997) The structure of mammalian 15-lipoxygenase reveals similarity to the lipases and the determinants of substrate specificity. Nat Struct Biol 4:1003–1009
112. Choi J, Chon JK, Kim S, Shin W (2008) Conformational flexibility in mammalian 15S-lipoxygenase: reinterpretation of the crystallographic data. Proteins 70:1023–1032
113. Walther M, Anton M, Wiedmann M, Fletterick R, Kuhn H (2002) The N-terminal domain of the reticulocyte-type 15-lipoxygenase is not essential for enzymatic activity but contains determinants for membrane binding. J Biol Chem 277:27360–27366
114. Walther M, Hofheinz K, Vogel R, Roffeis J, Kühn H (2011) The N-terminal β-barrel domain of mammalian lipoxygenases including mouse 5-lipoxygenase is not essential for catalytic activity and membrane binding but exhibits regulatory functions. Arch Biochem Biophys 516:1–9
115. Romanov S, Wiesner R, Myagkova G, Kuhn H, Ivanov I (2006) Affinity labeling of the rabbit 12/15-lipoxygenase using azido derivatives of arachidonic acid. Biochemistry 45:3554–3562
116. Hammel M, Walther M, Prassl R, Kuhn H (2004) Structural flexibility of the N-terminal beta-barrel domain of 15-lipoxygenase-1 probed by small angle X-ray scattering. Functional consequences for activity regulation and membrane binding. J Mol Biol 343:917–929

117. Shang W, Ivanov I, Svergun DI, Borbulevych OY, Aleem AM, Stehling S, Jankun J, Kuhn H, Skrzypczak-Jankun E (2011) Probing dimerization and structural flexibility of mammalian lipoxygenases by small-angle X-ray scattering. J Mol Biol 409:654–668
118. Di Venere A, Horn T, Stehling S, Mei G, Masgrau L, Gonzalez-Lafont A, Kuhn H, Ivanov I (2013) Role of Arg403 for thermostability and catalytic activity of rabbit 12/15-lipoxygenase. Biochim Biophys Acta 1831:1079–1088
119. Suardiaz R, Masgrau L, Lluch JM, Gonzalez-Lafont A (2014) Regio- and stereospecificity in the oxygenation of arachidonic acid catalyzed by Leu597 mutants of rabbit 15-lipoxygenase: a QM/MM study. Chemphyschem 15:2303–2310
120. Toledo L, Masgrau L, Lluch JM, Gonzalez-Lafont A (2011) Substrate binding to mammalian 15-lipoxygenase. J Comput Aided Mol Des 25:825–835
121. Xu S, Mueser TC, Marnett LJ, Funk MO Jr (2012) Crystal structure of 12-lipoxygenase catalytic-domain-inhibitor complex identifies a substrate-binding channel for catalysis. Structure 20:1490–1497
122. Putnam CD, Hammel M, Hura GL, Tainer JA (2007) X-ray solution scattering (SAXS) combined with crystallography and computation: defining accurate macromolecular structures, conformations and assemblies in solution. Q Rev Biophys 40:191–285
123. Dainese E, Sabatucci A, van Zadelhoff G, Angelucci CB, Vachette P, Veldink GA, Agro AF, Maccarrone M (2005) Structural stability of soybean lipoxygenase-1 in solution as probed by small angle X-ray scattering. J Mol Biol 349:143–152
124. Moin ST, Hofer TS, Sattar R, Ul-Haq Z (2011) Molecular dynamics simulation of mammalian 15S-lipoxygenase with AMBER force field. Eur Biophys J 40:715–726
125. Ivanov I, Shang W, Toledo L, Masgrau L, Svergun DI, Stehling S, Gomez H, Di Venere A, Mei G, Lluch JM et al (2012) Ligand-induced formation of transient dimers of mammalian 12/15-lipoxygenase: a key to allosteric behavior of this class of enzymes? Proteins 80:703–712
126. Schewe T (2002) 15-lipoxygenase-1: a prooxidant enzyme. Biol Chem 383:365–374
127. Bryant RW, Schewe T, Rapoport SM, Bailey JM (1985) Leukotriene formation by a purified reticulocyte lipoxygenase enzyme. Conversion of arachidonic acid and 15-hydroperoxyeicosatetraenoic acid to 14, 15-leukotriene A4. J Biol Chem 260:3548–3555
128. Brash AR, Yokoyama C, Oates JA, Yamamoto S (1989) Mechanistic studies of the dioxygenase and leukotriene synthase activities of the porcine leukocyte 12S-lipoxygenase. Arch Biochem Biophys 273:414–422
129. Ivanov I, Saam J, Kuhn H, Holzhutter HG (2005) Dual role of oxygen during lipoxygenase reactions. FEBS J 272:2523–2535
130. Zheng Y, Brash AR (2010) On the role of molecular oxygen in lipoxygenase activation: comparison and contrast of epidermal lipoxygenase-3 with soybean lipoxygenase-1. J Biol Chem 285:39876–39887
131. Toledo L, Masgrau L, Maréchal JD, Lluch JM, González-Lafont A (2010) Insights into the mechanism of binding of arachidonic acid to mammalian 15-lipoxygenases. J Phys Chem B 114:7037–7046
132. Schwarz K, Borngraber S, Anton M, Kuhn H (1998) Probing the substrate alignment at the active site of 15-lipoxygenases by targeted substrate modification and site-directed mutagenesis. Evidence for an inverse substrate orientation. Biochemistry 37:15327–15335
133. Kuhn H, Schewe T, Rapoport SM (1986) The stereochemistry of the reactions of lipoxygenases and their metabolites. Proposed nomenclature of lipoxygenases and related enzymes. Adv Enzymol Relat Areas Mol Biol 58:273–311
134. Van Os CP, Rijke-Schilder GP, Van Halbeek H, Verhagen J, Vliegenthart JF (1981) Double dioxygenation of arachidonic acid by soybean lipoxygenase-1. Kinetics and regio-stereo specificities of the reaction steps. Biochim Biophys Acta 663:177–193
135. Serhan CN, Petasis NA (2011) Resolvins and protectins in inflammation resolution. Chem Rev 111:5922–5943

136. Yamamoto S, Ueda N, Yokoyama C, Fitzsimmons BJ, Rokach J, Oates JA, Brash AR (1988) Lipoxin syntheses by arachidonate 12- and 5-lipoxygenases purified from porcine leukocytes. Adv Exp Med Biol 229:15–26
137. Belkner J, Wiesner R, Kuhn H, Lankin VZ (1991) The oxygenation of cholesterol esters by the reticulocyte lipoxygenase. FEBS Lett 279:110–114
138. Kuhn H, Brash AR (1990) Occurrence of lipoxygenase products in membranes of rabbit reticulocytes. Evidence for a role of the reticulocyte lipoxygenase in the maturation of red cells. J Biol Chem 265:1454–1458
139. van Leyen K, Duvoisin RM, Engelhardt H, Wiedmann M (1998) A function for lipoxygenase in programmed organelle degradation. Nature 395:392–395
140. Belkner J, Stender H, Kuhn H (1998) The rabbit 15-lipoxygenase preferentially oxygenates LDL cholesterol esters, and this reaction does not require vitamin E. J Biol Chem 273:23225–23232
141. Upston JM, Neuzil J, Witting PK, Alleva R, Stocker R (1997) Oxidation of free fatty acids in low density lipoprotein by 15-lipoxygenase stimulates nonenzymic, alpha-tocopherol-mediated peroxidation of cholesteryl esters. J Biol Chem 272:30067–30074
142. Cyrus T, Witztum JL, Rader DJ, Tangirala R, Fazio S, Linton MF, Funk CD (1999) Disruption of the 12/15-lipoxygenase gene diminishes atherosclerosis in apo E-deficient mice. J Clin Invest 103:1597–1604
143. George J, Afek A, Shaish A, Levkovitz H, Bloom N, Cyrus T, Zhao L, Funk CD, Sigal E, Harats D (2001) 12/15-lipoxygenase gene disruption attenuates atherogenesis in LDL receptor-deficient mice. Circulation 104:1646–1650
144. Zhao L, Pratico D, Rader DJ, Funk CD (2005) 12/15-lipoxygenase gene disruption and vitamin E administration diminish atherosclerosis and oxidative stress in apolipoprotein E deficient mice through a final common pathway. Prostaglandins Other Lipid Mediat 78:185–193
145. Garssen GJ, Vliegenthart JF, Boldingh J (1971) An anaerobic reaction between lipoxygenase, linoleic acid and its hydroperoxides. Biochem J 122:327–332
146. de Groot JJ, Garssen GJ, Vliegenthart JF, Boldingh J (1973) The detection of linoleic acid radicals in the anaerobic reaction of lipoxygenase. Biochim Biophys Acta 326:279–284
147. Garssen GJ, Vliegenthart JF, Boldingh J (1972) The origin and structures of dimeric fatty acids from the anaerobic reaction between soya-bean lipoxygenase, linoleic acid and its hydroperoxide. Biochem J 130:435–442
148. Streckert G, Stan HJ (1975) Conversion of linoleic acid hydroperoxide by soybean lipoxygenase in the presence of guaiacol: identification of the reaction products. Lipids 10:847–854
149. Zheng Y, Brash AR (2010) Dioxygenase activity of epidermal lipoxygenase-3 unveiled: typical and atypical features of its catalytic activity with natural and synthetic polyunsaturated fatty acids. J Biol Chem 285:39866–39875
150. Munoz-Garcia A, Thomas CP, Keeney DS, Zheng Y, Brash AR (2014) The importance of the lipoxygenase-hepoxilin pathway in the mammalian epidermal barrier. Biochim Biophys Acta 1841:401–408
151. Salzmann U, Kuhn H, Schewe T, Rapoport SM (1984) Pentane formation during the anaerobic reactions of reticulocyte lipoxygenase. Comparison with lipoxygenases from soybeans and green pea seeds. Biochim Biophys Acta 795:535–542
152. Belkner J, Kuhn H, Wiesner R (1990) Oxygenation of biological membranes by the reticulocyte lipoxygenase. Lack of stoichiometry between oxygen uptake and product formation. Biomed Biochim Acta 49:S31–34
153. Kuhn H, Salzmann-Reinhardt U, Ludwig P, Ponicke K, Schewe T, Rapoport S (1986) The stoichiometry of oxygen uptake and conjugated diene formation during the dioxygenation of linoleic acid by the pure reticulocyte lipoxygenase. Evidence for aerobic hydroperoxidase activity. Biochim Biophys Acta 876:187–193

154. Maas RL, Brash AR (1983) Evidence for a lipoxygenase mechanism in the biosynthesis of epoxide and dihydroxy leukotrienes from 15(S)-hydroperoxyicosatetraenoic acid by human platelets and porcine leukocytes. Proc Natl Acad Sci USA 80:2884–2888
155. Yamamoto S, Ueda N, Yokoyama C, Kaneko S, Shinjo F, Yoshimoto T, Oates JA, Brash AR, Fitzsimmons BJ, Rokach J (1987) Dioxygenase and leukotriene A synthase activities of arachidonate 5- and 12-lipoxygenases purified from porcine leukocytes. Adv Prostaglandin Thromboxane Leukot Res 17A:55–59
156. Brash AR, Ingram CD, Harris TM (1987) Analysis of a specific oxygenation reaction of soybean lipoxygenase-1 with fatty acids esterified in phospholipids. Biochemistry 26:5465–5471
157. Saam J, Ivanov I, Walther M, Holzhutter HG, Kuhn H (2007) Molecular dioxygen enters the active site of 12/15-lipoxygenase via dynamic oxygen access channels. Proc Natl Acad Sci USA 104:13319–13324
158. Knapp MJ, Klinman JP (2003) Kinetic studies of oxygen reactivity in soybean lipoxygenase-1. Biochemistry 42:11466–11475
159. Knapp MJ, Seebeck FP, Klinman JP (2001) Steric control of oxygenation regiochemistry in soybean lipoxygenase-1. J Am Chem Soc 123:2931–2932
160. Cohen J, Arkhipov A, Braun R, Schulten K (2006) Imaging the migration pathways for O2, CO, NO, and Xe inside myoglobin. Biophys J 91:1844–1857
161. Svensson-Ek M, Abramson J, Larsson G, Tornroth S, Brzezinski P, Iwata S (2002) The X-ray crystal structures of wild-type and EQ(I-286) mutant cytochrome c oxidases from Rhodobacter sphaeroides. J Mol Biol 321:329–339

Platelets and Lipoxygenases

Michael Holinstat, Katrin Niisuke, and Benjamin E. Tourdot

Abstract Platelets are a cellular component of blood whose primary function is to maintain hemostasis in response to vessel insult. Beyond their traditional role in hemostasis, platelets are also known to regulate inflammation. These small anucleated cells express a single lipoxygenase, platelet 12(S)-LOX, which is capable of oxidizing a number of fatty acids. Depending on the fatty acid, the metabolite(s) produced by 12(S)-LOX have often been shown to impact platelet and vessel function in distinct ways and therefore are important regulators of platelet reactivity and inflammation. This chapter reviews the roles of 12(S)-LOX and its metabolites on platelet functions in thrombosis, hemostasis and inflammation.

Platelets, the progeny of bone marrow derived megakaryocytes, are small (2–3 μM) anucleated cells that circulate in the blood at a high concentration (human 150–400,000 platelets/μL) [1]. Traditionally, platelets or thrombocytes (in lower vertebrates) have been recognized for their role in regulation of hemostasis and thrombosis (Fig. 1). During vascular injury platelets become activated and form a clot, a platelet plug that functions as a physical barrier to prevent blood loss. Specifically, an injury to the vascular system results in damage to the endothelial cells leading to subsequent exposure of the extracellular matrix, and initiation of the coagulation cascade and platelet activation. Following initial activation steps, soluble mediators released by activated platelets function to reinforce the initial platelet clot through recruitment of naïve platelets from circulation thus enhancing the formation of an intravascular thrombus. Signaling through the platelet integrin $\alpha_{IIb}\beta_3$ results in the formation of a stable clot which stops blood from exiting the vasculature. Failure to adequately regulate clot formation may lead to an occlusive thrombus, while failure to adequately form a thrombus may lead to a bleeding diathesis; both of which can lead to life threatening complications [2].

Beyond their traditional hemostatic role, platelets have been shown to participate in diseases that include an inflammatory component, such as infectious diseases [3], cancer [4, 5], atherosclerosis [6, 7] and arthritis [1, 8, 9]. Indeed, platelets express

M. Holinstat (✉) • K. Niisuke • B.E. Tourdot
University of Michigan Medical School, 1150 West Medical Center Drive, 2240 Medical Sciences Research Building III (MSRB III), Ann Arbor, MI 48109-5632, USA
e-mail: mholinst@med.umich.edu

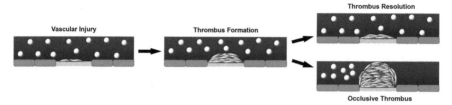

Fig. 1 Upon platelet stimulation, free polyunsaturated fatty acids (PUFAs) are released from the plasma membrane and metabolized by platelet 12-lipoxygenase (12(S)-LOX) into 12-LOX oxylipins. 12-LOX derived oxylipins have been shown to bind to GPCRs, become esterified into the plasma membrane and may function to bind intracellular signaling complexes. Additional work is required to determine how 12(S)-LOX oxylipins exert there functions in platelets

immune receptors and secrete an array of soluble mediators in response to activation that allows them to participate in the inflammatory process through direct or indirect interaction with noxious agents and pathogens, or by eliciting a response from other immune cells [7, 10]. Depending on the type of inflammatory challenge, platelets have been purported to play either protective or pathogenic roles in circulation. In response to an inflammatory challenge, platelets can directly neutralize the noxious agent or release soluble mediators such as cytokines that recruit leukocytes to the site of inflammation. These functions of platelets in response to infection appear to play a protective role early in the inflammatory response pathway [3]. However, as the infection persists platelets can have deleterious consequences due to increased tissue damage [3]. Consistent with a pro-inflammatory role platelets have been shown to exacerbate disease progression to chronic inflammatory diseases. As with hemostasis, inflammation must be carefully regulated to avoid deleterious consequences.

Both hemostatic and immune functions of platelets rely on the generation of soluble molecules to coordinate the initiation and resolution of these complex processes. One group of such soluble mediators, oxylipins are generated by oxygenases such as cyclooxygenase (COX) and lipoxygenase (LOX) acting on polyunsaturated fatty acids (PUFAs) during cellular activation. The clinical importance of COX-derived oxylipins in hemostasis and inflammation is highlighted by the effectiveness of aspirin, a COX inhibitor, to prolong bleeding, prevent the occurrence of thrombotic diseases, and prevent inflammation. Unlike COX the role of lipoxygenases and their oxylipins on platelet activation is less well defined.

1 Lipoxygenases in Platelets

The only LOX expressed in human platelets, and their precursors the megakaryocyte, is 12(S)-LOX [11]. Additionally, it was the first LOX described in humans, as well as among mammals in general. Several studies have shown all mammalian platelets investigated to date solely express 12(S)-LOX including bovine [12], sheep [12], mouse [13, 14], rabbit [15, 16], rat [17, 18], and dog [19].

Similar to other members of the LOX family, platelet 12(S)-LOX (12(S)-LOX) catalyzes a peroxidation of PUFAs that is site-specific and stereochemically

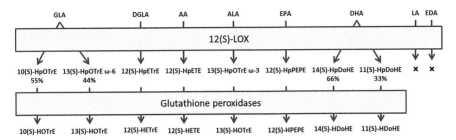

Fig. 2 The hydroperoxides that are formed by 12-LOX from PUFAs, such as GLA, DGLA, AA, ALA, EPA, and DHA are quickly reduced to corresponding hydroxyl products by glutathione peroxidases. 12-LOX cannot peroxidize LA and EDA

controlled. 12(S)-LOX can oxygenate arachidonic acid (AA), dihomo-γ-linolenic acid (DGLA), eicosapentaenoic acid (EPA), α-linolenic acid (ALA), docosahexaenoic acid (DHA) and γ-linolenic acid (GLA) [20]. Four of these PUFAs (AA, DGLA, EPA, and, ALA) will be converted to a single lipid metabolite, 12(S)-hydroperoxyicosa-5,8,10,14-tetraenoic acid (12(S)-HpETE), 12(S)-hydroperoxyicosa-5,10,14-trienoic acid (12(S)-HpETrE), 12(S)-hydroperoxyicosa-5,8,10,14,17-pentaenoic acid (12(S)-HpEPE) and 13(S)-hydroperoxyoctadeca-9,11,15-trienoic acid (13(S)-HpOTrE ω-3), respectively, while two of the PUFAs (DHA and GLA) produce two different products (Fig. 2). The major product derived from DHA is 14(S)-hydroperoxydocosa-4,7,10,12,16,19-hexaenoic acid (14(S)-HpDoHE, 66 %) and the minor 11(S)-hydroperoxydocosa-4,7,9,13,16,19-hexaenoic acid (11(S)-HpDoHE, 33 %). GLA oxidation by 12(S)-LOX produces a mixture of 10 (S)-hydroperoxyoctadeca-6,8,12-trienoic acid (10(S)-HpOTrE, 55 %) and 13(S)-hydroperoxyoctadeca-6,9,11-trienoic acid (13(S)-HpOTrE ω-6, 44 %) [20]. Two other tested fatty acids, eicosadienoic acid (EDA) and linoleic acid (LA), show no reactivity when incubated with 12(S)-LOX, suggesting that these two fatty acids are not substrates for the enzyme *in vivo* [20]. That observation may be explained by the fact that substrates for 12(S)-LOX enter the cavity methyl-end first [21] in which case the hydrogen abstraction site for EDA and LA are not positioned properly to yield product. Other possible explanations however exist which could play an important role in substrate activity such as conjugation and ongoing investigations in the field will help to further identify what determines substrate specificity for the 12(S)-LOX enzyme. The hydroperoxides that are formed from 12(S)-LOX oxidation of PUFAs are labile and quickly reduced to their corresponding hydroxylated products by glutathione peroxidases that are highly expressed in platelets [22].

The majority of substrates for 12(S)-LOX are esterified at the sn-2 position in the phospholipids located in the cell membrane. Under basal conditions however, the majority of 12(S)-LOX is located in the cytosol where it translocates to the lipid membrane during platelet activation in a Ca^{2+}-dependent manner [23]. Once at the membrane, 12(S)-LOX associates with either cPLA$_2$ or sPLA$_2$, enzymes which hydrolyze PUFAs at the sn-2 position of phospholipids, providing the substrates for LOX and COX [24, 25]. The importance of cPLA$_2$ in liberating PUFAs was recently highlighted in a patient expressing a mutation in cPLA$_{2\alpha}$ rendering it non-functional [26] and whose oxylipin production was reduced by 95 % compared to healthy controls following blood coagulation [27]. Similarly, deleting $cPLA_2$ in mice results in an 80 % reduction in serum 12-HETE levels [28].

2 Effect of LOX Oxylipins on Thrombosis and Hemostasis

In human and animal studies, the role of 12(S)-LOX and its oxylipin 12(S)-HETE have been controversial. In human studies, numerous labs have shown that 12(S)-LOX, primarily through production of 12(S)-HETE, can function in a prothrombotic manner, however the exact nature of this mechanism is not fully elucidated. The prothrombotic role of 12(S)-LOX is supported by evidence from patients with myeloproliferative disorders [29–31] who have decreased 12(S)-LOX expression level and show increased bleeding and less thrombotic complications compared to myeloproliferative patients with normal levels of 12(S)-LOX. While these results are consistent with a pro-thrombotic role for 12(S)-LOX, caution should be taken with interpretation of these studies as they were made on a limited sample size of patients with other complications. In further support of a prothrombotic role of 12(S)-LOX in humans, ex vivo pharmacological inhibition of 12(S)-LOX in platelets attenuates platelet aggregation in response to a myriad of agonists including U46619 [32], ADP [33], Thrombin [32], PAR4-AP [33], (canine/platelets; baicalein/OPC) [19], collagen [33, 34], and FcγRIIa [35]. While earlier studies used 12(S)-LOX inhibitors, such as baicalein and NDGA that later have been shown to affect proteins that contribute to the decrease in platelet aggregation [36], newer more selective 12(S)-LOX inhibitors have produced similar abilities to attenuate platelet aggregation [37, 38].

In contrast to observed effects in humans, studies using platelets from mice deficient in 12(S)-LOX have produced conflicting results. The original characterization of platelets from platelet 12(S)-LOX deficient mice suggested an increase in ADP-mediated platelet aggregation compared to platelets from WT mice, but showed no difference in aggregation to other common platelet agonists including U46619, AA, collagen or (Gamma) thrombin [39]. Conversely, more recent reports from the same 12(S)-LOX deficient mouse line have shown that platelets from these mice have a bleeding diathesis [33] as well as attenuated aggregation in response to a number of agonists including GPVI, PAR4, and FcγRIIa compared to platelets from wild-type mice [33, 35, 40]. Discrepancies in these findings could have to do with the age of mice used in each experiment. Additionally, there is also the possibility that 12(S)-LOX plays a unique role in regulation of platelet reactivity and thrombosis in mouse and human platelets.

Studies on the impact of LOX derived oxylipins on hemostasis and thrombosis are controversial. However, it should be noted that all PUFAs are not equal and can exert opposite effects on platelet activity when metabolized by 12(S)-LOX. Based on available 12(S)-LOX kinetics data with PUFAs, such as AA, DGLA, EPA and ALA, we can divide those substrates into two groups based on their carbon-chain length. The longer substrates, with 20 carbon atoms (AA, DGLA and EPA) have over tenfold greater k_{cat} and k_{cat}/K_M values than the shorter substrate, with

18 carbon atoms (ALA), suggesting that length is a key factor in the rate of substrate capture and product release [20]. Even if platelet 12(S)-LOX processes AA, DGLA and EPA at similar rates, the 12(S)-HpETE and its reduced form 12(S)-HETE are found to be the predominant metabolites produced, simply because AA is the most abundant PUFA in the cell membranes.

The mechanism by which 12(S)-LOX exerts a prothrombotic influence on platelet reactivity is not entirely understood, however 12(S)-HETE is generally thought to play an important role in activating/reinforcing platelets in both the ex vivo and *in vivo* environments. As previously stated, 12-HETE is produced in response to platelet stimulation by a myriad of agonists including ADP [19], collagen [24], U46619 [19], thrombin [32, 41–43] and FcγRIIa cross-linking [35]. The general agreement in the field is that exogenous 12-HETE as well as mono-HETEs derived from other cells including 5 and 15-HETE does not directly cause platelet aggregation. However, controversy exists in the literature as to the role 12-HETE plays in agonist induced platelet aggregation with studies reporting that 12-HETE potentiates [44, 45], inhibits [46–48], or plays no role [34] in agonist induced platelet aggregation. To address a potential role for 12(S)-LOX oxylipins in regulation of platelet function exogenous addition of 12-HpETE or 12-HETE has been shown to potentiate agonist induced platelet activation using either thrombin [44] or AA [45]. One mechanism by which 12-HETE is believed to enhance agonist induced platelet aggregation is by activating PKC [40] and increasing dense granule secretion as shown in thrombin stimulated platelets [33, 49], presumably through a 12-HETE receptor.

In contrast to the prothrombotic observations described above, other studies have demonstrated an inhibitory effect of 12-HpETE or 12-HETE on platelet aggregation and dense granule secretion after stimulation with collagen [46], AA [46] and U-46619 [47, 48]. In mice lacking 12(S)-LOX, exogenous addition of super-physiological concentrations of 12-HETE attenuated the ADP hyper-activation induced aggregation [39]. The inhibitory action of 12-HETE at these pharmacologically elevated levels on platelet aggregation has been purported to disrupt AA signaling either through the inhibition of cPLA$_2$ [50] or by acting as antagonist to the TPα receptor [51]. Further, it has also been reported that addition of excess 12-HpETE, the labile 12(S)-LOX product, stimulates its own product production by increasing 12(S)-LOX activity [52]. That observation is supported by experiments with EPA in which case a number of metabolic 12(S)-LOX oxidized EPA products are formed following incubation with AA and the process is dependent of AA conversion by 12(S)-LOX to 12(S)-HpETE [53]. Pre-incubation with reduced product, 12-HETE, however, did not induce EPA metabolism [53]. In other experiments pre-incubation with 12(S)-HpETE has shown to increase the amount of non-esterified AA in collagen-stimulated platelets and significantly changing platelet aggregation and the formation of TXB$_2$ [54]. The true effect of 12(S)-HpETE and 12(S)-HETE on platelet activation during inflammatory states due to physiological levels of these oxylipins has yet to be determined in human or animal models.

The discrepancies in these studies are an area of ongoing investigation in the field. The purification and storage of these oxidized lipids is not a trivial matter. Firstly, they are easily oxidized further or can undergo modifications at molecular

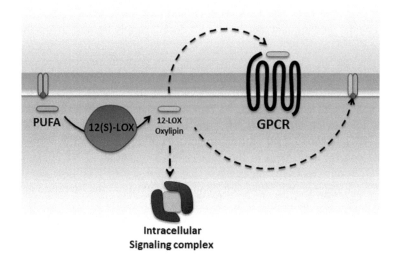

Fig. 3 Upon vascular injury, subendothelial matrix components including VWF, and collagen are exposed to platelets resulting in there activation. These activated platelets spread to cover the site of vascular injury and release soluble mediators to enhance the recruitment of other platelets

level. Secondly, the exogenous addition of mono-HETEs may not recapitulate the spatial or temporal effects of endogenously produced mono-HETEs. Finally, the amount of 12-HETE used in these experiments has varied from 250 nM to 200 µM [39] leading to a potentiation or inhibition of platelet activation, respectively. This data suggests that some of 12-HETEs functions may be concentration dependent. Confirming GPR31 as the GPCR for 12-HETE on platelets, as it has been shown in endothelial cells [55] and neurons [56] will significantly advance our understanding of the role of the 12-HETE-mediated pathway in modulating platelet function (Fig. 3).

An alternative mechanism for regulation of platelet function through 12(S)-LOX oxylipins is esterification of 12(S)-LOX products (Fig. 3). Generation of phospholipid-esterified eicosanoids in agonist-activated human platelets has been demonstrated by several labs and is dependent on sPLA2, but not on cPLA2 for its function. The liberated PUFA is believed to be oxidized by 12(S)-LOX and re-esterified shortly thereafter [57, 58]. 12-HETE, for example, has been shown to be esterified into phosphatidylethanolamine (PE) and phosphatidylcholine (PC) following formation and subsequently externalized. Upon externalization esterified 12-HETE has been shown to enhance thrombin generation on the platelet surface *in vitro* [57].

In addition to oxylipins derived from AA catalysis by 12(S)-LOX, 12(S)-LOX is capable of metabolizing other PUFAs which leads to the formation of a number of oxylipins; [20] some of which have been shown to inhibit hemostasis. For example, the 12(S)-LOX oxylipin derived from DGLA, 12-HETrE, has been shown to inhibit aggregation in response to multiple platelet agonists including PAR1-AP, ADP, and collagen [20]. The function of other 12(S)-LOX oxylipins such as 14-HDoHE remain to be elucidated.

In addition to oxylipin regulation of platelets and inflammation, published work suggests that 12(S)-LOX may regulate the inflammatory potential of the cell through direct or indirect interaction with a variety of signaling complexes. For example, 12(S)-LOX has been shown to act as a scaffold that may play an important role in regulation of platelet reactivity and inflammation [55], directly interact with integrins in cancer cells, and possibly interact with the ITAM signalosome in the platelet.

3 Effect of Platelet LOX Oxylipins on Inflammation

While the ability of platelets to modulate inflammation has been known for a long time the extent to which platelets regulate inflammation under various models is an area of active investigation. The recruitment and activation of platelets at sites of vascular injury positions them to initiate release inflammatory mediators and inflammation. Interestingly, platelets are also active in a number of chronic inflammatory conditions such as asthma. The ability of LOX oxylipins to modulate inflammation has been well studied with the primarily focus on LOX-derived oxylipins produced by leukocytes predominately 5 and 15-HETE's but less is known about LOX-derived oxylipins from platelets.

Platelet 12(S)-LOX is the major producer of 12-HETE in humans. 12-HETE has been shown to function as a pro-inflammatory molecule by causing vasodilation [59], neutrophil chemotaxis [60, 61], monocyte transendothelial cell migration (adhesion) [62], and cellular proliferation. However, the majority of the studies that demonstrated the pro-inflammatory functions of 12-HETE occurred in platelet independent cellular systems. In particular, mouse leukocytes express a leukocyte 12/15 LOX which predominately produces 12-HETE, while its homolog in humans predominately produces 15-HETE. Additionally in humans, other cells can produce 12-HETE in a platelet independent manner including epidermis, pancreatic β cells, Schwann cells, adipocytes, endothelial cells and leukocytes especially under inflammatory or stress conditions [63]. Whether or not 12-HETE from platelets has these inflammatory functions *in vivo* remains to be determined.

Recent data suggests that 12(S)-LOX plays an important role in platelet FcγRIIa signaling through a yet to be identified mechanism [35]. FcγRIIa signaling in platelets is required for the pathogenesis of heparin-induced thrombocytopenia, an iatrogenic disorder caused by the administration of heparin, leads to an immune mediated thrombocytopenia and in some case thrombosis. H1N1 infection, another disease that can generate immune complexes, has been associated with intravascular thrombotic complication [64]. These influenza A virus mediated immune complexes also activate platelets in an FcγRIIa-dependent manner and thrombin-dependent manner [65]. Further, 12-HETE was increased in platelets stimulated by influenza A immune complexes [65].

Helicobacter pylori, a bacteria detected in atherosclerotic plaque and associated with cardiovascular disease [66], has been shown to cause platelet aggregation in a strain dependent manner [67]. A study by Wassermann et al. demonstrated that one way H. pylori activates platelets is through the secretion of an urease that causes platelet aggregation in a 12(S)-LOX dependent manner [68]. The activation of platelets by H. pylori in atherosclerotic plaque could lead to the progression of the plaque.

Platelets contribute to chronic inflammatory diseases such as atherosclerosis and asthma by increasing the cellular proliferation and migration of smooth muscle cells [69]. In culture using cells from rats, 12-HETE was shown to cause aortic smooth muscle cell migration [70, 71] and pulmonary artery cell proliferation [72]. Additionally, airway smooth muscle cells (ASMC) proliferation contributes to airway remodeling, a hallmark of asthma. The co-culture of platelets with ASMCs has been shown to increase their proliferation in a 12(S)-LOX dependent manner [72, 73]. The increase in 12-HETE in exhaled breath condensate has been suggested to be useful as a biomarker for the diagnosis of Churg-Strauss syndrome, an airway inflammatory disorder. However, the source of the 12-HETE being produced has not been delineated and could be from eosinophil instead of platelets [36].

4 Trans-cellular Regulation

Numerous cell types are activated during hemostasis and inflammation in the vascular system, each capable of producing unique LOX oxylipins that have the ability to influence platelet activation and respond to inflammatory conditions. LOX products are known to be released from leukocytes following recruitment to the site of vascular injury [74]. Specifically, the leukocyte 5-LOX oxylipin 5-HETE and the macrophage derived-oxylipin product 15-HETE both have been shown to potentiate thrombin induced platelet aggregation and ADP release [49, 75]. Clinically, 15-HETE can also be produced by macrophage laden atherosclerotic plaque in the carotid artery and thus potentially exacerbate atherothrombotic events.

Experimental work suggests that there are trans-cellular routes for the biosynthesis of some of the bioactive lipids: a PUFA that is released from one cell can be taken up by another cell for synthesis of eicosanoids or a synthesized eicosanoid can be transferred to another cell and further used for synthesis of other eicosanoids (Fig. 4) [76, 77]. This inter-cellular communication through PUFA oxylipins has been routinely observed in the platelet. Within the area of leukotrienes (LT) and lipoxins (LX) it is almost a prerequisite, especially if participating cells are lacking one of the enzymes necessary for the synthesis of named compounds [78, 79]. Lipoxins, such as LXA_4 and LXB_4, are tetraene-containing eicosanoids

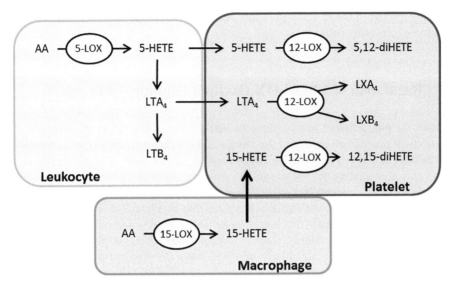

Fig. 4 Transcellular biosynthesis of oxylipins mediated by platelet 12-LOX

generated from LTA_4 [80–83] and the biosynthesis of oxylipins requires a coordinated action of two lipoxygenases—5-LOX to form LTA_4 in the macrophage and 12(S)-LOX to convert it to a lipoxin in the platelet [84]. Other examples of transcellular regulation of platelet function through 12(S)-LOX oxylipins includes leukocyte-type 5-LOX in neutrophils and reticulocyte-type 15-LOX in monocytes and macrophages, which participate in the synthesis of different platelet-derived oxylipins. These coordinated signaling schemes involving the transfer of oxylipins between cells in circulation must be taken into consideration when studying inflammatory processes in animal models due to the difference in LOX expression profiles between mammalian models. In contrast to humans, there is no reticulocyte-type 15-LOX in mice, but the leukocyte-type 12(S)-LOX (possesses 12S/15S-LOX activity) appears to be its functional equivalent [85]. This difference needs to be considered if enzymes exhibit their biological activities via the formation of bioactive AA oxygenation products (12(S)-HETE in mice versus 15(S)-HETE in humans). In rabbits the situation is even more complicated by having two separate genes encoding for reticulocyte-type 15-LOX and a leukocyte-type 12(S)-LOX, both of which have 12(S)/15(S)-LOX activity [86, 87].

In addition to lipoxins, platelet 12(S)-LOX has shown the ability to generate 5S,12S-dihydroxyeicosatetraenoic acid (5S,12S-diHETE) from the 5-LOX reduced product 5-HETE and 14,15-diHETE from the 15-LOX reduced product 15-HETE [88]. Furthermore, 11,12-diHETE generation has been reported to be dependent on platelet 12(S)-LOX activity. However, the metabolic pathway for this lipid product differs from other diHETEs since the formation of this compound probably proceeds via oxygenation by the platelet 12(S)-LOX followed by epoxide formation and hydrolysis of the epoxide [89] and therefore is not dependent on lipid modification by the other cells

5 Regulation of 12(S)-LOX Oxylipin Exposure to Platelets

While the production of trans-cellular bioactive lipids is dependent on coordinated activity of several cell types and the mechanism requires its transport from one cell to another in a highly regulated spatial–temporal pattern, another factor that comes into play in regulating the rate of synthesis for these oxylipins is albumin. Albumin, which is highly concentrated in the blood and represents over 40 % of protein in plasma, can bind fatty acids and their oxylipins. For example, it has been shown that albumin binds TxA_2, stabilizes it and, as a result, prolongs its half-life in aqueous media [90]. Similarly, highly reactive intermediate, LTA_4, is stabilized by the addition of low concentration of albumin [91]. It therefore presents a way to provide cells with these reactive substrates over longer period of time. For example albumin reduces the formation of 5(S),12(S)-diHETE from 5(S)-HETE and 12(S)-HETE from exogenous AA, while having no effect on 12(S)-HETE production from endogenous AA [88].

The previously mentioned glutathione peroxidases are enzymes play a role in the defense against oxidative damage to cells by regulating hydroperoxide tone, which is the sum of steady-state concentrations of all hydroperoxides, and is determined by the balance of hydroperoxide-generating and hydroperoxide-consuming processes [92]. Hydroperoxydated PUFAs contribute to this hydroperoxide tone to a large extent. At low concentrations hydroperoxy-PUFAs activate cyclooxygenases [93] and lipoxygenases [94, 95], but at higher concentrations promote the suicide inactivation of these enzymes and upregulate hydroperoxidase activities [96]. It has been shown that inactivation of glutathione peroxidases in platelets results in inhibition of 12-HpETE conversion to 12-HETE and the metabolism of the hydroperoxide is switched towards the formation of hepoxilins (HXA_3 and HXB_3) by 12 (S)-LOX. However, with functional glutathione hydroperoxides, 12-HpETE is primarily reduced to 12-HETE. Therefore, we can speculate that the function of glutathione peroxidases is to limit specific oxylipin products and acts as an effective enzymatic system for controlling hepoxilin synthesis [22]. While both 12-HETE and hepoxilins are biologically active eicosanoids, they are known to have distinct arrays of biological actions [97–99]. For example, while there is controversy in scientific literature about 12-HETE action on platelets, HXA_3 has been reported to inhibit ADP-induced aggregation and to mediate a regulatory volume decrease [100]. HXA_3 has also been reported to induce the heat-shock protein HSP72 in human neutrophils [100], which have been suggested to protect cells from reactive oxygen species and inflammation [101]—the latter effect may be implicated in the compensatory role of hepoxilin formation under oxidative stress conditions. Therefore, it is reasonable to assume that the diversion of the 12(S)-LOX pathway metabolism from a reduction route to isomerization route in the platelet under severe oxidative stress conditions could constitute a compensatory mechanism to maintain the functional integrity of platelets under those conditions.

6 Influencing Matters for 12(S)-LOX Signal Outcome

Bioactive lipids produced through lipoxygenation can differentially influence hemostasis as well as the inflammatory state of the vessel depending on which PUFAs are used as substrates. A number of studies have shown that the cell membrane where most of the substrate is located is a dynamic system and the lipid content on the membrane of cells in circulation can quickly change due to consumed diet of essential fatty acids reflecting the relative lipid ratio in the bloodstream. Research groups have taken advantage of this fact to study these physiological processes, which are mediated either by dietary fatty acids or the metabolic products produced following enzymatic oxidation of the free fatty acids in the membrane.

A Change in dietary intake of certain omega-3 and omega-6 PUFAs have shown to modulate their content in the lipid bilayer of platelets and the type of metabolic products generated by 12(S)-LOX, which in whole has an impact on platelet [102–105].

Large prospective studies with omega-3 fatty acids have shown that there is a positive relationship between increased dietary intake of these PUFAs and reduced CVD [46, 106–109]. These observations are supported by a study that showed consumption of higher content of mackerel for 1 week lead to increased EPA/AA ratio and reduced platelet aggregation and thromboxane synthesis after low-dose collagen stimulation [110].

On the omega-6 fatty acid side, supplementing people with GLA has shown to end up with an increased content of DGLA and in some cases also GLA, but with no significant changes in AA in platelet membrane [104, 105]. While it has been shown that DGLA inhibits platelet aggregation and ADP secretion ex vivo [20], the above-mentioned clinical studies were unable to show any significant effect on platelet function.

Thus, larger amount of specific PUFAs in the diet may result in an increased pool of fatty acids made available to 12(S)-LOX in platelet. Depending on the specific fatty acid substrates available, oxidation by 12(S)-LOX generates several pools of eicosanoids, which may in turn act as one of the key regulators of platelet reactivity and play a role both within the platelet itself as well as at distal tissues and organ systems.

7 Conclusions

The role of lipoxygenases in regulating the normal hemostasis and inflammatory processes of platelets has received increased attention as of late. While a number of important questions regarding the regulatory role of 12(S)-LOX and its oxylipins in mediating platelet reactivity, clot formation, and occlusive thrombi have recently been addressed, there is still uncertainty regarding the mechanism by which 12(S)-LOX exerts these effects. The main effects of 12(S)-LOX under physiologic conditions appear to be due to the production of 12-HETE; however, the mechanism by which 12-HETE exerts its effects remains to be elucidated. Interestingly, other 12(S)-LOX products appear to have inhibitory effects on platelet function and hemostasis. Advances in lipidomics will go a long way in answering what other LOX oxylipins are present in platelets and how they help to regulate overall platelet function. A more complete understanding of platelet LOX activity and its oxylipin products in inflammation and hemostasis may allow for improved therapeutic intervention to regulate the production of pro- and anti-thrombotic oxylipin products which play a role in inflammation in the vessel.

References

1. Semple JW, Italiano JE Jr, Freedman J (2011) Platelets and the immune continuum. Nat Rev Immunol 11(4):264–274
2. Stalker TJ, Newman DK, Ma P, Wannemacher KM, Brass LF (2012) Platelet signaling. Handb Exp Pharmacol 210:59–85
3. Morrell CN, Aggrey AA, Chapman LM, Modjeski KL (2014) Emerging roles for platelets as immune and inflammatory cells. Blood 123(18):2759–2767
4. Gay LJ, Felding-Habermann B (2011) Contribution of platelets to tumour metastasis. Nat Rev Cancer 11(2):123–134
5. Bambace NM, Holmes CE (2011) The platelet contribution to cancer progression. J Thromb Haemost 9(2):237–249
6. Huo Y, Ley KF (2004) Role of platelets in the development of atherosclerosis. Trends Cardiovasc Med 14(1):18–22
7. Lievens D, von Hundelshausen P (2011) Platelets in atherosclerosis. Thromb Haemost 106(5):827–838
8. Boilard E, Nigrovic PA, Larabee K et al (2010) Platelets amplify inflammation in arthritis via collagen-dependent microparticle production. Science 327(5965):580–583
9. Michou L, Cornelis F, Baron M et al (2013) Association study of the platelet collagen receptor glycoprotein VI gene with rheumatoid arthritis. Clin Exp Rheumatol 31(5):770–772
10. Aslam R, Speck ER, Kim M et al (2006) Platelet Toll-like receptor expression modulates lipopolysaccharide-induced thrombocytopenia and tumor necrosis factor-alpha production in vivo. Blood 107(2):637–641
11. Hamberg M, Samuelsson B (1974) Prostaglandin endoperoxides. Novel transformations of arachidonic acid in human platelets. Proc Natl Acad Sci USA 71(9):3400–3404
12. Nugteren DH (1975) Arachidonate lipoxygenase in blood platelets. Biochim Biophys Acta 380(2):299–307

13. Burger F, Krieg P, Marks F, Furstenberger G (2000) Positional- and stereo-selectivity of fatty acid oxygenation catalysed by mouse (12S)-lipoxygenase isoenzymes. Biochem J 348 (Pt 2):329–335
14. Chen X, Kurre U, Jenkins N, Copeland N, Funk C (1994) cDNA cloning, expression, mutagenesis of C-terminal isoleucine, genomic structure, and chromosomal localizations of murine 12-lipoxygenases. J Biol Chem 269(19):13979–13987
15. Michibayashi T (2005) Platelet aggregation and vasoconstriction related to platelet cyclooxygenase and 12-lipoxygenase pathways. J Atheroscler Thromb 12(3):154–162
16. Yu JY, Lee JJ, Jung JK et al (2012) Anti-platelet activity of diacetylated obovatol through regulating cyclooxygenase and lipoxygenase activities. Arch Pharm Res 35(12):2191–2198
17. Baba A, Sakuma S, Okamoto H, Inoue T, Iwata H (1989) Calcium induces membrane translocation of 12-lipoxygenase in rat platelets. J Biol Chem 264(27):15790–15795
18. Stern N, Kisch ES, Knoll E (1996) Platelet lipoxygenase in spontaneously hypertensive rats. Hypertension 27(5):1149–1152
19. Katoh A, Ikeda H, Murohara T, Haramaki N, Ito H, Imaizumi T (1998) Platelet-derived 12-hydroxyeicosatetraenoic acid plays an important role in mediating canine coronary thrombosis by regulating platelet glycoprotein IIb/IIIa activation. Circulation 98 (25):2891–2898
20. Ikei KN, Yeung J, Apopa PL et al (2012) Investigations of human platelet-type 12-lipoxygenase: role of lipoxygenase products in platelet activation. J Lipid Res 53 (12):2546–2559
21. Ivanov I, Heydeck D, Hofheinz K et al (2010) Molecular enzymology of lipoxygenases. Arch Biochem Biophys 503(2):161–174
22. Sutherland M, Shankaranarayanan P, Schewe T, Nigam S (2001) Evidence for the presence of phospholipid hydroperoxide glutathione peroxidase in human platelets: implications for its involvement in the regulatory network of the 12-lipoxygenase pathway of arachidonic acid metabolism. Biochem J 353(Pt 1):91–100
23. Ozeki Y, Nagamura Y, Ito H et al (1999) An anti-platelet agent, OPC-29030, inhibits translocation of 12-lipoxygenase and 12-hydroxyeicosatetraenoic acid production in human platelets. Br J Pharmacol 128(8):1699–1704
24. Coffey MJ, Jarvis GE, Gibbins JM et al (2004) Platelet 12-lipoxygenase activation via glycoprotein VI: involvement of multiple signaling pathways in agonist control of H(P) ETE synthesis. Circ Res 94(12):1598–1605
25. Morgan LT, Thomas CP, Kuhn H, O'Donnell VB (2010) Thrombin-activated human platelets acutely generate oxidized docosahexaenoic-acid-containing phospholipids via 12-lipoxygenase. Biochem J 431(1):141–148
26. Reed K, Tucker D, Aloulou A et al (2011) Functional characterization of mutations in inherited human cPLA,ÇÇ deficiency. Biochemistry 50(10):1731–1738
27. Adler DH, Cogan JD, Phillips JA 3rd et al (2008) Inherited human cPLA(2alpha) deficiency is associated with impaired eicosanoid biosynthesis, small intestinal ulceration, and platelet dysfunction. J Clin Invest 118(6):2121–2131
28. Wong DA, Kita Y, Uozumi N, Shimizu T (2002) Discrete role for cytosolic phospholipase A (2)alpha in platelets: studies using single and double mutant mice of cytosolic and group IIA secretory phospholipase A(2). J Exp Med 196(3):349–357
29. Schafer A (1982) Deficiency of platelet lipoxygenase activity in myeloproliferative disorders. N Engl J Med 306(7):381–386
30. Okuma M, Kanaji K, Ushikubi F et al (1989) Reduced 12-lipoxygenase activity in platelets of patients with myeloproliferative disorders. Adv Prostaglandin Thromboxane Leukot Res 19:148–151
31. Matsuda S, Murakami J, Yamamoto Y et al (1993) Decreased messenger RNA of arachidonate 12-lipoxygenase in platelets of patients with myeloproliferative disorders. Biochim Biophys Acta 1180(3):243–249

32. Nyby MD, Sasaki M, Ideguchi Y et al (1996) Platelet lipoxygenase inhibitors attenuate thrombin- and thromboxane mimetic-induced intracellular calcium mobilization and platelet aggregation. J Pharmacol Exp Ther 278(2):503–509
33. Yeung J, Apopa PL, Vesci J et al (2013) 12-lipoxygenase activity plays an important role in PAR4 and GPVI-mediated platelet reactivity. Thromb Haemost 110(3):569–581
34. Svensson Holm AC, Grenegard M, Ollinger K, Lindstrom EG (2014) Inhibition of 12-lipoxygenase reduces platelet activation and prevents their mitogenic function. Platelets 25(2):111–117
35. Yeung J, Tourdot BE, Fernandez-Perez P et al (2014) Platelet 12-LOX is essential for FcgammaRIIa-mediated platelet activation. Blood 124:2271–2279
36. Yeung J, Holinstat M (2011) 12-lipoxygenase: a potential target for novel anti-platelet therapeutics. Cardiovasc Hematol Agents Med Chem 9(3):154–164
37. Kenyon V, Rai G, Jadhav A et al (2011) Discovery of potent and selective inhibitors of human platelet-type 12-lipoxygenase. J Med Chem 54(15):5485–5497
38. Luci DK, Jameson JB 2nd, Yasgar A et al (2014) Synthesis and structure-activity relationship studies of 4-((2-hydroxy-3-methoxybenzyl)amino)benzenesulfonamide derivatives as potent and selective inhibitors of 12-lipoxygenase. J Med Chem 57(2):495–506
39. Johnson EN, Brass LF, Funk CD (1998) Increased platelet sensitivity to ADP in mice lacking platelet-type 12-lipoxygenase. Proc Natl Acad Sci USA 95(6):3100–3105
40. Yeung J, Apopa P, Vesci J et al (2012) Protein kinase C regulation of 12-lipoxygenase-mediated human platelet activation. Mol Pharmacol 81(3):420–430
41. Ozeki Y, Ito H, Nagamura Y, Unemi F, Igawa T (1998) 12(S)-HETE plays a role as a mediator of expression of platelet CD62 (P-selectin). Platelets 9(5):297–302
42. Holinstat M, Boutaud O, Apopa P et al (2011) Protease-activated receptor signaling in platelets activates cytosolic phospholipase A2α differently for cyclooxygenase-1 and 12-lipoxygenase catalysis. Arterioscler Thromb Vasc Biol 31(2):435–442
43. Burzaco J, Conde M, Parada LA, Zugaza JL, Dehaye JP, Marino A (2013) ATP antagonizes thrombin-induced signal transduction through 12(S)-HETE and cAMP. PLoS One 8(6): e67117
44. Sekiya F, Takagi J, Usui T et al (1991) 12S-hydroxyeicosatetraenoic acid plays a central role in the regulation of platelet activation. Biochem Biophys Res Commun 179(1):345–351
45. Calzada C, Vericel E, Lagarde M (1997) Low concentrations of lipid hydroperoxides prime human platelet aggregation specifically via cyclo-oxygenase activation. Biochem J 325 (Pt 2):495–500
46. Takenaga M, Hirai A, Terano T, Tamura Y, Kitagawa H, Yoshida S (1986) Comparison of the in vitro effect of eicosapentaenoic acid (EPA)-derived lipoxygenase metabolites on human platelet function with those of arachidonic acid. Thromb Res 41(3):373–384
47. Aharony D, Smith JB, Silver MJ (1982) Regulation of arachidonate-induced platelet aggregation by the lipoxygenase product, 12-hydroperoxyeicosatetraenoic acid. Biochim Biophys Acta 718(2):193–200
48. Sekiya F, Takagi J, Sasaki K et al (1990) Feedback regulation of platelet function by 12S-hydroxyeicosatetraenoic acid: inhibition of arachidonic acid liberation from phospholipids. Biochim Biophys Acta 1044(1):165–168
49. Setty BN, Werner MH, Hannun YA, Stuart MJ (1992) 15-hydroxyeicosatetraenoic acid-mediated potentiation of thrombin-induced platelet functions occurs via enhanced production of phosphoinositide-derived second messengers—sn-1,2-diacylglycerol and inositol-1,4,5-trisphosphate. Blood 80(11):2765–2773
50. Chang J, Blazek E, Kreft AF, Lewis AJ (1985) Inhibition of platelet and neutrophil phospholipase A2 by hydroxyeicosatetraenoic acids (HETES). A novel pharmacological mechanism for regulating free fatty acid release. Biochem Pharmacol 34(9):1571–1575
51. Fonlupt P, Croset M, Lagarde M (1991) 12-HETE inhibits the binding of PGH2/TXA2 receptor ligands in human platelets. Thromb Res 63(2):239–248

52. Siegel MI, McConnell RT, Abrahams SL, Porter NA, Cuatrecasas P (1979) Regulation of arachidonate metabolism via lipoxygenase and cyclo-oxygenase by 12-HPETE, the product of human platelet lipoxygenase. Biochem Biophys Res Commun 89(4):1273–1280
53. Morita I, Takahashi R, Saito Y, Murota S (1983) Stimulation of eicosapentaenoic acid metabolism in washed human platelets by 12-hydroperoxyeicosatetraenoic acid. J Biol Chem 258(17):10197–10199
54. Calzada C, Vericel E, Mitel B, Coulon L, Lagarde M (2001) 12(S)-hydroperoxy-eicosatetraenoic acid increases arachidonic acid availability in collagen-primed platelets. J Lipid Res 42(9):1467–1473
55. Guo Y, Zhang W, Giroux C et al (2011) Identification of the orphan G protein-coupled receptor GPR31 as a receptor for 12-(S)-hydroxyeicosatetraenoic acid. J Biol Chem 286 (39):33832–33840
56. Hampson AJ, Grimaldi M (2002) 12-hydroxyeicosatetrenoate (12-HETE) attenuates AMPA receptor-mediated neurotoxicity: evidence for a G-protein-coupled HETE receptor. J Neurosci 22(1):257–264
57. Thomas CP, Morgan LT, Maskrey BH et al (2010) Phospholipid-esterified eicosanoids are generated in agonist-activated human platelets and enhance tissue factor-dependent thrombin generation. J Biol Chem 285(10):6891–6903
58. Morgan A, Dioszeghy V, Maskrey B et al (2009) Phosphatidylethanolamine-esterified eicosanoids in the mouse: tissue localization and inflammation-dependent formation in Th-2 disease. J Biol Chem 284(32):21185–21191
59. Faraci FM, Sobey CG, Chrissobolis S, Lund DD, Heistad DD, Weintraub NL (2001) Arachidonate dilates basilar artery by lipoxygenase-dependent mechanism and activation of K(+) channels. Am J Physiol Regul Integr Comp Physiol 281(1):R246–253
60. Turner SR, Tainer JA, Lynn WS (1975) Biogenesis of chemotactic molecules by the arachidonate lipoxygenase system of platelets. Nature 257(5528):680–681
61. Palmer RM, Stepney RJ, Higgs GA, Eakins KE (1980) Chemokinetic activity of arachidonic and lipoxygenase products on leuocyctes of different species. Prostaglandins 20(2):411–418
62. Sultana C, Shen Y, Rattan V, Kalra VK (1996) Lipoxygenase metabolites induced expression of adhesion molecules and transendothelial migration of monocyte-like HL-60 cells is linked to protein kinase C activation. J Cell Physiol 167(3):477–487
63. Dobrian AD, Lieb DC, Cole BK, Taylor-Fishwick DA, Chakrabarti SK, Nadler JL (2011) Functional and pathological roles of the 12- and 15-lipoxygenases. Prog Lipid Res 50 (1):115–131
64. Bunce PE, High SM, Nadjafi M, Stanley K, Liles WC, Christian MD (2011) Pandemic H1N1 influenza infection and vascular thrombosis. Clin Infect Dis 52(2):e14–17
65. Boilard E, Pare G, Rousseau M et al (2014) Influenza virus H1N1 activates platelets through FcgammaRIIA signaling and thrombin generation. Blood 123(18):2854–2863
66. Manolakis A, Kapsoritakis AN, Potamianos SP (2007) A review of the postulated mechanisms concerning the association of Helicobacter pylori with ischemic heart disease. Helicobacter 12(4):287–297
67. Corcoran PA, Atherton JC, Kerrigan SW et al (2007) The effect of different strains of Helicobacter pylori on platelet aggregation. Can J Gastroenterol 21(6):367–370
68. Wassermann GE, Olivera-Severo D, Uberti AF, Carlini CR (2010) Helicobacter pylori urease activates blood platelets through a lipoxygenase-mediated pathway. J Cell Mol Med 14 (7):2025–2034
69. Ross R (1999) Atherosclerosis—an inflammatory disease. N Engl J Med 340(2):115–126
70. Nakao J, Ito H, Chang WC, Koshihara Y, Murota S (1983) Aortic smooth muscle cell migration caused by platelet-derived growth factor is mediated by lipoxygenase product (s) of arachidonic acid. Biochem Biophys Res Commun 112(3):866–871
71. Nakao J, Ooyama T, Ito H, Chang WC, Murota S (1982) Comparative effect of lipoxygenase products of arachidonic acid on rat aortic smooth muscle cell migration. Atherosclerosis 44 (3):339–342

72. Preston IR, Hill NS, Warburton RR, Fanburg BL (2006) Role of 12-lipoxygenase in hypoxia-induced rat pulmonary artery smooth muscle cell proliferation. Am J Physiol Lung Cell Mol Physiol 290(2):L367–374
73. Nieves D, Moreno JJ (2008) Enantioselective effect of 12(S)-hydroxyeicosatetraenoic acid on 3T6 fibroblast growth through ERK 1/2 and p38 MAPK pathways and cyclin D1 activation. Biochem Pharmacol 76(5):654–661
74. Walenga RW, Boone S, Stuart MJ (1987) Analysis of blood HETE levels by selected ion monitoring with ricinoleic acid as the internal standard. Prostaglandins 34(5):733–748
75. Vijil C, Hermansson C, Jeppsson A, Bergstrom G, Hulten LM (2014) Arachidonate 15-lipoxygenase enzyme products increase platelet aggregation and thrombin generation. PLoS One 9(2):e88546
76. Maclouf J (1993) Transcellular biosynthesis of arachidonic acid metabolites: from in vitro investigations to in vivo reality. Baillieres Clin Haematol 6(3):593–608
77. Marcus AJ, Hajjar DP (1993) Vascular transcellular signaling. J Lipid Res 34(12):2017–2031
78. Brady HR, Papayianni A, Serhan CN (1994) Leukocyte adhesion promotes biosynthesis of lipoxygenase products by transcellular routes. Kidney Int Suppl 45:S90–97
79. Lindgren JA, Edenius C (1993) Transcellular biosynthesis of leukotrienes and lipoxins via leukotriene A4 transfer. Trends Pharmacol Sci 14(10):351–354
80. Fiore S, Ryeom SW, Weller PF, Serhan CN (1992) Lipoxin recognition sites. Specific binding of labeled lipoxin A4 with human neutrophils. J Biol Chem 267(23):16168–16176
81. Romano M, Chen XS, Takahashi Y, Yamamoto S, Funk CD, Serhan CN (1993) Lipoxin synthase activity of human platelet 12-lipoxygenase. Biochem J 296(Pt 1):127–133
82. Serhan CN, Sheppard KA (1990) Lipoxin formation during human neutrophil-platelet inter-actions. Evidence for the transformation of leukotriene A4 by platelet 12-lipoxygenase in vitro. J Clin Invest 85(3):772–780
83. Weber PC, Fischer S (1984) Arachidonic acid and eicosapentaenoic acid metabolism in platelets and vessel walls. Med Biol 62(2):129
84. Serhan CN, Romano M (1995) Lipoxin biosynthesis and actions: role of the human platelet LX-synthase. J Lipid Mediat Cell Signal 12(2–3):293–306
85. Freire-Moar J, Alavi-Nassab A, Ng M, Mulkins M, Sigal E (1995) Cloning and characteri-zation of a murine macrophage lipoxygenase. Biochim Biophys Acta 1254(1):112–116
86. Berger M, Schwarz K, Thiele H et al (1998) Simultaneous expression of leukocyte-type 12-lipoxygenase and reticulocyte-type 15-lipoxygenase in rabbits. J Mol Biol 278(5):935–948
87. Fleming J, Thiele BJ, Chester J et al (1989) The complete sequence of the rabbit erythroid cell-specific 15-lipoxygenase mRNA: comparison of the predicted amino acid sequence of the erythrocyte lipoxygenase with other lipoxygenases. Gene 79(1):181–188
88. Dadaian M, Westlund P (1999) Albumin modifies the metabolism of hydroxyeicosatetraenoic acids via 12-lipoxygenase in human platelets. J Lipid Res 40(5):940–947
89. Westlund P, Palmblad J, Falck JR, Lumin S (1991) Synthesis, structural identification and biological activity of 11,12-dihydroxyeicosatetraenoic acids formed in human platelets. Biochim Biophys Acta 1081(3):301–307
90. Maclouf J, Kindahl H, Granstrom E, Samuelsson B (1980) Interactions of prostaglandin H2 and thromboxane A2 with human serum albumin. Eur J Biochem 109(2):561–566
91. Fitzpatrick FA, Morton DR, Wynalda MA (1982) Albumin stabilizes leukotriene A4. J Biol Chem 257(9):4680–4683
92. Flohé L (1989) Glutathione: chemical, biochemical, and medical aspects (Dolphin, David.; Avramovi, Olga.; Poulson, Rozanne.). Wiley, New York
93. Marshall PJ, Kulmacz RJ, Lands WE (1987) Constraints on prostaglandin biosynthesis in tissues. J Biol Chem 262(8):3510–3517
94. Ludwig P, Holzhutter HG, Colosimo A, Silvestrini MC, Schewe T, Rapoport SM (1987) A kinetic model for lipoxygenases based on experimental data with the lipoxygenase of reticulocytes. Eur J Biochem 168(2):325–337

95. Hecker G, Utz J, Kupferschmidt RJ, Ullrich V (1991) Low levels of hydrogen peroxide enhance platelet aggregation by cyclooxygenase activation. Eicosanoids 4(2):107–113
96. Schewe T, Rapoport SM, Kuhn H (1986) Enzymology and physiology of reticulocyte lipoxygenase: comparison with other lipoxygenases. Adv Enzymol Relat Areas Mol Biol 58:191–272
97. Yamamoto S, Suzuki H, Ueda N (1997) Arachidonate 12-lipoxygenases. Prog Lipid Res 36 (1):23–41
98. Honn KV, Tang DG, Gao X et al (1994) 12-lipoxygenases and 12(S)-HETE: role in cancer metastasis. Cancer Metastasis Rev 13(3–4):365–396
99. Pace-Asciak CR (1994) Hepoxilins: a review on their cellular actions. Biochim Biophys Acta 1215(1–2):1–8
100. Lin Z, Laneuville O, Pace-Asciak CR (1991) Hepoxilin A3 induces heat shock protein (HSP72) expression in human neutrophils. Biochem Biophys Res Commun 179(1):52–56
101. Jacquier-Sarlin MR, Fuller K, Dinh-Xuan AT, Richard MJ, Polla BS (1994) Protective effects of hsp70 in inflammation. Experientia 50(11–12):1031–1038
102. Larson MK, Shearer GC, Ashmore JH et al (2011) Omega-3 fatty acids modulate collagen signaling in human platelets. Prostaglandins Leukot Essent Fatty Acids 84(3–4):93–98
103. von Schacky C, Fischer S, Weber PC (1985) Long-term effects of dietary marine omega-3 fatty acids upon plasma and cellular lipids, platelet function, and eicosanoid formation in humans. J Clin Invest 76(4):1626–1631
104. Boberg M, Vessby B, Selinus I (1986) Effects of dietary supplementation with n-6 and n-3 long-chain polyunsaturated fatty acids on serum lipoproteins and platelet function in hypertriglyceridaemic patients. Acta Med Scand 220(2):153–160
105. Marra F, Riccardi D, Melani L et al (1998) Effects of supplementation with unsaturated fatty acids on plasma and membrane lipid composition and platelet function in patients with cirrhosis and defective aggregation. J Hepatol 28(4):654–661
106. Dyerberg J, Bang HO (1979) Haemostatic function and platelet polyunsaturated fatty acids in Eskimos. Lancet 2(8140):433–435
107. Knauss HJ, Sheffner AL (1967) Effect of unsaturated fatty acid supplements upon mortality and clotting parameters in rats fed thrombogenic diets. J Nutr 93(3):393–400
108. Tamura Y, Hirai A, Terano T et al (1986) Clinical and epidemiological studies of eicosapentaenoic acid (EPA) in Japan. Prog Lipid Res 25(1–4):461–466
109. Lorenz R, Spengler U, Fischer S, Duhm J, Weber PC (1983) Platelet function, thromboxane formation and blood pressure control during supplementation of the Western diet with cod liver oil. Circulation 67(3):504–511
110. Siess W, Roth P, Scherer B, Kurzmann I, Bohlig B, Weber PC (1980) Platelet-membrane fatty acids, platelet aggregation, and thromboxane formation during a mackerel diet. Lancet 1 (8166):441–444

Lipoxygenases and Cardiovascular Diseases

Andrés Laguna-Fernández, Marcelo H. Petri, Silke Thul, and Magnus Bäck

Abstract The lipoxygenase (LO) family of enzymes metabolize fatty acids into bioactive lipid mediators that exert potent actions on inflammatory reactions related to several cardiovascular diseases, such as atherosclerosis. The polyunsaturated omega-6 fatty acid arachidonic acid serves as a substrate for 5-LO, 12- and 15-LO, which catalyzes the formation of several bioactive lipid mediators. For example, 5-LO-derived leukotrienes transduce pro-inflammatory signaling in leukocytes and within the vascular wall. Targeting leukotriene receptors reduces experimental atherosclerosis, and pharmacoepidemiological studies indicate that leukotriene receptor antagonism is associated with a decreased cardiovascular risk. In contrast, sequential lipoxygenation of arachidonic acid yields lipoxins, which are anti-inflammatory and transduce the resolution of inflammation. The FPR2/ALX receptor is activated by both lipoxins and peptide agonists, and in murine models of atherosclerosis, FPR2/ALX deletion decreases atherosclerotic lesion size but increases atherosclerotic plaque instability. Finally, omega-3 essential fatty acids may serve as substrate for the LO enzymes yielding mediators that promote inflammation resolution, and omega-3 supplementation reduces experimental atherosclerosis. In conclusion, it is important to fully consider and explore all possible pathways of LO metabolism and their downstream metabolites when considering the role of the 5-, 12- and 15-LO pathways in cardiovascular disease.

A. Laguna-Fernández • M.H. Petri • S. Thul
Translational Cardiology, Center for Molecular Medicine, L8:03, Karolinska University Hospital, 171 76 Stockholm, Sweden

Department of Medicine, Karolinska Institutet, 171 76 Stockholm, Sweden

M. Bäck (✉)
Translational Cardiology, Center for Molecular Medicine, L8:03, Karolinska University Hospital, 171 76 Stockholm, Sweden

Department of Medicine, Karolinska Institutet, 171 76 Stockholm, Sweden

Department of Cardiology, Karolinska University Hospital, Stockholm, Sweden
e-mail: Magnus.Back@ki.se

1 Introduction

The lipoxygenase (LO) family of enzymes metabolize fatty acids into bioactive lipid mediators that exert potent actions on inflammatory reactions related to several cardiovascular diseases, such as atherosclerosis [1, 2], abdominal aortic aneurysms [3], aortic valve stenosis [4] and hypertension [5]. The present chapter will briefly review the biosynthetic pathways involving 5-LO, 12-LO and 15-LO in the cardiovascular system, and then focus on the potential role of LO-derived lipid mediators in atherosclerosis-related cardiovascular diseases.

2 LO Metabolism

2.1 LO Metabolism of Arachidonic Acid

Arachidonic acid (AA) is a polyunsaturated omega-6 fatty acid that serves as a substrate for 5-LO, which together with the 5-LO activating protein (FLAP), catalyzes the formation of the unstable leukotriene (LT) precursor LTA_4. Through subsequent enzymatic steps, LTA_4 is either hydrolyzed into LTB_4 or conjugated with glutathione to form the cysteinyl-LTs, i.e. LTC_4, LTD_4, and LTE_4. The structure of four double bonds is common for LTB_4 and the cysteinyl-LTs, and all together they constitute the 4-Series LTs (Fig. 1). 5-LO is in general restricted to myeloid cells, such as granulocytes, macrophages and mast cells, which also express downstream enzymes to synthesize LTs from AA [6]. Cells lacking or expressing low levels of 5-LO, such as platelets, endothelial cells and smooth muscle cells, may however participate in LT biosynthesis by means of accepting LTA_4 donated from other 5-LO expressing cells; a process referred to as transcellular metabolism [7]. It should also be considered that 5-LO expression is regulated by promoter methylation [8], and that 5-LO expression can be induced in non-myeloid cells by epigenetic mechanisms [9, 10].

AA lipoxygenation by 12-LO leads to the synthesis of a variety of products; of which for example 12(S)-hydroxyeicosatetraenoic acid (12(S)HETE; Fig. 1) and its effects on endothelial cells and smooth muscle cells have been widely studied. There are several isoforms of 12-LO, and in the cardiovascular system, platelets represent the most abundant source.

The final LO addressed in this chapter is 15-LO, for which there are two isoforms, referred to as reticulocyte type (15-LO-1) and epidermis type (15-LO-2). The main product derived from 15-LO metabolism of AA is 15(S)-HETE, which exerts effects on neutrophil migration [11], and modulates vascular tone through activation of both endothelial and smooth muscle cells [12].

Whereas AA metabolism by the individual LOs hence directly yields bioactive mediators with potent actions in the cardiovascular systems, dual lipoxygenation of AA by sequential action of either 15-LO and 5-LO on the one hand, or 5-LO and

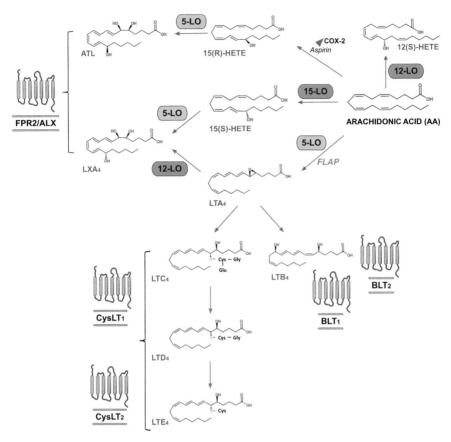

Fig. 1 Lipid mediators derived from lipoxygenase metabolism of arachidonic acid in human cells. The different lipoxygenases (5-, 12- and 15-LO) act either alone or by sequential metabolism to generate bioactive lipids. The specific receptors activated by the respective mediators are also indicated

12-LO on the other, yields a group of lipid mediators referred to as lipoxins (LXs; Fig. 1). This family of lipid mediators is composed of LXA_4, LXB_4 and the 15-epimer of LXA_4 (15-epi LXA_4). The latter mediator is synthesized through 5-LO metabolism of 15(R)HETE, a metabolite synthesized from AA by acetylated cyclooxygenase (COX)-2, as depicted in Fig. 1, and is therefore also referred to as aspirin-triggered lipoxin (ATL). In contrast to other AA-derived LO metabolites, LXs exert multiple anti-inflammatory actions, and participate in the resolution of inflammation [13] with possible implications for cardiovascular disease, as will be discussed in detail in the following sections.

2.2 LO Metabolism of Omega 3 Fatty Acids

In addition to AA, also omega-3 essential polyunsaturated fatty acids, such as eicosapentanoic acid (EPA) and docosahexaenoic acid (DHA), serve as substrates for LOs in human cells (Fig. 2), which has several implications for cardiovascular inflammatory circuits. First, when metabolized by 5-LO, EPA generates LTs with five double bonds (5-series LTs, for example LTB_5; see Fig. 2) which are less biologically active compared with the 4-series LTs but compete with LT ligation at the LT receptors, suggesting that 5-LO metabolites of omega-3 fatty acids may act as inhibitors of inflammation [14]. As depicted in Fig. 2, omega-3 fatty acids can follow multiple biosynthetic pathways involving both different LOs and COX-2. As described above for ATL biosynthesis, after aspirin-induced acetylation, COX-2 can introduce a 18R hydroperoxy-group into the EPA molecule leading to formation of 18-Hydroxyeicosapentaenoic acid (18-HEPE), which serves as substrate for 5-LO to generate Resolvin E1 (RvE1), where the E stands for EPA-derived (Fig. 2). Furthermore, with the omega-3 fatty acid DHA as substrate, either acetylated COX-2 followed by 5-LO, or dual lipoxygenation by 5- and 15-LO will yield the D-series of Resolvins (D for DHA-derived). The action of the three LOs on DHA, by means of dual lipoxygenation or in combination with the action of additional enzymes, also yields other bioactive lipids referred to as protectins and maresins. Importantly, all the LO-derived metabolites of omega-3 fatty acids share the characteristics of being mediators of the resolution of inflammation, as has been described by Serhan et al. [13, 15].

3 Receptors for LO Products

3.1 LT Receptors

LTs exert their actions via 7-transmembrane G-Protein-Coupled Receptors (GPCRs) divided into two subclasses: BLT receptors, activated by LTB_4; and CysLT receptors, activated by the cysteinyl-LTs. These LT receptor classes are further subdivided into BLT_1 and BLT_2, as well as $CysLT_1$ and $CysLT_2$, respectively [16] (Fig. 1). BLT receptors were initially identified on phagocytic leukocytes. Subsequent studies revealed BLT also transduced responses in lymphocyte populations, suggesting LTB_4 as a mediator of both innate and adaptive immune responses. Chemotaxis, one of the principal effects of LTB_4, occurs via activation of the BLT_1 receptor subtype, which is the high affinity LTB_4 receptor. The gene encoding the second subtype of BLT receptor, BLT_2, was identified during the analysis of the BLT_1 promoter [17]. This receptor exhibits lower affinity for LTB_4 compared with the BLT_1 receptor, and in addition responds to other LO products, such as 12-epi LTB_4, 12(S)-HETE and 15(S)-HETE. [16]. Importantly, BLT_1 receptor expression is not limited to leukocytes, since also non-myeloid cells,

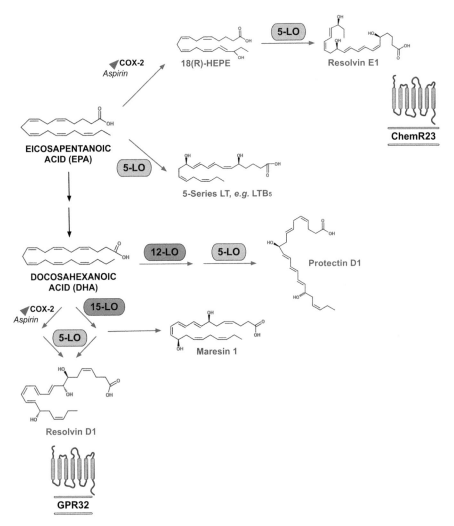

Fig. 2 Lipid mediators derived from lipoxygenase metabolism of omega-3 fatty acids in human cells. The essential omega-3 fatty acids eicosapentanoic acid (EPA) and docosahexaenoic acid (DHA), serve as substrates for LOs in human cells. When metabolized by 5-LO, EPA generates leukotrienes with five double bonds (5-series LTs, for example LTB_5) which inhibit leukotriene-induced responses. The omega-3 fatty acids can also follow multiple biosynthetic pathways involving both different lipoxygenases (5-, 12- and 15-LO) and cyclooxygenase (COX)-2. The specific receptors activated by the respective mediators are also indicated

such as vascular smooth muscle and endothelial cells are activated by LTB_4 by means of BLT_1 signaling, associated with vascular smooth muscle cell migration and proliferation [18], as well as endothelium-dependent responses [19].

The potent bronchoconstrictive effects of cysteinyl-LTs in asthma are mediated through the $CysLT_1$ receptor, which is the target for the clinically introduced anti-asthmatic LT receptor antagonists [20]. The second CysLT receptor, $CysLT_2$, exhibits a wider expression pattern, extending also to cardiovascular and cerebral tissues. In addition, CysLT receptors are also expressed on several leukocyte populations, mediating for example cell adhesion and migration, as well as release of cytokines and other inflammatory mediators. A recent observation is the nuclear/perinuclear expression pattern of CysLT receptors, which may be coupled to nuclear calcium signalling, cell survival, mitogenesis, and alterations of gene expression in different cell populations in the cardiovascular system [4, 21, 22].

3.2 FPR2/ALX

LXA_4 and ATL, as well as RvD1 and aspirin-triggered RvD1, transduce anti-inflammatory and pro-resolution responses by means of the FPR2/ALX receptor. This receptor in addition binds prokaryotic peptides initiated with an N-formylmethionine as well as a number of endogenous ligands, such as amyloidogenic and antibacterial peptides transducing pro-inflammatory responses [23]. Due to the particularity of responding both to LO products and to a number of peptide/protein agonists, this receptor has been denoted FPR2/ALX (formyl peptide receptor 2 and A type lipoxin receptor) [23] (Fig. 1). FPR2/ALX is mainly expressed by phagocytic leukocytes, endothelial and vascular smooth muscle cells, neurons, astrocytes and hepatocytes.

When activated by the lipid agonists, FPR2/ALX reduces neutrophil infiltration/activation and promotes the resolution of inflammation [24]. On the other hand, leukocyte chemotaxis has been described following FPR2/ALX activation by prokaryotic peptides [23]. These observations point to a dual role of FPR2/ALX in inflammation, and the implication of this duality in atherosclerosis will be further discussed below

3.3 GPR32

In addition to FPR2/ALX, the actions of RvD1 and LXA_4 in humans may also be mediated by G-protein coupled receptor 32 (GPR32) [25] (Fig. 1). GPR32 is in contrast a pseudogene in mice and rats, and so far no murine ortholog for GPR32 has been described [26]. Activation of GPR32 does not induce Ca^{2+} mobilization or modify cAMP levels [25], but RvD1 signaling through by GPR32 may inhibit the MAPK and NF-κB pathways leading to decreased expression of pro-inflammatory cytokines in for example human airway epithelial cells [27]. In addition to monocytes, macrophages and granulocytes, GPR32 expression has also been detected in the vascular wall [28].

3.4 ChemR23

The receptor activated by RvE1 is referred to as ChemR23 (Fig. 1) and also termed Chemokine-like receptor 1 [29]. In humans, ChemR23 expression is detected in the cardiovascular system, brain, kidney, gastrointestinal tissues and myeloid tissues. Notably, ChemR23 is highly expressed in adipocytes, monocytes, macrophages, dendritic cells and endothelial cells. In the latter cell type, pro-inflammatory cytokines, such as TNF-α, IL-1β, and IL-6, up-regulate ChemR23 expression [30]. ChemR23 was originally described as the receptor of chemerin, an adipokine that regulates adipogenesis and adipocyte metabolism [31]. Interestingly, chemerin plays a dual role on inflammation, promoting chemotaxis of immature dendritic cells and macrophages [32] on the one hand, and preventing endothelial cells activation by means of NF-κB and p38 inhibition [33] on the other hand. However, the role of ChemR23 in several chemerin-induced responses remains to be established. RvE1 activation of ChemR23 in macrophages reduces the activation of NF-κB and down-regulates cytokine expression [34]. In different in vitro models, RvE1 signals through phosphorylation of Akt and its downstream targets, thus enhancing macrophage phagocytosis [35].

Pharmacokinetic experiments show that although the extent of ChemR23 activation evoked by chemerin was three times higher than the one obtained with RvE1, the latter agonist was one order of magnitude more potent than chemerin at inhibiting NF-κB activation. These results suggest that RvE1 may be the preferential agonist for ChemR23, and that RvE1 and chemerin may differentially interact with ChemR23 [36].

4 Atherosclerosis

4.1 Pathophysiology and Clinical Context

Atherosclerosis is a chronic inflammatory disease of the vascular wall. Early atherosclerotic lesions are characterized by a thickening of the intima layer, lipid accumulation and macrophage infiltration [37]. Activated endothelial cells participate in the recruitment of immune cells, which secrete cytokines (such as TNF-α, IL-1β and interferon [IFN]-γ) and lipid mediators, including LO-derived mediators. The modification of lipids inside the atherosclerotic plaque, in particular oxidation, plays a critical role in the development of the lesion [37]. 15-LO (or its ortholog 12/15-LO in mice) may play a key role in the oxidation of low density lipoproteins (LDL). The latter notion was raised based on the colocalization of 15-LO with epitopes of oxidized LDL in macrophage-rich areas of atherosclerotic lesions [38]. In addition, transfer of human 15-LO gene into rabbit iliac arteries produces the appearance of oxidation-specific lipid-protein adducts characteristic of oxidized

LDL [39]. Further studies in mice have revealed that LDL uptake in macrophages through the LDL receptor-related protein (LRP) is necessary for the 12/15-LO oxidation of LDL [40].

As the atherosclerotic plaque develops, smooth muscle cells and collagen form a cap to stabilize the lesion. Degradation of this fibrous cap surrounding the core of the atherosclerotic lesion, induces plaque rupture provoking acute thrombosis and occlusion of the blood vessel, leading to for example myocardial infarction or stroke.

4.2 LO Expression and Metabolism in Human Atherosclerotic Lesions

The release of LTB_4 from human vessels is higher in atherosclerotic lesions compared with healthy arteries [41, 42]. Indeed, the enzymatic machinery for LT biosynthesis is present in human atherosclerotic lesions, as detected by immunohistochemistry [43, 44], Western blot [45, 46] and mRNA analysis [47]. Some of the latter studies associated 5-LO expression levels with increased atherosclerotic lesion instability. Clinical observations and correlations with protease activity suggest that an increase in 5-LO expression and LT production could potentially induce plaque instability, which precedes rupture and thus causes the clinical event. However, a recent study in human carotid atherosclerotic lesions did not reveal any significant correlation between LTB_4 content and histological features in terms of stable/unstable characteristics [42]. A possible interpretation of the latter findings could be that cysteinyl-LTs are the main drivers of plaque instability. It should be taken into account that each method of associating enzyme and lipid levels with plaque characteristics has its limitations and should be interpreted with care in the absence of a mechanistic evaluation. The possible signaling pathways linking LT signaling to atherosclerotic lesion development, progression and eventually plaque rupture will be further discussed below.

In comparison with 5-LO, the expression levels of 15-LO-1 in the aortas, as well as in coronary and carotid arteries are several orders of magnitude lower, and are not related to the degree of lesion pathology [44]. Nevertheless, a subsequent immunohistological study revealed that 15-LO-2 is highly expressed in human carotid plaques, and colocalizes with macrophages [48]. Also analyses of mRNA levels have confirmed that 15-LO-2 is the dominant subtype in atherosclerotic lesions [49]. In the latter study, 15-LO-2 mRNA levels were higher in carotid atherosclerotic lesions derived from patients with clinical manifestations of cerebral ischemia compared with asymptomatic lesions [49], suggesting a possible link between 15-LO pathways and plaque instability.

Finally, analyses of mRNA levels have failed to detect significant levels of the 12-LO subtypes in human atherosclerotic lesions [49]. It should however be taken into consideration that platelets may represent a significant source of 12-LO in atherosclerotic pathologies. Since platelets contribution to whole lesion mRNA levels is likely to be negligible, the role of 12-LO may be underestimated in whole tissue transcriptional analyses.

In support of significant 5-, 12-, and 15-LO activity in atherosclerosis, selective lipidomic analyses have characterized the metabolites of all three LOs as the predominant eicosanoids produced by human atherosclerotic lesions [50]. Interestingly, and in support of a role of 12-LO in atherosclerosis, 12-HETE (Fig. 1), is the most abundant LO metabolite produced in atherosclerotic lesions [50].

Taken together, the above-mentioned studies have provided convincing evidence for an active LO metabolism in atherosclerosis, and that this process takes place locally within atherosclerotic lesions. In the following sections, the potential role of these mediators in atherosclerosis will be discussed.

4.3   LT Signaling in Atherosclerosis

The local LTB_4 production leads to activation of BLT_1 and BLT_2 receptors, which have been detected in macrophage-rich areas within human carotid atherosclerotic lesions [18]. Since macrophages represent a major source of 5-LO, LTB_4-induced monocyte recruitment could potentially further increase the inflammatory activity at sites of atherosclerotic lesions. As mentioned above, LTB_4 also induces chemotaxis of T-lymphocytes, which accordingly accumulate in the vicinity of 5-LO-positive macrophages in atherosclerotic lesions [51]. The immunological signaling pathways transducing the atherogenic effects of LTs might also be mediated through enhancement of other pro-inflammatory pathways. For example both LTB_4 and cysteinyl-LTs up-regulate the expression of monocyte chemoattractant protein (MCP)-1, promoting subsequent MCP-1-induced chemotaxis and further amplification of the inflammatory response [52].

In addition to immune cells, immunohistochemical analysis of human atherosclerotic lesions have also revealed that smooth muscle cells express both BLT [18] and CysLT [21] receptors, which transduce LT-induced effects with implications for intimal hyperplasia, as will be further discussed below. Furthermore, LT receptor activation is associated with release of matrix metalloproteinases (MMPs) [53], leading to collagen degradation and weakening of the fibrous cap, with potential implications for unstable atherosclerotic lesions and eventually plaque rupture.

4.4   LXs Signaling: Implications for Atherosclerosis

LXA_4 is locally produced in coronary atherosclerotic lesions, and the levels increase after aspirin treatment [54]. Furthermore, plasma levels of 15-epi LXA_4

are significantly lower in patients with symptomatic peripheral artery disease compared with healthy volunteers [55], suggesting that decreased levels of these lipid mediators may indicate a failure in the resolution of inflammation. Although no study has specifically addressed the LX-induced effects on cells derived from subjects with atherosclerosis, several of the established effects of LXA_4 on inflammatory cells could potentially have beneficial effects in atherosclerosis. For example, LXA_4 decreases pro-inflammatory cytokines production in monocytic cells [56]. A potential anti-inflammatory role of LXA_4 in atherosclerosis may also include effects on adaptive immunological circuits, as suggested by a reduced TNF-α production after CD3 stimulation of T lymphocytes in vitro [57].

In human carotid atherosclerotic lesions, the expression of the lipoxin receptor FPR2/ALX has been histologically detected co-localizing with markers of macrophages, smooth muscle cells and endothelial cells [58]. mRNA levels of FPR2/ALX were significantly up-regulated in atherosclerotic compared with healthy vessels and correlated with cyto- and chemokine expression both within atherosclerotic lesions and in leukocytes derived from subjects with atherosclerosis [58]. The latter findings are consistent with a possible inflammation-resolution deficit in atherosclerosis, with decreased LX generation [55] and a preferential FPR2/ALX signaling by means of pro-inflammatory ligands [58]. Nevertheless, in a cohort of 127 subjects, expression levels of FPR2/ALX in carotid atherosclerotic lesions were inversely and independently associated with clinical signs of cerebral ischemia [58], supporting that signaling through this receptor may transduce increased atherosclerotic plaque stability.

4.5 D-Series Resolvins Signaling: Implications for Atherosclerosis

Although the role of D-resolvins in atherosclerosis has remained largely unexplored, the pro-resolving and anti-inflammatory effects that these mediators exert on inflammatory cells could potentially inhibit and reverse the development of chronic inflammation in atherosclerotic lesions and thus be of therapeutic value. In macrophages, RvD1 reduces pro-inflammatory cytokines, such as IL-6, IL-8, TNF-α and MCP-1 [59–62], whereas anti-inflammatory mediators like IL-10 are increased [60, 63], hence providing the rationale for a pro-resolving function of D-resolvins during inflammation. The resolution of an inflammatory response further depends on the uptake of both apoptotic cells and excess bacteria by macrophages. Important for this process, RvD1 and RvD3 facilitate the phagocytotic activity of macrophages toward apoptotic thymocytes and neutrophils [61, 64, 65]. In addition, RvD1, RvD2 and RvD5 enhance phagocytosis of macrophages and polymorphonuclear neutrophils (PMNs), thus promoting clearance of bacterial infections in murine models of sepsis [59, 62].

Further anti-inflammatory and pro-resolving properties of D-series Rvs derive from the inhibition of murine and human PMN transendothelial cell migration [63, 66], which is important in the initial phase of atherosclerotic inflammation. In particular, RvD1 impairs the interaction between PMNs and endothelial cells by attenuating actin polymerization and β2-integrin CD11b surface expression in PMNs [25, 66]. Moreover, also RvD2 reduces the interaction between activated endothelial cells and PMNs and promotes nitric oxide (NO) release via endothelial NO-Synthase [62, 63]. Taken together, the anti-inflammatory and pro-resolving effects of D-series Rvs on leukocyte recruitment, phagocytosis and cytokine production provide increasing evidence for a potential beneficial role of D-series Rvs to counteract inflammation in the pathogenesis of atherosclerosis.

4.6 E-Series Resolvin Signaling: Implications for Atherosclerosis

There are hitherto no studies exploring the role of E-resolvin signaling in atherosclerosis. In periodontal inflammation, RvE1 reduces the levels of IL-1β and C-reactive protein (CRP) [67], which may indicate potential roles for this mediator in reducing atherosclerosis. In terms of cellular effects, RvE1 decreases the migration of PDGF-stimulated endothelial cells [55], blocks TNF-signaling, and inhibits migration and IL-12 production in dendritic cells [36]. In macrophages, RvE1 down-regulates cytokine expression [34], and stimulates the phosphorylation of Akt and its downstream targets, thus enhancing the phagocytic activities of this specific cell type [35]. Nanomolar concentrations of RvE1 acting on PMNs reduce their transendothelial migration, and their capacity to generate superoxides [68, 69] may also be potentially beneficial in atherosclerosis. Furthermore, RvE1 reduces platelet activation and aggregation [70]. Finally, RvE1 interact with the BLT1 receptor [71] and may hence counteract the pro-atherogenic actions of LTB_4 described above.

5 Animal Models of Atherosclerosis

5.1 Targeting 5-LO/FLAP

Hyperlipidemic mice, such as apolipoprotein E ($ApoE^{-/-}$) and LDL receptor ($LDLR^{-/-}$) knock-out mice represent established models of atherosclerosis, in which either genetic or pharmacological targeting of the different lipoxygenases has been explored. While the initial study suggested 95 % inhibition in $5LO^{+/-}$

Table 1 Effects of 5-LO knock-out on murine atherosclerosis

Genotype	Diet	Age or time on diet	Aortic root	En face	Comments	References
LDLR$^{-/-}$	HFD	4–6 months	↓	NA	5-LO$^{+/-}$	[72]
LDLR$^{-/-}$	HFD	12 w	↔	↔	–	[51]
ApoE$^{-/-}$	Chow	6–12 months	↔	↔	–	[51]
ApoE$^{-/-}$	HFD	8 w	↔	↓	–	[51]
ApoE$^{-/-}$	HFD	8 w	↔	↔	"Paigen" diet	[73]
ApoE$^{-/-}$	HFD	6 months	↔	↔	"Western" diet	[73]

w weeks, *HFD* high fat diet, *NA* not available

LDLR$^{-/-}$ mice [72], subsequent studies failed to repeat those findings after either homozygous or heterozygous 5-LO knock-out in both ApoE$^{-/-}$ and LDLR$^{-/-}$ mice [51, 73], as indicated in Table 1. In contrast, other models have shown that targeting FLAP significantly reduces murine atherosclerosis, but only under certain conditions. More specifically, the following murine models exhibit decreased atherosclerosis after either pharmacological or genetic targeting of FLAP; namely ApoE$^{-/-}$ and LDLR$^{-/-}$ double knock-out mice [74], transgenic ApoE$^{-/-}$ mice with a dominant negative TGFβ type II receptor under the CD4 promoter [75], and most recently ApoE$^{-/-}$ mice with postnatal COX2 deletion [76]. Importantly, those animal models all exhibit an up-regulated leukotriene pathway compared with single ApoE$^{-/-}$ mice [1, 76], suggesting that further activation of the LT pathway is necessary to see effects of its inhibition.

In addition, certain differential effects of targeting 5-LO versus FLAP may also be expected in terms of altered AA metabolism. Whereas direct targeting of 5-LO will disrupt the formation of all 5-LO-derived lipid mediators, FLAP is not obligatory in the 5-LO-induced enzymatic conversion of either 15-LO derived 15(S)-HETE or acetylated COX-2-derived 15(R)HETE into LXA$_4$ and 15-epi LXA$_4$, respectively. This potentially makes FLAP a specific target of LT biosynthesis while leaving other biosynthetic pathways for LXs unaffected. The latter assumption has received support from in vivo studies of other pathologies, such as liver injury, in which the FLAP antagonist BAYx1005 inhibited LT synthesis but actually increased LX formation [77]. Likewise, a recent study showed that RvD1 inhibited LTB$_4$ generation while increasing LXA$_4$ levels in macrophages by means of preventing the calcium-dependent translocation of 5-LO to the nuclear membrane, which is an obligatory step for 5-LO/FLAP-mediated metabolism of AA into 5-HPETE and subsequently LTs, but not for the LX formation by means of 5-LO activity subsequent to AA metabolism by 15-LO (Fig. 1; [78]).

Taken together, the somewhat conflicting results on murine atherosclerosis obtained by either 5-LO or FLAP targeting may depend on the models and timepoints studied, but although hitherto not specifically addressed, it cannot be excluded that targeting this enzymatic step may affect both pro-inflammatory (e.g. LTs) and pro-resolving (e.g. LX) lipid mediators.

5.2 Targeting 12/15-LO

Genetic deletion of 12/15-LO, which is the murine homologue of the human 12- and 15-LOs, has been evaluated in hyperlipidemic mice. Since 12/15 LO is necessary for LXA_4 formation from AA in mice (Fig. 1), genetic targeting of this enzyme blocks LXs formation, which has been shown in murine arthritis [79]. However, as mentioned above, 12/15-LO may also participate in LDL oxidation, which further complicates its role in atherosclerosis. Several studies have shown atheroprotective effects following 12/15-LO disruption [80–88]. The effects in $12/15\text{-LO}^{-/-}$ and $\text{ApoE}^{-/-}$ double knock-out mice were similar but not additive to those induced by anti-oxidant vitamin E treatment [88], supporting that the observed atheroprotection was related to 12/15-LO oxidative activity. In contrast, another study reported accelerated atherosclerosis in $\text{ApoE}^{-/-}$ and $12/15\text{-LO}^{-/-}$ double knock-out mice, which was mimicked by bone-marrow transplantation, and suggested that macrophage-derived pro-resolution mediators, such as LXA_4 and RvD1, were increased [61].

Overexpession of human 15-LO under the preproendothelin promoter, to generate a vascular wall-specific transgene, increased atherosclerosis in $\text{LDLR}^{-/-}$ mice [89]. Likewise, overexpression of murine 12/15-LO accelerates atherosclerosis by means of an increased production of endothelial adhesion molecules [90]. In contrast, macrophage-specific overexpression of either human 15-LO in hyperlipidemic rabbits or murine 12/15-LO in $\text{ApoE}^{-/-}$ mice is atheroprotective and associated with increased LXA_4 formation [61, 91].

Taken together, these studies further highlight the complexity of studying the 12/15-LO pathway in experimental atherosclerosis. The specific role of 12/15-LO in atherosclerosis may also vary depending on the time-points studied, the animal models used and their dietary conditions, as shown in Table 2.

5.3 Targeting LT Receptors

Given the variable results obtained by altered LO activity in hyperlipidemic mice, targeting lipid mediator receptors could potentially provide insight into the specific role of each of the different LO metabolites in atherosclerosis. For example, the BLT_1 receptor antagonist CP105696 significantly reduces atherosclerosis lesion size and macrophage content in $\text{ApoE}^{-/-}$ mice [92], whereas the BLT_2 receptor antagonist LY255283 did not alter atherosclerotic lesion size [93]. In further support of a dominant role of the $LTB_4\text{-}BLT_1$ pathways in atherosclerosis development, several studies have demonstrated decreased lesion size in $\text{ApoE}^{-/-}$ and $BLT_1^{-/-}$ double knock-out mice [94–96]. In addition to LTB_4, also cysteinyl-LTs appear to mediate pro-atherogenic signaling through the $CysLT_1$ receptor subtype, as suggested from the decreased atherosclerotic lesion size observed after

Table 2 Effects of 12/15-LO knock-out on murine atherosclerosis

Gender	Genotype	Diet	Age or time on diet	Aortic root	En face	Comments	References
NA	ApoE$^{-/-}$	Chow	15 w and 52 w	↓	↓	–	[81]
M and F	LDLR$^{-/-}$	HFD	3, 6, 9, 12 and 18 w	↓	↓	–	[82]
NA	ApoE$^{-/-}$	Chow	10, 32, 52 and 64 w	NA	↓	–	[80]
M and F	LDLR$^{-/-}$	Chow	15 w and 32 w	NA	↓	Reduced in M and no difference in F	[87]
NA	BMT → ApoE$^{-/-}$	HFD	12 w	NA	↓	–	[83]
M and F	ApoE$^{-/-}$	Chow + vit E	12 w	NA	↓	–	[88]
M	ApoE$^{-/-}$	Chow + TP antagonist	12 w	NA	↓	–	[86]
M and F	ApoE$^{-/-}$	Chow	25 w	↓	↓	Reduced in F and no differences in M	[84]
M	LDLR$^{-/-}$/ BMT → LDLR$^{-/-}$	HFD	16 w	NA	↓	M as recipients and bone marrow from F	[85]
M and F	ApoE$^{-/-}$	Chow	22 and 27 w	↑	↑	–	[61]

M male, *F* female, *w* weeks, *HFD* high fat diet, *BMT* bone marrow transplantation, *NA* not available

administration of the clinically used anti-asthmatic CysLT$_1$ receptor antagonist montelukast to different hyperlipidemic mice [97, 98].

5.4 Targeting the LXA$_4$ Receptor

As mentioned above, FPR2/ALX, exhibits a certain duality, defined as the particularity of this receptor to transduce pro-inflammation in response to peptide agonists, and anti-inflammatory or pro-resolution responses induced by LXA$_4$, ATL and RvD1. In line with the findings of increased FPR2/ALX mRNA levels in human atherosclerotic compared with healthy vessels, and an association with pro-inflammatory pathways in inflammatory cells (cf. supra), LDLR$^{-/-}$ and FPR2$^{-/-}$ double knock-out mice exhibit decreased atherosclerosis [99]. The latter finding was mimicked by bone marrow transplantation, suggesting that leukocyte-driven pro-atherogenic effects of FPR2/ALX signaling may be caused by higher levels of pro-inflammatory (compared with pro-resolution) FPR2/ALX agonists in this particular model [99]. This notion is supported by the atherosclerosis protection offered by genetic deletion of the FPR2/ALX agonist cathelicidin-related anti-microbial peptide (CRAMP) [100], and highlights the complexity of FPR2/ALX signaling in atherosclerosis.

Several in vitro studies of cells with endogenous and recombinant FPR2/ALX expression have generated contradictory results in terms of lipoxin signaling [25, 101–103], which have been recently reviewed [104]. Smooth muscle cell migration is inhibited by LXA$_4$, whereas LXA$_4$ is inactive as an agonist in murine aortic smooth muscle cells derived from FPR2 knock-out mice, supporting that LXA$_4$ signals through FPR2/ALX in these cells. Furthermore, atherosclerotic lesions in FPR2/ALX knock-out mice had lower collagen content, and smooth muscle cells derived from these mice exhibited transcriptional profiles indicating decreased collagen crosslinking and increased collagenase expression [99]. Taken together with the increased association of FPR2/ALX expression and plaque instability in human atherosclerotic lesions, the latter findings provide an initial suggestion of differential effects of FPR2/ALX signaling in atherosclerosis, promoting atherosclerosis progression while increasing plaque stability. The exact role of other LO-derived mediators that bind FPR2/ALX in atherosclerosis remains to be established.

5.5 Omega-3 Supplementation in Animal Models of Atherosclerosis

In contrast to AA, which is a ubiquitous LO substrate, DHA and EPA are essential fatty acids. Dietary supplementation with omega-3 rich fish oil in atherosclerosis-prone mice leads to increased incorporation of DHA and EPA in the aorta and the heart, whereas the AA content is decreased [105], hence providing a rationale for increased cardiovascular formation of LO-derived resolvins. The main studies evaluating the effects of dietary omega-3 fatty acids supplementation on atherosclerosis development in mice are summarized in Table 3 [105–113]. Despite somewhat variable results, the main conclusion that emerges from those studies is a beneficial effect of omega-3 supplementation in terms of atherosclerosis development reduction. Since the studies listed in Table 3 display a certain heterogeneity in critical aspects such as diet composition and exposure time, and in terms of the mouse strains used, it cannot be excluded that methodological differences may explain the differential results obtained.

Table 3 Effects of omega-3 supplementation on murine atherosclerosis

Gender	Genotype	Diet	Treatment	Time	Aortic root	En face	Comments	Reference
M and F	ApoE$^{-/-}$	HFD	1 % DHA ethyl ester	8 w	↔	↔	–	[106]
M	ApoE$^{-/-}$	Chow	1 % fish oil	14 w	↔	↔	–	[107]
M	ApoE$^{-/-}$	Chow	0.5 % fish oil	8 w	→	NA	Atherosclerosis ↓ in intermittent hypoxia group	[105]
F	ApoE$^{-/-}$ or LDLR$^{-/-}$	HFD	10 % of palm oil, echium oil or fish oil	16 w LDLR$^{-/-}$ 12 ApoE$^{-/-}$	→	NA	Atherosclerosis ↓ LDLR$^{-/-}$, but not in ApoE$^{-/-}$ mice	[108]
M	ApoE$^{-/-}$ or LDLR$^{-/-}$	HFD	5 % EPA	12–13 w	→	→	More collagen and less macrophages	[109]
M and F	ApoE$^{-/-}$ or LDLR$^{-/-}$	ApoE$^{-/-}$ on chow LDLR$^{-/-}$ on HFD	1 % fish oil	20 w	→	→	Atherosclerosis ↓ in male LDLR$^{-/-}$ but not in female ApoE$^{-/-}$ mice	[110]
M	ApoE$^{-/-}$ LDLR$^{-/-}$	HFD	Different n-6/n-3 ratios	8 and 16 w	NA	→	Atherosclerosis ↓ with lower n-6/n-3 ratio	[111]
M	LDLR$^{-/-}$	HFD	Different n-6/n-3 ratios	32 w	NA	→	Atherosclerosis ↓ with lower n-6/n-3 ratio	[112]
M	LDLR$^{-/-}$	HFD	Different n-6/n-3 ratios	12 w	→	NA	Atherosclerosis ↓ with lower n-6/n-3 ratio	[113]

M male, *F* female, *HFD* high fat diet, *w* weeks, *n* omega, *NA* not available

In addition to lesion size, also plaque stability may be altered by omega-3 supplementation. For example, ApoE$^{-/-}$ mice fed a relatively high concentration of EPA have more stable lesions compared with untreated animals [109]. In the same strain, fish oil supplementation prevents the rise in aortic levels of MMP-2 after exposure to chronic intermittent hypoxia [105], supporting a potential plaque stabilizing effect of omega-3 supplementation.

It should however be pointed out that none of the studies listed in Table 3 specifically measured the alterations of LO metabolism towards resolvins after omega-3 supplementation, and that the potential effects of resolvins on murine atherosclerosis remains to be established.

6 Intimal Hyperplasia

6.1 Pathophysiology and Clinical Context

In addition to effects on inflammatory circuits, atherosclerotic cardiovascular disease also includes direct effects on structural cells of the vessel wall, such as vascular smooth muscle cells. Migration and proliferation of smooth muscle cells are part of the process of vascular wall remodeling in early atherosclerosis. As outlined above, smooth muscle cells also participate in the formation of a fibrous cap stabilizing the plaque, a process which involves extracellular matrix production. Smooth muscle cells are in addition crucial in the response to vascular injury, for example during percutaneous coronary interventions (PCI). The intimal hyperplasia of a vessel in response to injury and PCI is characterized by proliferation and migration of smooth muscle cells, which may lead to either re-occlusion or restenosis of a vessel.

6.2 LO in Intimal Hyperplasia

Different animal models have been established to stimulate the migration and proliferation of vascular smooth muscle cells, leading to hypertrophy of the intimal layer of the vascular wall. The notion of a role of LO activation in this pathophysiological process originates in studies of the non-selective LO inhibitor phenidone, which reduced intimal hyperplasia following rodent carotid artery balloon injury [114]. Early studies also showed that the FLAP antagonist MK886 decreased neutrophil deposition at sites of carotid arterial injury in pigs [115] and inhibited intimal hyperplasia after femoral artery photochemical injury in rats [116]. The importance of leukocytes as a source of 5-LO metabolites acting on the vascular wall was recently supported by the observation that the protection against intimal hyperplasia observed after arterial injury in mice lacking FLAP was mimicked by transplantation of FLAP-deficient bone marrow into wild type mice [117].

In addition to the 5-LO/FLAP pathway, also 12-LO has been implicated in smooth muscle cell proliferation and vascular hypertrophy in several animal models for intimal hyperplasia following vascular injury. A rat carotid artery model of balloon injury and restenosis identified a strong 12-LO expression in neointimal vascular smooth muscle cells, endothelial cells as well as leukocytes [118]. This effect was induced 4–14 days after injury and could be reversed by using the LO inhibitor phenidone, leading to a greater lumen as well as reduced intimal proliferation and thickening of the vessel wall. These findings implied that 12-LO may be involved in the development of intimal hyperplasia and were supported by use of a specific ribozyme inhibitor cleaving leukocyte-type 12-LO in the same model of balloon-injured rat carotid arteries [119]. The ribozyme inhibitor further attenuated migration and fibronectin deposition in an in vitro approach using rat aortic vascular smooth muscle cells.

The levels of 12/15-LO correlates with intima formation after carotid balloon-injury in mice [120]. Interestingly, 12/15-LO overexpressing mice had a more pronounced neointimal hyperplasia compared to wild type mice, whereas 12/15 deficient mice exhibited a reduced intima formation [120], suggesting that downstream metabolites of 12/15-LO stimulate intimal hyperplasia. Although the 12/15-LO metabolite mediating this effect remains to be established, it is interesting that 15(S)HETE, a major 15-LO metabolite in vascular smooth muscle cells, induces smooth muscle cell migration and pro-inflammatory activation [121]. In line with those findings, balloon injury in rat carotid arteries followed by adenoviral-induced expression of 15-LO increases 15(S)HETE production and intimal hyperplasia, associated with an elevated number of infiltrating CD68-positive macrophages in balloon injured arteries [121]. Furthermore, smooth muscle cells derived from transgenic 12/15-LO overexpressing mice exhibit an increased proliferation rate compared with cells derived from wild type mice, whereas vascular smooth muscle cells derived from 12/15-LO deficient mice exhibit a slower proliferation [122, 123]. Taken together, those studies suggest that products of 5-, 12- and 15-LO may drive intimal hyperplasia. However, also in this context there appears to exist a balance between metabolites both stimulating and inhibiting smooth muscle cell responses, as revealed by studies targeting the different receptors activated by LO products, as will be addressed in the next section.

6.3 LT Receptors and Intimal Hyperplasia

BLT receptor antagonism reduced intimal hyperplasia in different models of rodent vascular injury [18, 116], consistent with LTB_4-induced vascular smooth muscle cell migration and proliferation observed in vitro [18]. Extrapolation of those results to atherosclerosis may however be limited by the normolipidemia in the latter models. Nevertheless, also atherosclerotic lesions of $ApoE^{-/-}$ mice display a reduced smooth muscle cell content after BLT_1 receptor disruption [94]. Likewise, cholesterol-fed rabbits subjected to carotid angioplasty with balloon dilatation and stent implantation as a model of PCI. In the latter model, the BLT receptor antagonist amebulant (BIIL284) significantly reduced in-stent intimal hyperplasia, which was correlated with decreased macrophage content and lower MMP activities in the stenotic lesions [53].

Targeting the cysteinyl-LT signaling after vascular injury in rats have however generated contradictory results, with both neutral effects [116] and reduced intimal hyperplasia [124] having been reported in different models and using different $CysLT_1$ receptor antagonists. In support of a protective effect of inhibiting this pathway, montelukast administration by osmotic pumps inhibited the formation of intimal hyperplasia after wire injury to the femoral artery in mice [125].

6.4 LX-Signaling in Smooth Muscle Cells and Intimal Hyperplasia

Both human and murine vascular smooth muscle cells express FPR2/ALX [58]. In vitro experiments have revealed that ATL reduces migration of smooth muscle cells derived both from human veins [55] and murine aortas [126], an effect which is abolished in cells derived from FPR2/ALX-deficient mice [126], hence supporting that ATL inhibits smooth muscle cell migration through FPR2/ALX receptor signaling.

In vivo administration of either ATL [126] or RvD2 [28] in murine models of vascular injury decreases intimal hyperplasia. Interestingly, ATL failed to induce any significant effects when administered to knock-out mice [126]. Globally, these findings point towards a crucial role of LO-derived mediators and their signaling through FPR2/ALX in the response to vascular injury.

6.5 Resolvin-Signaling in Smooth Muscle Cells and Intimal Hyperplasia

In a rabbit model of vascular injury, RvD1 and RvD2 inhibited vascular smooth muscle cell proliferation, migration and pro-inflammatory cytokine gene expression [28]. Following balloon injury of femoral arteries in rabbits, a broad variety of pro-resolving lipid mediators, such as RvD1 and RvD5 could be detected, suggesting that vascular injury in addition to LTs also induces endogenous pathways towards pro-resolution lipid mediators. Furthermore, expression of both RvD1 receptors FPR2/ALX and GPR32 was found in the rabbit artery wall. In particular, GPR32 expression in the medial layer of the vessel was increased following injury and removal of endothelium [28]. When RvD2 was administered locally directly after the induction of acute vascular injury, cell proliferation and leukocyte recruitment was reduced, as well as a decrease in early pro-inflammatory cytokine expression of TNF-α and IL-1α. On the long term, the anti-inflammatory effects of Rvs led to a reduction in neointimal hyperplasia and might therefore prove a novel approach in the treatment of vascular diseases [28].

7 Clinical Implications

7.1 Anti-LTs

Despite a certain heterogeneity of experimental results, the notion which emerges from the studies discussed above is that a protective effect could be anticipated in cardiovascular prevention, if pro-inflammatory LT signaling could be inhibited and/or pro-resolution pathways could be promoted. The translation of the latter assumption from bench to bedside has received support from both epidemiological studies and clinical trials. Two clinical trials of LT pathway modifiers have specifically evaluated their effects in the context of atherosclerosis. First, the FLAP antagonist veliflapon (DG031/BAYx1005) was evaluated for effects on atherosclerosis-associated biomarkers in a randomized controlled trial of subjects being carriers of either FLAP or LTA4 hydrolase haplotypes (which had previously been associated with an increased risk for myocardial infarction), showing a discrete reduction in CRP levels compared with placebo [127]. Second, the direct 5-LO inhibitor atreleuton (ABT-761/VIA2291) was shown to inhibit LT synthesis in patients with acute coronary syndromes and in addition reduced atherosclerotic plaque volume measured by coronary CT [128].

LT receptor antagonists have also been associated with decreased pro-inflammatory biomarkers in asthma studies [129]. Recently, a nation-wide observational study revealed a decreased cardiovascular risk associated with LT receptor antagonist use [130]. In summary, the $CysLT_1$ receptor antagonist montelukast exhibited a borderline significant prevention of recurrent ischemic stroke [130]. In a post hoc analysis it was shown that montelukast had most pronounced and significant protective effects in subjects not taking angiotensin-modifying drugs (ACE inhibitors and/or angiotensin receptor blockers) [130]. In addition, montelukast use was associated with a decreased risk of recurrent myocardial infarction in males but not in females [130]. Taken together, these observations suggested a generalizability of the anti-inflammatory effects of LT modifiers beyond pulmonary disease.

7.2 Omega 3 Fatty Acids

As discussed above, dietary supplementation with omega-3 fatty acids may skew LO metabolism toward pro-resolution rather than pro-inflammation. A link between omega-3 fatty acids and decreased cardiovascular inflammation was first observed in Greenland Inuits. The higher levels of EPA and DHA found in the plasma and platelets of Inuits compared with other Scandinavians were inversely related to population rates of acute myocardial infarction [131]. Epidemiological and clinical trial evidence subsequently accumulated in further support of anti-inflammatory effects for omega-3 fatty acids, associated with cardioprotection and reduced mortality [132]. However, these findings have not been consequently replicated [133]. Importantly, the recent structural elucidation of the active pro-resolving omega-3-derived mediators, such as the Rvs (Fig. 1) implies that also Aspirin is needed for the biosynthesis of these mediators, which may have been overlooked in clinical trials of omega-3 supplementation.

8 Summary and Conclusions

There is today convincing evidence that LOs are present and active in human atherosclerotic lesions. The lipid metabolism catalyzed by the human LOs lead to the biosynthesis of several lipid mediators, which together make up an intricate network with in part opposing and balancing effects on inflammation and atherosclerosis. At several occasions, dual effects are observed, making the physiological response to altered LO metabolism sometimes difficult to predict. For example, whereas 5-LO generates pro-inflammatory LTs from AA, this enzyme also participates in the biosynthesis of anti-inflammatory and pro-resolving LXs. Likewise, although inhibition of 15-LO can decrease LX formation, this enzyme may also be pro-atherogenic by means of LDL oxidation. Finally, changing the substrate availability from omega-6 (AA) to omega-3 (DHA or EPA) fatty acids may skew LO metabolism from being pro-inflammatory into pro-resolving signaling.

Further knowledge of the specific actions of the different LO metabolites has been obtained through targeting of the specific receptors. However, also in this context, dual actions exist, with the FPR2/ALX receptor mediating pro-inflammation when activated by peptide agonists, whereas lipid agonists transduce the resolution of inflammation through this receptor.

Recent studies providing translational implications for the therapeutic value of targeting the LO pathways point to beneficial effects of anti-leukotrienes. It is also important to fully consider and explore all possible pathways of LO metabolism and their downstream metabolites (Figs. 1 and 2) during different conditions and time points of disease when considering the 5-, 12- and 15-LO pathways as therapeutic targets for cardiovascular disease

Acknowledgements M.B. is supported by the Swedish Research Council (grant number 2014–2312); the Swedish Heart and Lung Foundation (grant numbers 20120474 and 20120827) and the Stockholm County Council (grant number 20140222). A.L.F. is a post-doctoral fellow within the CERIC Linnaeus Program. S.T. is supported by a post-doctoral fellowship from the Deutsche Forschungsgemeinschaft.

References

1. Bäck M (2008) Inflammatory signaling through leukotriene receptors in atherosclerosis. Curr Atheroscler Rep 10(3):244–251
2. Bäck M (2009) Leukotriene signaling in atherosclerosis and ischemia. Cardiovasc Drugs Ther 23(1):41–48. doi:10.1007/s10557-008-6140-9
3. Houard X, Ollivier V, Louedec L, Michel JB, Bäck M (2009) Differential inflammatory activity across human abdominal aortic aneurysms reveals neutrophil-derived leukotriene B4 as a major chemotactic factor released from the intraluminal thrombus. FASEB J 23 (5):1376–1383. doi:10.1096/fj.08-116202
4. Nagy E, Andersson DC, Caidahl K, Eriksson MJ, Eriksson P, Franco-Cereceda A, Hansson GK, Bäck M (2011) Upregulation of the 5-lipoxygenase pathway in human aortic valves correlates with severity of stenosis and leads to leukotriene-induced effects on valvular myofibroblasts. Circulation 123(12):1316–1325. doi:10.1161/CIRCULATIONAHA.110.966846
5. Labat C, Temmar M, Nagy E, Bean K, Brink C, Benetos A, Back M (2013) Inflammatory mediators in saliva associated with arterial stiffness and subclinical atherosclerosis. J Hypertens 31(11):2251–2258. doi:10.1097/HJH.0b013e328363dccc, discussion 2258
6. Samuelsson B, Dahlén SE, Lindgren JÅ, Rouzer CA, Serhan CN (1987) Leukotrienes and lipoxins: structures, biosynthesis, and biological effects. Science 237(4819):1171–1176
7. Peters-Golden M, Henderson WR Jr (2007) Leukotrienes. N Engl J Med 357(18):1841–1854. doi:10.1056/NEJMra071371
8. Katryniok C, Schnur N, Gillis A, von Knethen A, Sorg BL, Looijenga L, Radmark O, Steinhilber D (2010) Role of DNA methylation and methyl-DNA binding proteins in the repression of 5-lipoxygenase promoter activity. Biochim Biophys Acta 1801(1):49–57. doi:10.1016/j.bbalip.2009.09.003
9. Nagy E, Bäck M (2012) Epigenetic regulation of 5-lipoxygenase in the phenotypic plasticity of valvular interstitial cells associated with aortic valve stenosis. FEBS Lett 586 (9):1325–1329. doi:10.1016/j.febslet.2012.03.039
10. Uhl J, Klan N, Rose M, Entian KD, Werz O, Steinhilber D (2002) The 5-lipoxygenase promoter is regulated by DNA methylation. J Biol Chem 277(6):4374–4379. doi:10.1074/jbc.M107665200
11. Takata S, Papayianni A, Matsubara M, Jimenez W, Pronovost PH, Brady HR (1994) 15-Hydroxyeicosatetraenoic acid inhibits neutrophil migration across cytokine-activated endothelium. Am J Pathol 145(3):541–549
12. Matsuda H, Miyatake K, Dahlen SE (1995) Pharmacodynamics of 15(S)-hydroperoxyeicosatetraenoic (15-HPETE) and 15(S)-hydroxyeicosatetraenoic acid (15-HETE) in isolated arteries from guinea pig, rabbit, rat and human. J Pharmacol Exp Ther 273(3):1182–1189
13. Serhan CN (2014) Pro-resolving lipid mediators are leads for resolution physiology. Nature 510(7503):92–101. doi:10.1038/nature13479
14. Prescott SM (1984) The effect of eicosapentaenoic acid on leukotriene B production by human neutrophils. J Biol Chem 259(12):7615–7621

15. Serhan CN, Yacoubian S, Yang R (2008) Anti-inflammatory and proresolving lipid mediators. Annu Rev Pathol 3:279–312. doi:10.1146/annurev.pathmechdis.3.121806.151409
16. Bäck M, Dahlen SE, Drazen JM, Evans JF, Serhan CN, Shimizu T, Yokomizo T, Rovati GE (2011) International Union of Basic and Clinical Pharmacology. LXXXIV: leukotriene receptor nomenclature, distribution, and pathophysiological functions. Pharmacol Rev 63(3):539–584. doi:10.1124/pr.110.004184
17. Yokomizo T, Kato K, Terawaki K, Izumi T, Shimizu T (2000) A second leukotriene B(4) receptor, BLT2. A new therapeutic target in inflammation and immunological disorders. J Exp Med 192(3):421–432
18. Bäck M, Bu DX, Branstrom R, Sheikine Y, Yan ZQ, Hansson GK (2005) Leukotriene B4 signaling through NF-kappaB-dependent BLT1 receptors on vascular smooth muscle cells in atherosclerosis and intimal hyperplasia. Proc Natl Acad Sci U S A 102(48):17501–17506. doi:10.1073/pnas.0505845102
19. Bäck M, Qiu H, Haeggstrom JZ, Sakata K (2004) Leukotriene B4 is an indirectly acting vasoconstrictor in guinea pig aorta via an inducible type of BLT receptor. Am J Physiol Heart Circ Physiol 287(1):H419–424. doi:10.1152/ajpheart.00699.2003
20. Capra V, Bäck M, Barbieri SS, Camera M, Tremoli E, Rovati GE (2013) Eicosanoids and their drugs in cardiovascular diseases: focus on atherosclerosis and stroke. Med Res Rev 33(2):364–438. doi:10.1002/med.21251
21. Eaton A, Nagy E, Pacault M, Fauconnier J, Bäck M (2012) Cysteinyl leukotriene signaling through perinuclear CysLT(1) receptors on vascular smooth muscle cells transduces nuclear calcium signaling and alterations of gene expression. J Mol Med 90(10):1223–1231. doi:10.1007/s00109-012-0904-1
22. Nielsen CK, Campbell JI, Ohd JF, Morgelin M, Riesbeck K, Landberg G, Sjolander A (2005) A novel localization of the G-protein-coupled CysLT1 receptor in the nucleus of colorectal adenocarcinoma cells. Cancer Res 65(3):732–742
23. Ye RD, Boulay F, Wang JM, Dahlgren C, Gerard C, Parmentier M, Serhan CN, Murphy PM (2009) International Union of Basic and Clinical Pharmacology. LXXIII. Nomenclature for the formyl peptide receptor (FPR) family. Pharmacol Rev 61(2):119–161. doi:10.1124/pr.109.001578
24. Chiang N, Serhan CN, Dahlén SE, Drazen JM, Hay DW, Rovati GE, Shimizu T, Yokomizo T, Brink C (2006) The lipoxin receptor ALX: potent ligand-specific and stereoselective actions in vivo. Pharmacol Rev 58(3):463–487. doi:10.1124/pr.58.3.4
25. Krishnamoorthy S, Recchiuti A, Chiang N, Yacoubian S, Lee CH, Yang R, Petasis NA, Serhan CN (2010) Resolvin D1 binds human phagocytes with evidence for proresolving receptors. Proc Natl Acad Sci U S A 107(4):1660–1665. doi:10.1073/pnas.0907342107
26. Haitina T, Fredriksson R, Foord SM, Schioth HB, Gloriam DE (2009) The G protein-coupled receptor subset of the dog genome is more similar to that in humans than rodents. BMC Genomics 10:24. doi:10.1186/1471-2164-10-24
27. Hsiao HM, Thatcher TH, Levy EP, Fulton RA, Owens KM, Phipps RP, Sime PJ (2014) Resolvin D1 attenuates polyinosinic-polycytidylic acid-induced inflammatory signaling in human airway epithelial cells via TAK1. J Immunol 193(10):4980–4987. doi:10.4049/jimmunol.1400313
28. Miyahara T, Runge S, Chatterjee A, Chen M, Mottola G, Fitzgerald JM, Serhan CN, Conte MS (2013) D-series resolvin attenuates vascular smooth muscle cell activation and neointimal hyperplasia following vascular injury. FASEB J 27(6):2220–2232. doi:10.1096/fj.12-225615
29. Gantz I, Konda Y, Yang YK, Miller DE, Dierick HA, Yamada T (1996) Molecular cloning of a novel receptor (CMKLR1) with homology to the chemotactic factor receptors. Cytogenet Cell Genet 74(4):286–290
30. Kaur J, Adya R, Tan BK, Chen J, Randeva HS (2010) Identification of chemerin receptor (ChemR23) in human endothelial cells: chemerin-induced endothelial angiogenesis. Biochem Biophys Res Commun 391(4):1762–1768. doi:10.1016/j.bbrc.2009.12.150

31. Goralski KB, McCarthy TC, Hanniman EA, Zabel BA, Butcher EC, Parlee SD, Muruganandan S, Sinal CJ (2007) Chemerin, a novel adipokine that regulates adipogenesis and adipocyte metabolism. J Biol Chem 282(38):28175–28188. doi:10.1074/jbc. M700793200
32. Wittamer V, Franssen JD, Vulcano M, Mirjolet JF, Le Poul E, Migeotte I, Brezillon S, Tyldesley R, Blanpain C, Detheux M, Mantovani A, Sozzani S, Vassart G, Parmentier M, Communi D (2003) Specific recruitment of antigen-presenting cells by chemerin, a novel processed ligand from human inflammatory fluids. J Exp Med 198(7):977–985. doi:10.1084/jem.20030382
33. Yamawaki H, Kameshima S, Usui T, Okada M, Hara Y (2012) A novel adipocytokine, chemerin exerts anti-inflammatory roles in human vascular endothelial cells. Biochem Biophys Res Commun 423(1):152–157. doi:10.1016/j.bbrc.2012.05.103
34. Ishida T, Yoshida M, Arita M, Nishitani Y, Nishiumi S, Masuda A, Mizuno S, Takagawa T, Morita Y, Kutsumi H, Inokuchi H, Serhan CN, Blumberg RS, Azuma T (2010) Resolvin E1, an endogenous lipid mediator derived from eicosapentaenoic acid, prevents dextran sulfate sodium-induced colitis. Inflamm Bowel Dis 16(1):87–95. doi:10.1002/ibd.21029
35. Ohira T, Arita M, Omori K, Recchiuti A, Van Dyke TE, Serhan CN (2010) Resolvin E1 receptor activation signals phosphorylation and phagocytosis. J Biol Chem 285(5):3451–3461. doi:10.1074/jbc.M109.044131
36. Arita M, Bianchini F, Aliberti J, Sher A, Chiang N, Hong S, Yang R, Petasis NA, Serhan CN (2005) Stereochemical assignment, antiinflammatory properties, and receptor for the omega-3 lipid mediator resolvin E1. J Exp Med 201(5):713–722. doi:10.1084/jem.20042031
37. Hansson GK (2005) Inflammation, atherosclerosis, and coronary artery disease. N Engl J Med 352(16):1685–1695
38. Yla-Herttuala S, Rosenfeld ME, Parthasarathy S, Glass CK, Sigal E, Witztum JL, Steinberg D (1990) Colocalization of 15-lipoxygenase mRNA and protein with epitopes of oxidized low density lipoprotein in macrophage-rich areas of atherosclerotic lesions. Proc Natl Acad Sci U S A 87(18):6959–6963
39. Yla-Herttuala S, Luoma J, Viita H, Hiltunen T, Sisto T, Nikkari T (1995) Transfer of 15-lipoxygenase gene into rabbit iliac arteries results in the appearance of oxidation-specific lipid-protein adducts characteristic of oxidized low density lipoprotein. J Clin Invest 95(6):2692–2698. doi:10.1172/JCI117971
40. Zhu H, Takahashi Y, Xu W, Kawajiri H, Murakami T, Yamamoto M, Iseki S, Iwasaki T, Hattori H, Yoshimoto T (2003) Low density lipoprotein receptor-related protein-mediated membrane translocation of 12/15-lipoxygenase is required for oxidation of low density lipoprotein by macrophages. J Biol Chem 278(15):13350–13355. doi:10.1074/jbc. M212104200
41. De Caterina R, Mazzone A, Giannessi D, Sicari R, Pelosi W, Lazzerini G, Azzara A, Forder R, Carey F, Caruso D et al (1988) Leukotriene B4 production in human atherosclerotic plaques. Biomed Biochim Acta 47(10–11):S182–S185
42. van den Borne P, van der Laan SW, Bovens SM, Koole D, Kowala MC, Michael LF, Schoneveld AH, van de Weg SM, Velema E, de Vries JP, de Borst GJ, Moll FL, de Kleijn DP, Quax PH, Hoefer IE, Pasterkamp G (2014) Leukotriene B4 levels in human atherosclerotic plaques and abdominal aortic aneurysms. PLoS One 9(1), e86522. doi:10.1371/journal. pone.0086522
43. Allen S, Dashwood M, Morrison K, Yacoub M (1998) Differential leukotriene constrictor responses in human atherosclerotic coronary arteries. Circulation 97(24):2406–2413
44. Spanbroek R, Grabner R, Lotzer K, Hildner M, Urbach A, Ruhling K, Moos MP, Kaiser B, Cohnert TU, Wahlers T, Zieske A, Plenz G, Robenek H, Salbach P, Kuhn H, Radmark O, Samuelsson B, Habenicht AJ (2003) Expanding expression of the 5-lipoxygenase pathway within the arterial wall during human atherogenesis. Proc Natl Acad Sci U S A 100(3):1238–1243. doi:10.1073/pnas.242716099

45. Cipollone F, Mezzetti A, Fazia ML, Cuccurullo C, Iezzi A, Ucchino S, Spigonardo F, Bucci M, Cuccurullo F, Prescott SM, Stafforini DM (2005) Association between 5-lipoxygenase expression and plaque instability in humans. Arterioscler Thromb Vasc Biol 25(8):1665–1670. doi:10.1161/01.ATV.0000172632.96987.2d
46. Zhou YJ, Wang JH, Li L, Yang HW, de Wen L, He QC (2007) Expanding expression of the 5-lipoxygenase/leukotriene B4 pathway in atherosclerotic lesions of diabetic patients promotes plaque instability. Biochem Biophys Res Commun 363(1):30–36. doi:10.1016/j.bbrc. 2007.08.134
47. Qiu H, Gabrielsen A, Agardh HE, Wan M, Wetterholm A, Wong CH, Hedin U, Swedenborg J, Hansson GK, Samuelsson B, Paulsson-Berne G, Haeggstrom JZ (2006) Expression of 5-lipoxygenase and leukotriene A4 hydrolase in human atherosclerotic lesions correlates with symptoms of plaque instability. Proc Natl Acad Sci U S A 103 (21):8161–8166. doi:10.1073/pnas.0602414103
48. Hulten LM, Olson FJ, Aberg H, Carlsson J, Karlstrom L, Boren J, Fagerberg B, Wiklund O (2010) 15-Lipoxygenase-2 is expressed in macrophages in human carotid plaques and regulated by hypoxia-inducible factor-1alpha. Eur J Clin Invest 40(1):11–17. doi:10.1111/j. 1365-2362.2009.02223.x
49. Gertow K, Nobili E, Folkersen L, Newman JW, Pedersen TL, Ekstrand J, Swedenborg J, Kuhn H, Wheelock CE, Hansson GK, Hedin U, Haeggstrom JZ, Gabrielsen A (2011) 12- and 15-lipoxygenases in human carotid atherosclerotic lesions: associations with cerebrovascular symptoms. Atherosclerosis 215(2):411–416. doi:10.1016/j.atherosclerosis.2011.01.015
50. Liu HQ, Zhang XY, Edfeldt K, Nijhuis MO, Idborg H, Bäck M, Roy J, Hedin U, Jakobsson PJ, Laman JD, de Kleijn DP, Pasterkamp G, Hansson GK, Yan ZQ (2013) NOD2-mediated innate immune signaling regulates the eicosanoids in atherosclerosis. Arterioscler Thromb Vasc Biol 33(9):2193–2201. doi:10.1161/ATVBAHA.113.301715
51. Zhao L, Moos MP, Grabner R, Pedrono F, Fan J, Kaiser B, John N, Schmidt S, Spanbroek R, Lotzer K, Huang L, Cui J, Rader DJ, Evans JF, Habenicht AJ, Funk CD (2004) The 5-lipoxygenase pathway promotes pathogenesis of hyperlipidemia-dependent aortic aneurysm. Nat Med 10(9):966–973
52. Kim N, Luster AD (2007) Regulation of immune cells by eicosanoid receptors. TheScientificWorldJournal 7:1307–1328
53. Hlawaty H, Jacob MP, Louedec L, Letourneur D, Brink C, Michel JB, Feldman LJ, Bäck M (2009) Leukotriene receptor antagonism and the prevention of extracellular matrix degradation during atherosclerosis and in-stent stenosis. Arterioscler Thromb Vasc Biol 29 (4):518–524
54. Brezinski DA, Nesto RW, Serhan CN (1992) Angioplasty triggers intracoronary leukotrienes and lipoxin A4. Impact of aspirin therapy. Circulation 86(1):56–63
55. Ho KJ, Spite M, Owens CD, Lancero H, Kroemer AH, Pande R, Creager MA, Serhan CN, Conte MS (2010) Aspirin-triggered lipoxin and resolvin E1 modulate vascular smooth muscle phenotype and correlate with peripheral atherosclerosis. Am J Pathol 177 (4):2116–2123. doi:10.2353/ajpath.2010.091082
56. Petri MH, Ovchinnikova O, Bäck M (2015) Differential regulation of macrophage expression of leukotriene and lipoxin receptors. Prostaglandins Other Lipid Mediat. doi:10.1016/j. prostaglandins.2015.07.005, pii: S1098-8823(15)30006-X
57. Ariel A, Chiang N, Arita M, Petasis NA, Serhan CN (2003) Aspirin-triggered lipoxin A4 and B4 analogs block extracellular signal-regulated kinase-dependent TNF-alpha secretion from human T cells. J Immunol 170(12):6266–6272
58. Petri MH, Laguna-Fernandez A, Gonzalez-Diez M, Paulsson-Berne G, Hansson GK, Bäck M (2014) The role of the FPR2/ALX receptor in atherosclerosis development and plaque stability. Cardiovasc Res 105(1):65–74
59. Chiang N, Fredman G, Backhed F, Oh SF, Vickery T, Schmidt BA, Serhan CN (2012) Infection regulates pro-resolving mediators that lower antibiotic requirements. Nature 484 (7395):524–528. doi:10.1038/nature11042

60. Hsiao HM, Sapinoro RE, Thatcher TH, Croasdell A, Levy EP, Fulton RA, Olsen KC, Pollock SJ, Serhan CN, Phipps RP, Sime PJ (2013) A novel anti-inflammatory and pro-resolving role for resolvin D1 in acute cigarette smoke-induced lung inflammation. PLoS One 8(3), e58258. doi:10.1371/journal.pone.0058258
61. Merched AJ, Ko K, Gotlinger KH, Serhan CN, Chan L (2008) Atherosclerosis: evidence for impairment of resolution of vascular inflammation governed by specific lipid mediators. FASEB J 22(10):3595–3606. doi:10.1096/fj.08-112201
62. Spite M, Norling LV, Summers L, Yang R, Cooper D, Petasis NA, Flower RJ, Perretti M, Serhan CN (2009) Resolvin D2 is a potent regulator of leukocytes and controls microbial sepsis. Nature 461(7268):1287–1291. doi:10.1038/nature08541
63. Serhan CN, Petasis NA (2011) Resolvins and protectins in inflammation resolution. Chem Rev 111(10):5922–5943. doi:10.1021/cr100396c
64. Arnardottir HH, Dalli J, Colas RA, Shinohara M, Serhan CN (2014) Aging delays resolution of acute inflammation in mice: reprogramming the host response with novel nano-proresolving medicines. J Immunol 193(8):4235–4244. doi:10.4049/jimmunol.1401313
65. Krishnamoorthy S, Recchiuti A, Chiang N, Fredman G, Serhan CN (2012) Resolvin D1 receptor stereoselectivity and regulation of inflammation and proresolving microRNAs. Am J Pathol 180(5):2018–2027. doi:10.1016/j.ajpath.2012.01.028
66. Norling LV, Dalli J, Flower RJ, Serhan CN, Perretti M (2012) Resolvin D1 limits polymorphonuclear leukocyte recruitment to inflammatory loci: receptor-dependent actions. Arterioscler Thromb Vasc Biol 32(8):1970–1978. doi:10.1161/ATVBAHA.112.249508
67. Hasturk H, Kantarci A, Goguet-Surmenian E, Blackwood A, Andry C, Serhan CN, Van Dyke TE (2007) Resolvin E1 regulates inflammation at the cellular and tissue level and restores tissue homeostasis in vivo. J Immunol 179(10):7021–7029
68. Hasturk H, Kantarci A, Ohira T, Arita M, Ebrahimi N, Chiang N, Petasis NA, Levy BD, Serhan CN, Van Dyke TE (2006) RvE1 protects from local inflammation and osteoclast-mediated bone destruction in periodontitis. FASEB J 20(2):401–403. doi:10.1096/fj.05-4724fje
69. Serhan CN, Clish CB, Brannon J, Colgan SP, Chiang N, Gronert K (2000) Novel functional sets of lipid-derived mediators with antiinflammatory actions generated from omega-3 fatty acids via cyclooxygenase 2-nonsteroidal antiinflammatory drugs and transcellular processing. J Exp Med 192(8):1197–1204
70. Dona M, Fredman G, Schwab JM, Chiang N, Arita M, Goodarzi A, Cheng G, von Andrian UH, Serhan CN (2008) Resolvin E1, an EPA-derived mediator in whole blood, selectively counterregulates leukocytes and platelets. Blood 112(3):848–855. doi:10.1182/blood-2007-11-122598
71. Arita M, Ohira T, Sun YP, Elangovan S, Chiang N, Serhan CN (2007) Resolvin E1 selectively interacts with leukotriene B4 receptor BLT1 and ChemR23 to regulate inflammation. J Immunol 178(6):3912–3917
72. Mehrabian M, Allayee H, Wong J, Shi W, Wang XP, Shaposhnik Z, Funk CD, Lusis AJ (2002) Identification of 5-lipoxygenase as a major gene contributing to atherosclerosis susceptibility in mice. Circ Res 91(2):120–126
73. Cao RY, St Amand T, Grabner R, Habenicht AJ, Funk CD (2009) Genetic and pharmacological inhibition of the 5-lipoxygenase/leukotriene pathway in atherosclerotic lesion development in ApoE deficient mice. Atherosclerosis 203(2):395–400. doi:10.1016/j.atherosclerosis.2008.07.045
74. Jawien J, Gajda M, Rudling M, Mateuszuk L, Olszanecki R, Guzik TJ, Cichocki T, Chlopicki S, Korbut R (2006) Inhibition of five lipoxygenase activating protein (FLAP) by MK-886 decreases atherosclerosis in apoE/LDLR-double knockout mice. Eur J Clin Invest 36(3):141–146. doi:10.1111/j.1365-2362.2006.01606.x
75. Bäck M, Sultan A, Ovchinnikova O, Hansson GK (2007) 5-Lipoxygenase-activating protein: a potential link between innate and adaptive immunity in atherosclerosis and adipose tissue inflammation. Circ Res 100(7):946–949. doi:10.1161/01.RES.0000264498.60702.0d

76. Yu Z, Crichton I, Tang SY, Hui Y, Ricciotti E, Levin MD, Lawson JA, Pure E, FitzGerald GA (2012) Disruption of the 5-lipoxygenase pathway attenuates atherogenesis consequent to COX-2 deletion in mice. Proc Natl Acad Sci U S A 109(17):6727–6732. doi:10.1073/pnas.1115313109
77. Titos E, Claria J, Planaguma A, Lopez-Parra M, Gonzalez-Periz A, Gaya J, Miquel R, Arroyo V, Rodes J (2005) Inhibition of 5-lipoxygenase-activating protein abrogates experimental liver injury: role of Kupffer cells. J Leukoc Biol 78(4):871–878. doi:10.1189/jlb.1204747
78. Fredman G, Ozcan L, Spolitu S, Hellmann J, Spite M, Backs J, Tabas I (2014) Resolvin D1 limits 5-lipoxygenase nuclear localization and leukotriene B4 synthesis by inhibiting a calcium-activated kinase pathway. Proc Natl Acad Sci U S A 111(40):14530–14535. doi:10.1073/pnas.1410851111
79. Kronke G, Katzenbeisser J, Uderhardt S, Zaiss MM, Scholtysek C, Schabbauer G, Zarbock A, Koenders MI, Axmann R, Zwerina J, Baenckler HW, van den Berg W, Voll RE, Kuhn H, Joosten LA, Schett G (2009) 12/15-lipoxygenase counteracts inflammation and tissue damage in arthritis. J Immunol 183(5):3383–3389. doi:10.4049/jimmunol.0900327
80. Cyrus T, Pratico D, Zhao L, Witztum JL, Rader DJ, Rokach J, FitzGerald GA, Funk CD (2001) Absence of 12/15-lipoxygenase expression decreases lipid peroxidation and atherogenesis in apolipoprotein e-deficient mice. Circulation 103(18):2277–2282
81. Cyrus T, Witztum JL, Rader DJ, Tangirala R, Fazio S, Linton MF, Funk CD (1999) Disruption of the 12/15-lipoxygenase gene diminishes atherosclerosis in apo E-deficient mice. J Clin Invest 103(11):1597–1604. doi:10.1172/JCI5897
82. George J, Afek A, Shaish A, Levkovitz H, Bloom N, Cyrus T, Zhao L, Funk CD, Sigal E, Harats D (2001) 12/15-Lipoxygenase gene disruption attenuates atherogenesis in LDL receptor-deficient mice. Circulation 104(14):1646–1650
83. Huo Y, Zhao L, Hyman MC, Shashkin P, Harry BL, Burcin T, Forlow SB, Stark MA, Smith DF, Clarke S, Srinivasan S, Hedrick CC, Pratico D, Witztum JL, Nadler JL, Funk CD, Ley K (2004) Critical role of macrophage 12/15-lipoxygenase for atherosclerosis in apolipoprotein E-deficient mice. Circulation 110(14):2024–2031. doi:10.1161/01.CIR.0000143628.37680.F6
84. Poeckel D, Zemski Berry KA, Murphy RC, Funk CD (2009) Dual 12/15- and 5-lipoxygenase deficiency in macrophages alters arachidonic acid metabolism and attenuates peritonitis and atherosclerosis in ApoE knock-out mice. J Biol Chem 284(31):21077–21089. doi:10.1074/jbc.M109.000901
85. Rong S, Cao Q, Liu M, Seo J, Jia L, Boudyguina E, Gebre AK, Colvin PL, Smith TL, Murphy RC, Mishra N, Parks JS (2012) Macrophage 12/15 lipoxygenase expression increases plasma and hepatic lipid levels and exacerbates atherosclerosis. J Lipid Res 53(4):686–695. doi:10.1194/jlr.M022723
86. Tang L, Ding T, Pratico D (2008) Additive anti-atherogenic effect of thromboxane receptor antagonism with 12/15lipoxygenase gene disruption in apolipoprotein E-deficient mice. Atherosclerosis 199(2):265–270. doi:10.1016/j.atherosclerosis.2007.11.038
87. Zhao L, Cuff CA, Moss E, Wille U, Cyrus T, Klein EA, Pratico D, Rader DJ, Hunter CA, Pure E, Funk CD (2002) Selective interleukin-12 synthesis defect in 12/15-lipoxygenase-deficient macrophages associated with reduced atherosclerosis in a mouse model of familial hypercholesterolemia. J Biol Chem 277(38):35350–35356. doi:10.1074/jbc.M205738200
88. Zhao L, Pratico D, Rader DJ, Funk CD (2005) 12/15-Lipoxygenase gene disruption and vitamin E administration diminish atherosclerosis and oxidative stress in apolipoprotein E deficient mice through a final common pathway. Prostaglandins Other Lipid Mediat 78(1-4):185–193. doi:10.1016/j.prostaglandins.2005.07.003
89. Harats D, Shaish A, George J, Mulkins M, Kurihara H, Levkovitz H, Sigal E (2000) Overexpression of 15-lipoxygenase in vascular endothelium accelerates early atherosclerosis in LDL receptor-deficient mice. Arterioscler Thromb Vasc Biol 20(9):2100–2105

90. Reilly KB, Srinivasan S, Hatley ME, Patricia MK, Lannigan J, Bolick DT, Vandenhoff G, Pei H, Natarajan R, Nadler JL, Hedrick CC (2004) 12/15-Lipoxygenase activity mediates inflammatory monocyte/endothelial interactions and atherosclerosis in vivo. J Biol Chem 279 (10):9440–9450. doi:10.1074/jbc.M303857200
91. Shen J, Herderick E, Cornhill JF, Zsigmond E, Kim HS, Kuhn H, Guevara NV, Chan L (1996) Macrophage-mediated 15-lipoxygenase expression protects against atherosclerosis development. J Clin Invest 98(10):2201–2208. doi:10.1172/JCI119029
92. Aiello RJ, Bourassa PA, Lindsey S, Weng W, Freeman A, Showell HJ (2002) Leukotriene B4 receptor antagonism reduces monocytic foam cells in mice. Arterioscler Thromb Vasc Biol 22(3):443–449
93. Hoyer FF, Albrecht L, Nickenig G, Muller C (2012) Selective inhibition of leukotriene receptor BLT-2 reduces vascular oxidative stress and improves endothelial function in ApoE-/- mice. Mol Cell Biochem 359(1–2):25–31. doi:10.1007/s11010-011-0995-y
94. Heller EA, Liu E, Tager AM, Sinha S, Roberts JD, Koehn SL, Libby P, Aikawa ER, Chen JQ, Huang P, Freeman MW, Moore KJ, Luster AD, Gerszten RE (2005) Inhibition of atherogenesis in BLT1-deficient mice reveals a role for LTB4 and BLT1 in smooth muscle cell recruitment. Circulation 112(4):578–586. doi:10.1161/CIRCULATIONAHA.105.545616
95. Li RC, Haribabu B, Mathis SP, Kim J, Gozal D (2011) Leukotriene B4 receptor-1 mediates intermittent hypoxia-induced atherogenesis. Am J Respir Crit Care Med 184(1):124–131. doi:10.1164/rccm.201012-2039OC
96. Subbarao K, Jala VR, Mathis S, Suttles J, Zacharias W, Ahamed J, Ali H, Tseng MT, Haribabu B (2004) Role of leukotriene B4 receptors in the development of atherosclerosis: potential mechanisms. Arterioscler Thromb Vasc Biol 24(2):369–375
97. Jawien J, Gajda M, Wolkow P, Zuranska J, Olszanecki R, Korbut R (2008) The effect of montelukast on atherogenesis in apoE/LDLR-double knockout mice. J Physiol Pharmacol 59 (3):633–639
98. Mueller CF, Wassmann K, Widder JD, Wassmann S, Chen CH, Keuler B, Kudin A, Kunz WS, Nickenig G (2008) Multidrug resistance protein-1 affects oxidative stress, endothelial dysfunction, and atherogenesis via leukotriene C4 export. Circulation 117(22):2912–2918. doi:10.1161/CIRCULATIONAHA.107.747667
99. Petri MH, Tellier C, Michiels C, Ellertsen I, Dogne JM, Bäck M (2013) Effects of the dual TP receptor antagonist and thromboxane synthase inhibitor EV-077 on human endothelial and vascular smooth muscle cells. Biochem Biophys Res Commun 441(2):393–398. doi:10.1016/j.bbrc.2013.10.078
100. Doring Y, Drechsler M, Wantha S, Kemmerich K, Lievens D, Vijayan S, Gallo RL, Weber C, Soehnlein O (2012) Lack of neutrophil-derived CRAMP reduces atherosclerosis in mice. Circ Res 110(8):1052–1056. doi:10.1161/CIRCRESAHA.112.265868
101. Dalli J, Consalvo AP, Ray V, Di Filippo C, D'Amico M, Mehta N, Perretti M (2013) Proresolving and tissue-protective actions of annexin A1-based cleavage-resistant peptides are mediated by formyl peptide receptor 2/lipoxin A4 receptor. J Immunol 190 (12):6478–6487. doi:10.4049/jimmunol.1203000
102. Forsman H, Onnheim K, Andreasson E, Dahlgren C (2011) What formyl peptide receptors, if any, are triggered by compound 43 and lipoxin A4? Scand J Immunol 74(3):227–234. doi:10.1111/j.1365-3083.2011.02570.x
103. Hanson J, Ferreiros N, Pirotte B, Geisslinger G, Offermanns S (2013) Heterologously expressed formyl peptide receptor 2 (FPR2/ALX) does not respond to lipoxin A(4). Biochem Pharmacol 85(12):1795–1802. doi:10.1016/j.bcp.2013.04.019
104. Bäck M, Powell WS, Dahlen SE, Drazen JM, Evans JF, Serhan CN, Shimizu T, Yokomizo T, Rovati GE (2014) Update on leukotriene, lipoxin and oxoeicosanoid receptors: IUPHAR review 7. Br J Pharmacol 171(15):3551–3574. doi:10.1111/bph.12665
105. Van Noolen L, Bäck M, Arnaud C, Rey A, Petri MH, Levy P, Faure P, Stanke-Labesque F (2014) Docosahexaenoic acid supplementation modifies fatty acid incorporation in tissues

and prevents hypoxia induced-atherosclerosis progression in apolipoprotein-E deficient mice. Prostaglandins Leukot Essent Fatty Acids 91(4):111–117. doi:10.1016/j.plefa.2014.07.016
106. Adan Y, Shibata K, Ni W, Tsuda Y, Sato M, Ikeda I, Imaizumi K (1999) Concentration of serum lipids and aortic lesion size in female and male apo E-deficient mice fed docosahexaenoic acid. Biosci Biotechnol Biochem 63(2):309–313. doi:10.1271/bbb.63.309
107. Xu Z, Riediger N, Innis S, Moghadasian MH (2007) Fish oil significantly alters fatty acid profiles in various lipid fractions but not atherogenesis in apo E-KO mice. Eur J Nutr 46 (2):103–110. doi:10.1007/s00394-006-0638-3
108. Brown AL, Zhu X, Rong S, Shewale S, Seo J, Boudyguina E, Gebre AK, Alexander-Miller MA, Parks JS (2012) Omega-3 fatty acids ameliorate atherosclerosis by favorably altering monocyte subsets and limiting monocyte recruitment to aortic lesions. Arterioscler Thromb Vasc Biol 32(9):2122–2130. doi:10.1161/ATVBAHA.112.253435
109. Matsumoto M, Sata M, Fukuda D, Tanaka K, Soma M, Hirata Y, Nagai R (2008) Orally administered eicosapentaenoic acid reduces and stabilizes atherosclerotic lesions in ApoE-deficient mice. Atherosclerosis 197(2):524–533. doi:10.1016/j.atherosclerosis.2007.07.023
110. Zampolli A, Bysted A, Leth T, Mortensen A, De Caterina R, Falk E (2006) Contrasting effect of fish oil supplementation on the development of atherosclerosis in murine models. Atherosclerosis 184(1):78–85. doi:10.1016/j.atherosclerosis.2005.04.018
111. Yamashita T, Oda E, Sano T, Yamashita T, Ijiru Y, Giddings JC, Yamamoto J (2005) Varying the ratio of dietary n-6/n-3 polyunsaturated fatty acid alters the tendency to thrombosis and progress of atherosclerosis in apoE-/- LDLR-/- double knockout mouse. Thromb Res 116(5):393–401. doi:10.1016/j.thromres.2005.01.011
112. Wang S, Wu D, Matthan NR, Lamon-Fava S, Lecker JL, Lichtenstein AH (2009) Reduction in dietary omega-6 polyunsaturated fatty acids: eicosapentaenoic acid plus docosahexaenoic acid ratio minimizes atherosclerotic lesion formation and inflammatory response in the LDL receptor null mouse. Atherosclerosis 204(1):147–155. doi:10.1016/j.atherosclerosis.2008.08.024
113. Chang CL, Torrejon C, Jung UJ, Graf K, Deckelbaum RJ (2014) Incremental replacement of saturated fats by n-3 fatty acids in high-fat, high-cholesterol diets reduces elevated plasma lipid levels and arterial lipoprotein lipase, macrophages and atherosclerosis in LDLR-/- mice. Atherosclerosis 234(2):401–409. doi:10.1016/j.atherosclerosis.2014.03.022
114. Fujita H, Saito F, Sawada T, Kushiro T, Yagi H, Kanmatsuse K (1999) Lipoxygenase inhibition decreases neointimal formation following vascular injury. Atherosclerosis 147 (1):69–75
115. Provost P, Borgeat P, Merhi Y (1998) Platelets, neutrophils, and vasoconstriction after arterial injury by angioplasty in pigs: effects of MK-886, a leukotriene biosynthesis inhibitor. Br J Pharmacol 123(2):251–258
116. Kondo K, Umemura K, Ohmura T, Hashimoto H, Nakashima M (1998) Suppression of intimal hyperplasia by a 5-lipoxygenase inhibitor, MK-886: studies with a photochemical model of endothelial injury. Thromb Haemost 79(3):635–639
117. Yu Z, Ricciotti E, Miwa T, Liu S, Ihida-Stansbury K, Landersberg G, Jones PL, Scalia R, Song WC, Assoian RK, FitzGerald GA (2013) Myeloid cell 5-lipoxygenase activating protein modulates the response to vascular injury. Circ Res 112(3):432–440. doi:10.1161/CIRCRESAHA.112.300755
118. Natarajan R, Pei H, Gu JL, Sarma JM, Nadler J (1999) Evidence for 12-lipoxygenase induction in the vessel wall following balloon injury. Cardiovasc Res 41(2):489–499
119. Gu JL, Pei H, Thomas L, Nadler JL, Rossi JJ, Lanting L, Natarajan R (2001) Ribozyme-mediated inhibition of rat leukocyte-type 12-lipoxygenase prevents intimal hyperplasia in balloon-injured rat carotid arteries. Circulation 103(10):1446–1452
120. Deliri H, Meller N, Kadakkal A, Malhotra R, Brewster J, Doran AC, Pei H, Oldham SN, Skaflen MD, Garmey JC, McNamara CA (2011) Increased 12/15-lipoxygenase enhances cell growth, fibronectin deposition, and neointimal formation in response to carotid injury. Arterioscler Thromb Vasc Biol 31(1):110–116. doi:10.1161/ATVBAHA.110.212068

121. Potula HS, Wang D, Quyen DV, Singh NK, Kundumani-Sridharan V, Karpurapu M, Park EA, Glasgow WC, Rao GN (2009) Src-dependent STAT-3-mediated expression of monocyte chemoattractant protein-1 is required for 15(S)-hydroxyeicosatetraenoic acid-induced vascular smooth muscle cell migration. J Biol Chem 284(45):31142–31155. doi:10.1074/jbc.M109.012526
122. Reddy MA, Kim YS, Lanting L, Natarajan R (2003) Reduced growth factor responses in vascular smooth muscle cells derived from 12/15-lipoxygenase-deficient mice. Hypertension 41(6):1294–1300. doi:10.1161/01.HYP.0000069011.18333.08
123. Taylor AM, Hanchett R, Natarajan R, Hedrick CC, Forrest S, Nadler JL, McNamara CA (2005) The effects of leukocyte-type 12/15-lipoxygenase on Id3-mediated vascular smooth muscle cell growth. Arterioscler Thromb Vasc Biol 25(10):2069–2074. doi:10.1161/01.ATV.0000178992.40088.f2
124. Porreca E, Di Febbo C, Di Sciullo A, Angelucci D, Nasuti M, Vitullo P, Reale M, Conti P, Cuccurullo F, Poggi A (1996) Cysteinyl leukotriene D4 induced vascular smooth muscle cell proliferation: a possible role in myointimal hyperplasia. Thromb Haemost 76(1):99–104
125. Kaetsu Y, Yamamoto Y, Sugihara S, Matsuura T, Igawa G, Matsubara K, Igawa O, Shigemasa C, Hisatome I (2007) Role of cysteinyl leukotrienes in the proliferation and the migration of murine vascular smooth muscle cells in vivo and in vitro. Cardiovasc Res 76(1):160–166
126. Petri M, Laguna-Fernandez A, Tseng C-N, Hedin U, Perretti M, Bäck M (2015) Aspirin-triggered 15-epi-lipoxin A4 signals through FPR2/ALX in vascular smooth muscle cells and protects against intimal hyperplasia after carotid ligation. Int J Cardiol 179:370–372
127. Hakonarson H, Thorvaldsson S, Helgadottir A, Gudbjartsson D, Zink F, Andresdottir M, Manolescu A, Arnar DO, Andersen K, Sigurdsson A, Thorgeirsson G, Jonsson A, Agnarsson U, Bjornsdottir H, Gottskalksson G, Einarsson A, Gudmundsdottir H, Adalsteinsdottir AE, Gudmundsson K, Kristjansson K, Hardarson T, Kristinsson A, Topol EJ, Gulcher J, Kong A, Gurney M, Thorgeirsson G, Stefansson K (2005) Effects of a 5-lipoxygenase-activating protein inhibitor on biomarkers associated with risk of myocardial infarction: a randomized trial. JAMA 293(18):2245–2256. doi:10.1001/jama.293.18.2245
128. Tardif JC, L'Allier PL, Ibrahim R, Gregoire JC, Nozza A, Cossette M, Kouz S, Lavoie MA, Paquin J, Brotz TM, Taub R, Pressacco J (2010) Treatment with 5-lipoxygenase inhibitor VIA-2291 (atreleuton) in patients with recent acute coronary syndrome. Circ Cardiovasc Imaging 3(3):298–307. doi:10.1161/CIRCIMAGING.110.937169
129. Allayee H, Hartiala J, Lee W, Mehrabian M, Irvin CG, Conti DV, Lima JJ (2007) The effect of montelukast and low-dose theophylline on cardiovascular disease risk factors in asthmatics. Chest 132(3):868–874. doi:10.1378/chest.07-0831
130. Ingelsson E, Yin L, Bäck M (2012) Nationwide cohort study of the leukotriene receptor antagonist montelukast and incident or recurrent cardiovascular disease. J Allergy Clin Immunol 129(3):702–707. doi:10.1016/j.jaci.2011.11.052, e702
131. Dyerberg J, Bang HO (1979) Haemostatic function and platelet polyunsaturated fatty acids in Eskimos. Lancet 2(8140):433–435
132. GISSItrial (1999) Dietary supplementation with n-3 polyunsaturated fatty acids and vitamin E after myocardial infarction: results of the GISSI-Prevenzione trial. Gruppo Italiano per lo Studio della Sopravvivenza nell'Infarto miocardico. Lancet 354(9177):447–455
133. OT Investigators, Bosch J, Gerstein HC, Dagenais GR, Diaz R, Dyal L, Jung H, Maggiono AP, Probstfield J, Ramachandran A, Riddle MC, Ryden LE, Yusuf S (2012) n-3 fatty acids and cardiovascular outcomes in patients with dysglycemia. N Engl J Med 367(4):309–318. doi:10.1056/NEJMoa1203859

Role of Lipoxygenases in Pathogenesis of Cancer

J. Roos, B. Kühn, J. Fettel, I.V. Maucher, M. Ruthardt, A. Kahnt,
T. Vorup-Jensen, C. Matrone, D. Steinhilber, and T.J. Maier

Abstract Increasing evidence in literature implicates lipoxygenases as players in the growth of several solid tumor types, including pancreatic, colorectal, prostate and breast cancer. Recently, a role of 5-lipoxygenase was reported in hemato-oncological tumors, such as chronic myeloid leukemia and acute myeloid leukemia. Whereas 5-lipoxygenase and 12-lipoxygenase are generally recognized as pro-carcinogenic, the related enzyme 15-lipoxygenase-2 is down-regulated in malignant tissues, considered to function as a tumor suppressor and to inhibit tumour growth. The role of 15-lipoxygenase-1 is still subject of controversy in the literature.

The following comprehensive review summarizes experimental findings arguing for a role of lipoxygenases in several most-concerned tumor types, including pancreatic cancer, breast cancer, prostate cancer, colorectal cancer and leukemia. Furthermore, we describe own experimental evidence raising question on the use of cytotoxic effects by lipoxygenase inhibitors or mitogenic effects by lipoxygenase-derived mediators in cell culture assays as argument for a tumorigenic role of lipoxygenases and their products in cancer.

1 Introduction

The transition from healthy to malignant cells requires several essential changes to normal cell physiology such as self-sufficiency in growth signals, insensitivity to growth inhibitory signals, suppression of programmed cell death and production of mediators to promote tissue invasion and metastasis [1]. Accordingly, a number of studies demonstrated alterations in the expression of lipoxygenases (LOs) or

---

J. Roos (✉) • B. Kühn • J. Fettel • I.V. Maucher • A. Kahnt • D. Steinhilber
Institute of Pharmaceutical Chemistry/ZAFES, Goethe-University, Max-von-Laue-Str. 9, 60438 Frankfurt/Main, Germany
e-mail: roos@pharmchem.uni-frankfurt.de

M. Ruthardt
Department of Hematology, Goethe University, 60590 Frankfurt, Germany

T. Vorup-Jensen • C. Matrone • T.J. Maier (✉)
Department of Biomedicine, Aarhus University, Bartholins Allé 6, 8000 Aarhus C, Denmark
e-mail: maier@biomed.au.dk

© Springer International Publishing Switzerland 2016
D. Steinhilber (ed.), *Lipoxygenases in Inflammation*, Progress in Inflammation Research, DOI 10.1007/978-3-319-27766-0_7

respective receptors for LO products in tissue samples of primary tumor cells as well as in established cancer cell lines [2–6]. Strongest evidence was reported for pancreatic, colorectal, prostate and breast cancer as well as hemato-oncological tumors as described in this review. However, some mechanistically related LO-dependent (anti-)tumorigenic effects were also seen in other tumor types, such as urological cancers, tumor of the central nervous system, kidney tumors and in skin carcinogenesis [7–10]. An overview about cell culture or animal studies assessing the role of lipoxygenases in most-relevant cancer types is presented in Fig. 1. Furthermore, LO products have been associated with several aspects of tumorigenesis, including promoting cancer cell proliferation and genotoxicity, prevention of apoptosis, and increase in tumor angiogenic and metastatic potential. The expression of LOs in tumor cells was found to be regulated by several well-recognized oncogenes and tumor suppressors, and several LO products have been shown to directly promote tumor cell proliferation and to contribute to cancer incidence, progression, and invasion [11]. Addition of 5-LO products to cultured tumor cells led to increased cell proliferation and activation of anti-apoptotic signaling pathways maintaining tumor cell viability [12, 13]. Furthermore, LO-antisense technology approaches were capable of modulating tumor cell growth [14] and pharmacological inhibition of 5-LO and 12-LO, but not of the 15-LOs, has

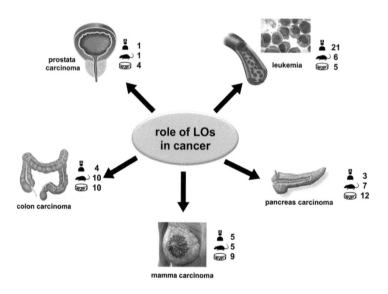

Fig. 1 Overview of the role of LOs in the most relevant tumors discussed in this chapter. Figure illustrates the respective number of studies of different types (using primary human material, cell lines and/or animal models) quoted to draw the connection between LOs and cancer. Only publications mentioned in this review have been listed. Notably, publications can contain more than one type of studies. For further information consult the respective section in this review

been shown to potently suppress tumor cell proliferation by inducing cell cycle arrest and triggering cell death via the intrinsic apoptotic pathway [15–17]. Based on these findings, drugs suppressing leukotriene (LT) biosynthesis or antagonizing LT signaling are considered a promising and novel pharmacological strategy for cancer prevention and therapy.

2 Studies Providing Evidence for a Role of LOs in Tumorigenesis

2.1 Role of LOs in Pancreatic Cancer

The first evidence for a potential role for 5-LO in pancreatic cancer growth was based on experiments with the FLAP (5-LO activating protein) inhibitor MK-886 which strongly induced programmed cell death in a pancreatic cancer cell line [18]. Further evidence for a role of 5-LO in pancreatic cancer cell growth was provided by Ding et al. who showed that (i) 5-LO was expressed in malignant cells but not in normal pancreatic ductal cells, (ii) 5-LO inhibitors and anti-sense approaches markedly suppressed tumor cell proliferation, and (iii) 5-LO products were capable of directly stimulating pancreatic tumor cell proliferation [15, 19]. These studies were supported by Hennig et al., who demonstrated an increased expression of 5-LO in a set of human pancreatic cancer cells and a decreased expression in normal pancreatic ductal cells [20]. In further support of this hypothesis, an elevated expression of the LTB_4 receptor was detected in human pancreatic cancer tissue [20]. Zhou et al. found that the anti-inflammatory and chemo-preventive herbal ingredient triptolide is able to suppress the growth of pancreas carcinoma cell lines by down regulating the expression of 5-LO and consequently LTB_4 synthesis [21].

The exact molecular mechanisms through which 5-LO products act on cancer cells in a growth-stimulatory manner remain, however, only incompletely understood. The first mechanistic information was provided by Tong et al. They described as a key event to explain 5-LO inhibitor induced cell death mitochondrial-mediated caspase activation via the intrinsic pathway of apoptosis [21]. In subsequent studies with the orally bioavailable LTB_4 receptor antagonist, LY293111, strong anti-pancreatic cancer effects were induced through tumor cell apoptosis in vitro and in vivo [22]. Also, triggering of a S-phase cell cycle arrest in pancreatic cancer cell lines were observed [23]. Thus, direct mitogenic and anti-apoptotic stimuli apparently play an important role. In line with these findings, 5-HETE 5-Hydroxyeicosatetraenoic acid and LTB_4 enhanced cell proliferation and viability of pancreatic cancer cells by activating the mitogenic and anti-apoptotic MAPK and AKT kinase signaling pathways [12].

Notably, only a limited number of studies are available to support a role of 5-LO in tumorigenesis in animal models or patients. Evidence for a role of 5-LO and its bioactive products in the pathogenesis of pancreatic cancer growth in vivo was provided by Henning et al. who investigated the expression of 5-LO in pancreatic cancer tissues extracted from animal models and patients. Here, 5-LO overexpression was shown to constitute an early pancreatic biomarker to indicate intraepithelial neoplastic lesions [24]. Furthermore, the only approved 5-LO inhibitor on the market, zileuton, was shown to reduce tumor growth of pancreatic tumors in a pancreatic cancer model in Syrian hamsters. Moreover, zileuton treatment reduced the incidence and size of pancreatic cancer tumors, both in monotherapy and in combination with the cyclooxygenase-2 (COX-2) inhibitor celecoxib [25, 26]. A few clinical trials have been conducted assessing the LTB_4 receptor antagonist LY293111 in patients with pancreatic cancer. Although a Phase I study found that LY293111 was well tolerated in combination with the chemotherapeutic drug irinotecan, significant therapeutic benefits were not observable in six-month survival or progression-free survival due to LY29311 [27, 28]. Another very recent study conducted in patients with non-small cell lung cancer receiving LY293111 and cisplatin/gemcitabine confirmed the lack of therapeutic benefit of the combination versus cisplatin/gemcitabine alone [29].

Less is known about the role of 12-LO and 15-LO in pancreatic cancer growth. Ding et al. observed overexpression of 12-LO in a series of humans pancreatic cancer lines, but not in normal pancreatic ductal cells. Apparently, the 12-LO metabolite 12-HETE directly stimulated pancreatic cancer cell proliferation [15, 19]. Mechanistic studies provided evidence that activation of the ERK and p38 MAPK pathways are crucially involved in the 12-HETE-induced pancreatic cancer cell proliferation [30]. Finally, the 12-LO inhibitor baicalein induced programmed cell death in pancreas carcinoma cells via the cytochrome c and caspase-3-dependent intrinsic pathway of apoptosis [31]. Henning et al. were among the first to demonstrate an enhanced pancreatic tumor cell proliferation due to downregulation of 15-LO-1 as compared to normal ductal pancreatic cells [32] suggesting that 15-LO-1 may function as a tumor suppressor like protein.

2.2 LOs in Breast Cancer

Over the last two decades evidence emerged in literature that the different types of LOs are also implicated in the growth of human breast cancer. Kort et al. [33] reported significant differences in the eicosanoid profile of tissues obtained from benign tumor material and malignant mammary tumors. In this study, no detectable quantities of LO metabolites (except 15-HETE) were present in the normal tissue, whereas in the malignant tumor material, substantial quantities of a number of metabolites of the LO pathway (5-/12-/15-HETE, LTB_4, LTC_4) were found [33]. Natarajan et al. [34] discovered an elevated expression of 12-LO in tumor tissue of breast cancer patients as well as in the two breast cancer cell lines MCF-7 and COH-BR1. By contrast, the non-tumorigenic breast epithelial cell line MCF-10F only expressed low levels of 12-LO [34]. These results are in line with findings by Jiang et al. [35] showing a significantly higher expression level of

12-LO in tumor tissues of a cohort of breast cancer patients compared to normal breast tissue. In the same study, it was found that overexpression of 12-LO was accompanied by a significantly reduced level of 15-LO expression in the same tumor tissue samples. Further investigation of the expression level of both 15-LO isoforms revealed a reduction of both 15-LO-1 and 15-LO-2 levels in breast tumor tissues, compared to normal tissues [36]. Interestingly, expression levels of both 12-LO and 5-LO were strikingly elevated in tissue samples from patients who died from breast cancer [35].

Studies regarding the expression level of FLAP, a protein which supplies the substrate arachidonic acid to 5-LO, disclosed an aberrant expression pattern of FLAP in human breast cancer tissues, especially in aggressive tumors [37]. The authors were able to show a correlation of high expression levels of FLAP and a poor prognosis and lower survival rate of the patients, indicating that FLAP may have a value as prognostic marker [37].

Another study [38] investigated the correlation of 5-LO and FLAP gene polymorphisms and the risk for breast cancer. Here, no significant correlation between different 5-LO or FLAP genotypes and breast cancer risk was found. Nevertheless, authors postulated a certain connection between high dietary linoleic acid intake and the FLAP -4900 A > G polymorphism predisposing for a significantly higher risk of breast cancer [38]. However, a subsequent population-based, multiethnic, but polymorphism-independent study could not confirm this finding for linoleic acid [39].

In addition to the reported 5- and 12-LO overexpression in breast cancer tissues, metabolites of arachidonic acid generated by the LO pathway have been shown to promote tumor cell proliferation. Exposure of five different breast tumor cell lines (MCF-7, ZR-75, T47D, SKBR3, MDA-MB-231) to exogenously added 5(S)-HETE increased tumor cell proliferation up to 45 % [2]. Additionally, LTD_4, which was tested only with the ZR-25 cell line, increased cell growth by 25 %. However, all other arachidonic acid metabolites such as 5(R)-HETE, 5-oxo-ETE, LTB_4, 12-/15-HETE did not display any significant mitogenic effects [2]. In other studies, proliferation enhancing effects of the 5-LO metabolites 5-oxo-ETE and 5-oxo-15-OH-ETE used at low concentrations < 10 µM [40] as well as of 12-HETE (at < 0.1 µM) were seen in the breast cancer cell lines MDA-MB-231 and MCF-7 [41]. Recently, stimulatory effects of the 15-LO metabolites 15-HETE and 13-HODE (13-hydroxyoctadecadienoic acid) on the proliferation of both cell lines were observed by cell growth assays [42], which was not observed before. Both, 15-HETE and 13-HODE are formed by 15-LO, however, 13-HODE is preferably generated by 15-LO-1 from linoleic acid [43]. These findings are in line with results provided by Reddy et al. [44] who showed an expression of 15-LO in human breast carcinoma cell line BT-20. Moreover, stimulation of BT-20 cells with epidermal (EGF) and transforming growth factor α (TGFα) led to increased DNA synthesis which was accompanied by an induction of 13-HODE formation. Finally, DNA synthesis as well as 13-HODE formation was attenuated by the unselective LO inhibitor nordihydroguaiaretic acid (NDGA) but not by the cyclooxygenase inhibitor indomethacin [44]. Authors concluded that 15-LO expression and 13-HODE formation as a result of high dietary levels of linoleic acid may enhance cell growth in cancerous breast tissue [44].

The above mentioned studies investigating either expression levels of LOs or metabolite formation provide evidence for an involvement of LOs in breast cancer tumorigenesis and growth. This initiated a number of studies investigating potential cytotoxic, anti-proliferative and pro-apoptotic effects of LO inhibitors in breast cancer cells.

Strong growth-inhibitory effects were seen with the two FLAP inhibitors MK-886 and MK-0591 used at comparatively high concentrations of 5 µM in five different breast cancer cell lines (MCF-7, ZR-75, T47D, SKBR3, MDA-MB-231) [2]. Addition of 5-HETE abolished the inhibitor-induced effects. Similar anti-proliferative effects in the same breast cancer cell lines were triggered by the unselective LO inhibitor NDGA at 5 µM, whereas only moderate effects on tumor cell growth were seen with AA-861 and zileuton (10 µM each) [2]. These findings were confirmed by Hammamieh et al. [45], showing that besides NDGA and AA-861 also curcumin, a major ingredient in the root of the curcuma plant and a dual 5-LO and COX-2 inhibitor [46] exerts anti-proliferative effects on MCF-7 breast cancer cells. Moreover, all inhibitors tested were also capable of effectively and potently blocking cell growth in the chemotherapeutic-resistant breast cancer cell line MCF-7 ADR [45].

In vivo studies using experimental animal carcinogenesis models investigated the effect of esculetin, a dual 5- and 12-LO inhibitor [47], on tumor development. In this study, rat mammary tumorigenesis and tumor proliferation induced by either 7,12-dimethylbenz[a]anthracene (DMBA) or N-methyl-N-nitrosourea (MNU) were inhibited by esculetin (0.03 % diet) [48, 49]. In addition, experimental animal carcinogenesis models were also conducted using the 5-LO inhibitor curcumin. The administration of dietary curcumin significantly reduced the incidence (28.0 %) of diethylstilbestrol (DES)-induced tumor promotion of rat mammary glands initiated with radiation [50]. In these investigations curcumin did not produce severe side-effects and is consequently likely to be an effective agent for chemoprevention acting at the radiation-induced initiation stage of mammary tumorigenesis [51].

Recent studies encouraged the design and testing of novel LO inhibitors regarding their potential as anti-proliferative anti-cancer agents. Hence, prenylated chalcone derivatives have been shown to inhibit 5-LO activity and to induce potent anti-proliferative effects in the human breast cancer cell line MCF-7 [52]. In another series of experiments, a peptide inhibitor of 12-LO was designed and analyzed in vitro and in vivo for its anti-tumorigenic effects [53]. The peptide YWCS significantly reduced cell viability in the breast cancer cell lines MCF-7 and MDA-MB-231. Moreover, the peptide increased the portion of apoptotic cells to 49.8 % and 20.8 % in MCF-7 and MDA-MB-231 cell lines, respectively. Finally, in vivo studies demonstrated that the parenteral administration of the peptide suppresses tumor growth in mice [53].

Together, a number of studies indicate for a certain involvement of the LO biosynthesis pathways in breast cancer development and progression. A considerable number of studies established a correlation of aberrant LO expression to tumor cell malignancy, poor prognosis and lower survival rate of the patients. Mitogenic effects of LO metabolites as well as anti-proliferative and pro-apoptotic actions of LO inhibitors provided a mechanistic basis to explain these effects.

2.3   Role of LOs in Prostate Cancer

Prostate cancer is the most common type of cancer with about 233,000 new cases in the United States in 2014. The American Cancer Society estimates that in 2014 approximately 30,000 men will die from prostate cancer in the United States (source: http://www.cancer.org). Development of new therapeutic strategies and identification of new targets for treatment of prostate cancer represents a still considerable unmet medical need. Several decades ago, a correlation between consumption of high fat food and development of prostate cancer was postulated [54]. Epidemiological studies and animal experiments both have strongly supported that incidence of prostate cancer was increased with a high fat diet [55].

Further studies report dietary fatty acids—particularly arachidonic acid, which is a major ingredient of "Western pattern diets"—and their 5-LO-derived metabolites as crucial players in promotion of the growth of prostate cancer cells [56].

From these studies 5-LO was considered as a potential therapeutic target for the treatment of prostate cancer. This hypothesis was strengthened by subsequent inhibitor studies showing that pharmacological suppression of 5-LO led to massive induction of apoptosis in prostate cancer cells [16].

This was followed by studies which discovered that prostate cancer cells are capable of constitutively producing 5-LO metabolites and that pharmacological inhibition of 5-LO and biosynthesis of 5-LO products triggers massive apoptosis in androgen-sensitive as well as androgen-independent cancer cells. This suggests a novel therapeutic option to treat androgen-independent cancer cells [57] and to prevent metastatic dissemination to other organ systems [58]. Furthermore, it was shown that exogenous 5(S)-HETE and more effectively its dehydrogenase-derivative 5-oxo-ETE prevented prostate cancer cells from 5-LO inhibitor induced apoptosis [57].

In 2001 Gupta et al. published that prostate cancer cells feature increased expression and activity of 5-LO whereas 5-LO protein expression was undetectable in healthy tissue [59]. Other studies could provide first evidence of a mechanistic linkage between 5-LO and the tumor cell viability, tumor cell proliferation and migration, metastasis as well as activation of apoptotic signaling cascades [60]. Furthermore it was shown that inhibition of 5-LO by the approved inhibitor Zileuton was able to suppress metastasis of prostate cancer cells [61]. According to Meng et al. this drug effect is probably mediated by an up-regulation of the calcium-regulated adhesion molecule E-cadherin and Paxillin. Notably, a number of previous cell culture studies already show a correlation of levels of E-cadherin and invasiveness [62]. Low E-cadherin expression has been associated with more aggressive tumors [63]. Paxillin has been identified as a key regulator of cellular migration mediating extra-nuclear and intra-nuclear signaling in prostate cancer proliferation [64]. Interestingly, zileuton-induced upregulation of Paxillin and E-cadherin significantly decreased the motility of the androgen-independent PC-3 prostate cancer cell line [61].

Meng et al. suggested zileuton as a beneficial and well tolerated short term treatment regime before and after prostatectomy [61]. To better understand 5-LO inhibitor-induced apoptosis in prostate cancer cells, Sarveswaran et al. investigated the effects of drugs on a series of tumorigenic enzymes, known to regulate growth and survival of a number of cancerous cells. These included phosphatidylinositol 3′-kinase-AKT/protein kinase B (PI3K/AKT), the mitogen- activated protein kinase- extracellular signal regulated kinase (MEK-ERK) and the protein kinase C-epsilon (PKCε).

When prostate cancer cells were treated with the FLAP inhibitor MK-0591 selectively suppressing LT biosynthesis, Sarveswaren et al. neither observed reduced phosphorylation of Akt at Ser^{473} or reduced enzymatic activity, nor inhibition of phosphorylation/activation of ERK [65]. These results suggest that 5-LO inhibitor mediated inhibition of prostate cancer cells proliferation is independent of Akt and ERK signaling [57].

After the treatment of prostate cancer cells with MK-0591, a significant and rapid loss of PKCε activity was detected, which was abolished by exogenous addition of 5-LO metabolites. These findings are consistent with the well-recognized function of PKCε as an oncogene overexpressed in several human cancers and involved in cell proliferation, migration, invasion, and survival [66].

In summary a number of studies found elevated expression of 5-LO and 5-LO products in prostate cancer cell lines or tumors indicating a tumorigenic role of 5-LO and its metabolites in survival of prostate cancer cells [57]. PKCε was identified as important player in 5-LO-mediated tumorigenesis. However, the exact molecular mechanisms how 5-LO metabolites lead to activation of the oncogenic PKCε still remains unknown.

2.4 LOs in Colorectal Cancer

Regarding incidence and mortality, colorectal cancer (CRC) still represents one of the most relevant human malignancies [67]. Considerable efforts have been made in the past to identify critical biochemical pathways involved in the pathogenesis of the disease. Besides the COX pathway, also dysfunctions in the LO signaling were related to the development of colorectal cancer. Already two decades ago researchers recognized the potential of 5-LO inhibitors to impair colorectal tumor growth. For example, BWA4C or BWB70C were found to inhibit the growth of murine adenocarcinomas (MACs). It was postulated that metabolites deriving from the LO pathway could play an important role in CRC [68]. In line with these findings, some structurally different 5-LO inhibitors, such as 2,3,5-trimethyl-6-(3-pyridylmethyl)-1,4-benzoquinone, were also found to induce anti-proliferative effects in MACs [69]. Öhd et al. [70] suggested that LT signaling could play an important role in pathogenesis of CRC as authors found increased expression levels of the LTD_4 receptor ($CysLT_1R$) in CRC tissue. This overexpression was significantly correlated to other tumorigenic factors like COX-2, Bcl-x_L and also 5-LO. Interestingly, male samples showed a higher expression level of $CysLT_1$ receptor than female samples. Furthermore, a high expression level was accompanied by a higher aggressiveness of the tumors [70]. Later on, it has been shown that 5-LO expression is up-regulated in CRC tumor epithelial cells compared to healthy mucosa. For this study, 91 CRC and adjacent normal mucosa samples from the same patients were evaluated based on polymerase chain reaction [71]. Another study showed that $CysLT_1R$ antagonists (used at µM concentrations) were able to induce apoptosis in intestinal epithelial cells, whereas there was no significant induction of cell death in CRC cell lines [72]. The authors considered $CysLT_1Rs$ localized in the nuclei of tumor cells as reason for this observation. Such nuclear receptors, first described by Nielsen et al. [73] may not be accessible by antagonists outside the cell. However, the FLAP inhibitor MK-886 caused apoptosis in both non tumor and tumor cells potentially by suppression of 5-LO product formation [72].

Another pro-inflammatory eicosanoid mediator with a documented effect on colon cancer cell proliferation is LTB_4. In human CRC tissues as well as in cultured Caco-2 and HT-29 cells, a strong expression of the LTB_4 receptor 1 (BLT1) was observed [74]. Treatment with the 5-LO inhibitor AA861 and the LTB_4-receptor antagonist U75302 markedly decreased the survival of both cell lines due to induction of apoptosis. Additional evidence for the importance of the LTB_4 receptor signaling pathway for cell survival and proliferation emerged from RNA silencing experiments demonstrating suppression of tumor cell proliferation by BLT1 RNAi approaches [74].

As colonic polyps are considered as precursors of colorectal cancer, several studies investigated a potential tumorigenic role of 5-LO and its products at early phases of the adenoma-carcinoma-sequence. An overexpression of 5-LO has been found in adenomatous polyps [75]. In the same study, the 5-LO inhibitor zileuton was able to inhibit the growth of human colon cancer cells in a xenograft model using athymic nude mice. Therefore, it was speculated that 5-LO may be a target candidate for CRC prevention and therapy [75]. In support of this hypothesis, genetic disruption of 5-LO led to a reduction in the number and size of intestinal polyps in an APC-driven mouse model of adenomatous polyposis [76].

Many factors are assumed to contribute to pathogenesis of CRC. Thus, cigarette smoking was found to be an important external trigger [77]. In 2004, a role of 5-LO in triggering CRC upon cigarette smoke exposure was observed in a mouse model [78]. Mice were exposed to cigarette smoke and co-treated with dextran sulfate sodium to stimulate cancer formation. This treatment significantly increased 5-LO protein expression (40 %) as well as vascular epithelial growth factor (VEGF) expression and activity of matrix metalloproteinase-2 (MMP-2) in the colonic tissue. VEGF and matrix metalloproteinases represent important tumorigenic angiogenesis regulators [79]. When animals were additionally treated with the 5-LO inhibitors AA861 or MK886, a significant reduction in MMP-2 activity, VEGF protein expression and number of blood vessels in the colonic tissue could be seen. Based on these findings authors proposed 5-LO inhibitors to be useful in the prevention of cigarette smoke-dependent adenoma formation in the colon [78].

Subsequently, Ye et al. [80] conducted studies on a xenograft model in which SW1116 cells were pretreated with cigarette smoke extract before inoculated into mice. This treatment provoked significantly upregulated LTB_4 levels in the tumor tissues. Interestingly, in the AA861 treated group of mice, LTB_4 levels were suppressed and tumor size was reduced by up to 50 % compared to controls [80]. Another group of animals received the COX-2 inhibitor celecoxib known to reduce incidence of adenomatous polyps [81]. Interestingly, treatment with celecoxib only led to reduction in tumor size of 30 %. The combination of both inhibitors synergistically caused the strongest reduction in tumor size and allowed reduced doses of both inhibitors. Additionally, a correlation between 5-LO DNA demethylation and exposure to cigarette smoke extract was found [80].

Whereas 5-LO was reported to exhibit tumorigenic effects, 15-LO was demonstrated to induce anticarcinogenic effects with relevance for progression of CRC. In human colorectal cancer cell lines it had been shown that the forced overexpression of 15-LO-1 not only reduced cell proliferation but was also able to induce apoptosis [82]. Furthermore, 15-LO-1-induced apoptosis was accompanied by an increase in caspase-3 activity [82]. Notably, an in vivo study with a transgenic mouse model confirmed the 15-LO-1-triggered reduction in proliferation [83]. Mice which expressed the human 15-LO-1 enzyme targeted to the intestinal epithelial cells showed a significantly inhibited number of azoxymethane (AOM) induced colonic tumors compared to control mice. As TNF-α-iNOS mRNA levels were inversely associated with 15-LO-1 expression, authors suggested a reduced cytokine signaling as explanation for the 15-LO-1 induced anti-carcinogenic effects [83]. Furthermore, restoration of expression of 15-LO-1 in colon cancer cell lines has been shown to attenuate the cells' hypoxia-induced response including induction of angiogenesis and VEGF expression and invasion of the tumor cells in cell culture experiments. Therefore, the authors recommend 15-LO-1 modulation as potential approach to suppress metastasis [84]. Comparatively less is known about the role of 12-LO in CRC. Exemplified, the 12-LO inhibitor cinnamyl-3-4-dihydroxy-a-cyanocinnamate was shown to inhibit ROS production, cell spreading and cell proliferation in colon cancer cell lines [85].

Some naturally occurring compounds were considered to be chemo-preventive against CRC. In rats treated with AOM to induce colon cancer, the plant phenolic compound curcumin was able to inhibit invasive and noninvasive adenocarcinomas. In addition, LO metabolites were significantly reduced in tumor tissue as well as in the normal mucosa. Based on these results, authors suggested the chemopreventive effect is partly due to LO inhibition [86]. In a later study, the 5-LO modulating effect of curcumin could be confirmed [87].

Furthermore, Rao et al. [88] found Phenylethyl-3-methylcaffeate (PEMC) to be another natural compound with properties similar to curcumin. PEMC is a caffeic acid ester derivative which is present in propolis, produced by honey bees. As true for curcumin, the oral administration of PEMC caused decreased LO metabolite levels in the tumors of AOM-induced CRC in rats. Compared to rats which received a control diet, this decrease was accompanied by a reduction in tumor volume of approx. 43 % [88].

Current CRC chemotherapy regimens mainly recommend use of 5-fluorouracil (5-FU). This pyrimidine analog can also be combined with other chemotherapeutic agents like irinotecan. Due to the still limited availability of effective and well-tolerated chemotherapeutics against CRC, the search for new pharmacological targets and pathways to induce cytotoxicity in colon cancer cells steadily increases in importance. Dual COX/FLAP inhibitors such as licofelone could be potential candidates. This compound was able to suppress colon tumor growth in APC mutant mice [89]. Another group of dually acting inhibitors addressing FLAP and 5-LO are the di-*tert*-butyl phenols. Well-characterized representatives of this group are darbufelone and derivatives which have been shown to inhibit cell proliferation and possess cytotoxic properties against colon cancer cell lines [90, 91].

In conclusion, a body of evidence accounts for clinical impact of lipid metabolites deriving from the arachidonic acid cascade on pathogenesis of CRC. Not only the COX pathway, but also the LO pathway is a player in development and progression of CRC. Compared to healthy tissue, a number of studies described alterations in LT receptor expression and LT biosynthesis in colon cancer cells and neoplastic tissues. Moreover, anti-proliferative effects of 5-LO inhibitors in cell culture assays are well-recognized. However, exact molecular mechanisms are still lacking to explain the 5-LO-mediated tumorigenic effects on CRC and warrant further investigations.

2.5 LOs as Players in Hemato-Oncological Malignancies

One of the first studies which indicated an involvement of 5-LO in hematological malignancies was performed by Tskuada et al. who studied the effect of the 5-LO inhibitor AA-861 in a human leukemic cell line [92]. AA-861 showed an anti-proliferative effect on these cells, which was accompanied by a reduction in cellular DNA, RNA and protein synthesis. From this study it was concluded that LT play an important role in cancer cell viability [92]. This role was further supported by a study in 1995, where primary human CML (chronic myeloid leukemia) blast cells in culture as well as leukemic cell lines showed reduced cell proliferation after treatment with different 5-LO inhibitors [93]. Similar results were obtained by Gillis et al. who showed that 5-LO inhibition by AA-861 increased apoptosis in eicosapentaenoic acid (EPA)- treated leukemic HL-60 cells [94]. Over these past years, only a small number of publications addressed the role of 5-LO in the pathogenesis of leukemia. Nevertheless, differential 5-LO expression was found in several types of primary leukemic cells from patients with pre-B cell acute lymphatic leukemia (ALL) [95, 96], chronic myeloid leukemia, [97, 98] acute myeloid leukemia (AML) [96, 99], B-cell chronic lymphocytic leukemia (B-CLL) [100–103] and mantle cell lymphoma (MCL) [104, 105]. In MCL cells, a rare subtype of non-Hodgkin lymphoma, 5-LO was up-regulated approximately sevenfold compared to normal B-cells. Inhibition of 5-LO and FLAP induced apoptosis in MCL cell lines and primary CLL cells suggesting an important role for the LT synthesis pathway in MCL and other B-cell malignancies [104]. This was in accordance with studies conducted by Runarsson et al., who showed that B-CLL cells produce LTB_4 and that the inhibition of the LT-synthesis by 5-LO and FLAP inhibitors (BWA4C and MK886) counteracted CD40-dependent activation of B-CLL cells by inhibiting CD40-induced DNA synthesis and CD40-induced expression of CD23, CD54 and CD150. Addition of exogenous LTB_4 almost completely reversed the observed effects [102]. Furthermore, overexpression of 5-LO correlated with a highly progressing disease in B-CLL patients [100]. The same was observed for the expression of $cPLA_{2\alpha}$ and BLT2. Although BLT1 and LTA_4 hydroxylase expression was also significantly raised in B-CLL cells compared to healthy B-cells, no statistically significant correlation was found with the stage of the disease [100]. Elevated expressions of $cPLA_{2\alpha}$ and increased synthesis of LTC_4 could also be found in granulocytes from AML patients compared to leukocytes from healthy donors. However in the same patient samples the formation of LTB_4 was also suppressed [106, 107]. In that context it should be noted that overexpression of one enzyme of the LT-synthesizing machinery in leukemic cells does inevitably account for an overexpression of the other enzyme [108].

Human CML microarray studies have revealed that 5-LO is differentially expressed in $CD34^+$ CML cells than in $CD34^+$ control cells [97, 98], suggesting likewise a role of 5-LO in human CML stem cells [109]. Furthermore a recent publication showed that 5-LO and BLTR1 levels were decreased both in blood samples and $CD34^+$ stem cells from CML patients compared to samples from healthy donors [110]. On the other hand, it was shown by Chen al. that 5-LO is up-regulated in BCR/ABL positive $GFP^+Lin^-ckit^+Sca1^+$ leukemic stem cells (LSCs) in comparison to non BCR/ABL expressing $GFP^+Lin^-ckit^+Sca1^+$ cells [109]. In a murine BCR/ABL leukemia induction model, recipients of BCR/ABL transduced bone marrow cells from $5\text{-LO}^{-/-}$ donor mice failed to develop CML, whereas recipients of BCR/ABL transduced bone marrow cells from wildtype donor mice developed and died within 4 weeks. They also observed a gradual disappearance of myeloid leukemia cells in CML mice in the absence of 5-LO. Additionally, 5-LO deficiency led to an impairment of LSCs but did not significantly cause a functional defect in normal hematopoietic stem cells (HSCs). More precisely, the differentiation and division of long-term (LT)-LSCs was affected, preventing these cells from developing CML. To further confirm their data, zileuton was used as a 5-LO inhibitor in their studies. Zileuton also led to an impairment of BCR/ABL positive LSCs in a similar manner. Pharmacological suppression of 5-LO in mice with a BCR/ABL-induced CML by zileuton prolonged their survival compared to placebo treated mice. The combination of imatinib, a BCR/ABL inhibitor, and zileuton had even a better therapeutic effect than with either zileuton or imatinib alone in prolonging survival of the mice. In another experiment loss of 5-LO did not affect BCR/ABL-induced lymphoid leukemia, suggesting that 5-LO is not required by ALL stem cells, but is specifically required by CML stem cells [109]. These results were followed by a clinical phase I study in which zileuton was tested in combination with imatinib for the treatment of CML (http://clinicaltrials.gov trial No. NCT01130688). The results remain to be published.

Recently it could be shown that 5-LO is strongly up-regulated in t(8;21)-positive AML [111]. The (8;21) translocation leads to the fusion protein AML1/ETO, which together with t(15;17) (PML/RARα), is one of the most prevalent translocations associated with AML. Gene expression microarray and ChIP-chip analysis identified 5-LO as a potential disease related target gene of AML1/ETO in an AML1/ETOex9 murine leukemia model [112]. Additionally, publically available microarray data of AML M2 patients with and without the (8;21) translocation was analyzed [99] and it was found that 5-LO expression was approximately 2.3-fold higher in t(8;21)-patients than in patients without the translocation. Both groups had elevated 5-LO levels relative to normal $CD34^+$ control samples [111]. AML1/ETO is able to up-regulate 5-LO via the transcription factor KLF6 (Kruppel-like factor 6), a protein critically required for early hematopoiesis [113]. Furthermore, KLF6 is also a cell-specific positive regulator of LTC_4 synthase [114]. In serial replating assays with wildtype and $5\text{-}LO^{-/-}$ murine total bone marrow cells, which were transduced with either control vector or AML1/ETOex9 fusion protein, loss of 5-LO decreased the aberrant replating efficiency of AML1/ETOex9 -positive cells. An impairment of in vitro self-renewal of other fusion proteins like PML/RARα and MLL/AF9 (translocation t(9;11)), was also seen. Nevertheless, AML1/ETOex9-expressing $5\text{-}LO^{-/-}$ cells were still capable of leukemia induction. These results led the authors hypothesize that in an AML1/ETOex9 leukemia induction mouse model other factors may exist that allow AML1/ETOex9 to overcome these defects and promote leukemia and that 5-LO is maybe more required by AML1/ETO than AML1/ETOex9 in leukemia development, especially given that AML1/ETO more strongly up-regulates KLF6. Another reason could be that in the in vitro and in vivo experiments conducted different cell populations with different ratios of HSCs were used (total bone marrow versus fetal liver cells), suggesting that different subtypes of HSCs respond differently to the introduction of AML1/ETOex9 and 5-LO deficiency [111].

However, in our recent work we showed that 5-LO is essential for the maintenance of LSC in a PML/RARα-positive stem cell model of AML [115]. As a LSC model we used $Sca1^+Lin^-$ murine hematopoietic stem and progenitor cells (HSPC), which were retrovirally transduced with PML/RARα. At first we studied the effects of 5-LO inhibition on LCS capacity with two potent 5-LO inhibitors, CJ-13,610 and zileuton. Treatment of PML/RARα-expressing HSPCs abolished the aberrant replating efficiency as well as LT- and ST(short-term)-stem cell capacity of these cells in a concentration-dependent manner. Treatment of the control non-PML/RARα-positive cells did not negatively affect these cells. On the contrary, LT- and

ST-HSC showed a higher frequency than DMSO treated cells, indicating that targeting 5-LO is not stem cell toxic, which is also in accordance with the work of Chen at al. [109]. Furthermore, suppression of stem cell capacity was accompanied by Wnt-signaling inhibition. Immunohistochemical staining of murine spleen colonies induced by PML/RARα against β-catenin showed a reversion of the PML/RARα-activation of Wnt-signaling by CJ-13,610. The oncogenic Wnt-signaling pathway is fundamental for an aberrant self-renewal and maintenance of cancer stem cells (CSC) in many cancers [116–119], but particularly in AML and CML [120–122]. Co-immunoprecipitation experiments in 293T cells as well as reporter gen assays and immunofluorescence staining in human leukemic U937 cells expressing inducible PML/RARα or NB4 cells, confirmed the involvement of 5-LO in regulation of the Wnt-signaling pathway. Furthermore, it was demonstrated that enzymatically inactive 5-LO directly interacts with β-catenin and inhibits the PML/RARα-induced activation of Wnt-signaling by preventing the translocation of β-catenin into the nucleus [115]. However, the exact molecular mechanism by which inactive 5-LO interacts with β-catenin and possible other Wnt components still remains unclear.

In summary, these studies mentioned above provide reasonable evidence for a crucial role of 5-LO in pathogenesis of leukemia and maintenance of LSCs in certain subtypes of leukemia. A pharmacological inhibition of 5-LO as a novel approach of stem cell therapy in leukemia is rendered feasible by the fact that 5-LO seems to be indispensable for the maintenance of LSCs in CML and AML. The recently finished clinical trial with zileuton and imatinib as treatment for CML will give further insight into this matter. Nevertheless, more in depth studies are needed to fully understand the role of 5-LO in the Wnt signaling pathway and maybe also other leukemia-related pathways such as JAK/STAT-, PI3K/AKT- or sonic hedgehog (SHH)- pathway.

In contrast to 5-LO, the role of other LOs in hemato-oncological malignancies is even less understood. Human CML microarray studies have shown that 15-LO is differentially expressed in $CD34^+$ CML cells [98,123]. In a DNA methylation study, including 127 AML patients, aberrant methylation of 12-LO was found to be associated with a negative outcome in AML patients, suggesting 12-LO methylation as a negative prognostic factor [124]. This is in accordance with the finding that the expression of 12-LO inversely correlates with the severity of CML [125]. It was also found that levels of 12-HETE are reduced in patients with CML compared with normal controls [126]. Furthermore, q-PCR analysis of AML and ALL patient samples revealed that leukemic blasts express both 12-LO and 15-LO transcripts in decreased levels compared to the level of healthy donor mononuclear cells [96]. Although it was reported that treatment with 12-HETE and 15-HETE stimulate the proliferation and differentiation of normal $CD34^+$ cells [127], addition of exogenous 12-HETE and 15-HETE (as 12-LO and 15-LO products) to freshly isolated leukemic blasts had no significant effect on the apoptosis and proliferation rate of these cells [96]. On the other hand, 15-LO metabolites (15-HPETE and 15-HETE) inhibited the growth of the CML cell line K562 by triggering apoptosis via activation of the intrinsic cell death pathway through cytochrome c release and caspase-3 activation. These effects were found to be mediated by reactive oxygen species (ROS), generated through the activation of NADPH oxidase [128]. Consistent with these findings, Middelton et al. showed that 15-LO-deficient mice develop a myelo-proliferative disorder (MPD) which is able to progress to transplantable leukemia [129]. Loss of 15-LO was associated with an enhanced PI3K/AKT-dependent phosphorylation of ICSBP (interferon consensus sequence binding protein) [129], a transcription factor known be involved in CML development [130–132].

However, in a recent publication it could be shown that 15-LO was essential for survival of LSC in a murine model of BCR/ABL-induced CML as well as in human $CD34^+$ CML cells [123]. This study was conducted by the same group who already showed the involvement of 5-LO in CML stem cell survival. 15-LO deficiency and pharmacological inhibition of this enzyme impaired LSC function by affecting proliferation and apoptosis of these cells, leading to an eventual depletion of LSCs. Additionally, loss of 15-LO prevented the initiation of BCR/ABL-induced CML in mice. This effect could be reversed by a knock down of P-selectin (SELP), a gene which is up-regulated in $15\text{-LO}^{-/-}$ mice and already known to play a suppressive role in hematopoiesis and CML development [133, 134]. Additionally, 15-LO deficiency altered the expression of β-catenin, PTEN, PI3K/AKT and ICSBP, which are known mediators in the pathogenesis of leukemia [122, 129, 135, 136]. However, in contrast to 5-LO, 15-LO deletion was also affecting normal stem cells, which led to a decreased number of HSCs in mice. This fact led the authors hypothesize that the MPD phenotype in $15\text{-LO}^{-/-}$ mice, which was observed by the group of Middelton et al., is likely a compensatory response to the reduced number of HSCs in the bone marrow of $15\text{-LO}^{-/-}$ mice [123, 129].

Taken together, the above mentioned studies indicate an anti-leukemic role at least for 12-LO, whereas the role of 15-LO in the pathogenesis of leukemia still remains unclear.

3 Studies Challenging the Role of LOs in Tumorigenesis

3.1 Cytotoxic Off-Target Effects of 5-LO Inhibitors and Mitogenic Effects of 5-LO Products

A number of publications have reported a crucial role of 5-LO and its products in tumorigenesis and there is no doubt that 5-LO inhibitors exert potent cytotoxic effects against 5-LO overexpressing tumor types and cultured tumor cells. These observations represent one of the most relevant experimental evidence for suggesting that 5-LO products directly stimulate tumor cell proliferation. From recent, carefully conducted studies we found that 5-LO inhibitors AA-861, Rev-5901, BWA4C, MK-886 and CJ-13,610, commonly used in cell culture studies for analyses of cytotoxicity, were able to potently suppress the viability of pancreatic cancer cells, cervix carcinoma cells and leukemic myeloid cells. However, surprisingly, this was independent of the suppression of 5-LO product formation [137]. This finding arguably derives from at least three different series of experimental approaches. First, 5-LO inhibitors markedly differed in their ability to reduce cell viability, to induce cytotoxic effects and to suppress the proliferation of cultured 5-LO-positive Capan-2 pancreas carcinoma cells. Though AA-861, MK-886 and Rev-5901 produced potent cytotoxic effects, other more selective and more potent 5-LO inhibitors, including CJ-13,610, BWA4C and zileuton failed to induce an anti-proliferative or cytotoxic response in all types of tumor cells employed. Second, IC_{50} values of reduction of cell viability of the cytotoxic 5-LO inhibitors highly exceeded the respective IC_{50} values for inhibition of 5-LO enzyme activity up to 5000-fold. Third, well-established 5-LO-negative tumor cell lines demonstrated a higher susceptibility towards the growth-inhibitory effects of 5-LO inhibitors than their morphologically related 5-LO-positive counterparts. These observations are in line with a report by Datta et al. in which MK-886 induced strong apoptotic effects independently of suppression of LT biosynthesis [137, 138]. Moreover, Sabirsh et al. recently described 5-LO-independent effects of various anti-LT drugs on Ca^{2+} signaling in 5-LO-deficient HeLa carcinoma cells [139]. Also some cytotoxic effects of licofelone, a dually acting COX/FLAP inhibitor, were related to off-target effects outside the arachidonic acid cascade [140].

Several studies reported enhanced proliferation of tumor cells and abolished anti-proliferative effects by 5-LO inhibitors if 5-LO products were added exogenously. Notably, in some studies extreme concentrations of 5-LO products were supplemented exceeding those in the medium of untreated cells by up to 10,000-fold [14, 16, 141]. This inconsistency represents an important caveat to these experiments.

3.2 Lacking Therapeutic Efficacy of 5-LO Inhibitors in Cancer Patients

As mentioned above a clinical double-blind phase II study with the LTB_4 antagonist LY293111 in patients suffering from advanced adenocarcinoma of the pancreas failed to show any therapeutic benefit [28]. Considering the remarkable susceptibility of pancreatic cancer cells towards anti-LT drugs in cell culture studies, a lack of correlation between the effects of 5-LO inhibitors in cultured cells and in patients becomes evident. Notably, mean C_{max} plasma concentrations of the drug achieved in patients after a dosage of 600 mg BID were 4.4 µM [142] and should almost fully antagonize LTB_4-induced signal transduction [143]. Thus, pleiotropic off-target effects including modulation of PPARγ (peroxisome proliferator-activated receptor gamma) activity may contribute to the observed cytotoxic effects in cell culture [144].

4 Conclusions and Future Direction

A number of studies constitute substantial rationale for a role of LOs, mainly 5-LO, in tumor development, by demonstrating overexpression of 5-LO in various types of malignant cells, mitogenic effects by 5-LO products, anti-proliferative activity by 5-LO inhibitors, impaired tumor cell growth by 5-LO gene silencing approaches as well genetic disruption in KO mice. However, strong cytotoxic off-target effects of the 5-LO inhibitors as well as the use of unphysiologically high concentrations of 5-LO products in cell culture add-back experiments raise question about the role of 5-LO products in tumorigenesis (see Fig. 2). Also, possible indirect tumorigenic effects of 5-LO products (e.g. promotion of angiogenesis) which are of relevance for tumor growth in-vivo but not in cell culture assays should be considered. Furthermore, overexpression of 5-LO in tumor tissues should not coercively indicate a crucial pro-tumorigenic role of 5-LO products. Non-enzymatic functions, including an interaction with cytoskeleton proteins or with the signal transduction adaptor protein Grb-2, involved in receptor tyrosine kinase (RTK)-dependent growth-factor signalling as well as interactions with components of the Wnt-signaling pathway have been reported for 5-LO [115, 145]. Because of the crucial

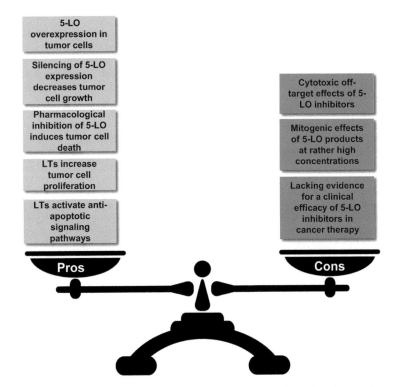

Fig. 2 Evidence arguing for (Pros) and against (Cons) a role of 5-LO products in tumorigenesis as discussed in detail in the chapter

role of oncogenic RTK and Wnt-signalling in cancer progression, a disrupted growth factor signalling may contribute to the reduction in tumor cell viability by 5-LO gene silencing approaches and 5-LO inhibitor experiments. Summarized, a body of evidence from the literature suggests a crucial, albeit poorly defined, role of LOs in tumorigenesis of several cancer types. Elucidation of the molecular mechanisms underlying these effects may include direct proliferative actions of LO products on tumor cells as well as indirect and so far neglected non-enzymatic effects of LOs and thereby draw novel connections between pathways that are currently regarded as unrelated.

Acknowledgement This work was supported by the German Research Foundation (DFG-MA-5825/1-1 and Sonderforschungsbereich SFB 1039) and the Else Kröner-Fresenius-Stiftung (stipend of the graduate school "Translational Research Innovation–Pharma"). T.J. Maier is recipient of a Heisenberg fellowship of the German Research Foundation (DFG-MA-5825/2-1).

References

1. Hanahan D, Weinberg RA (2011) Hallmarks of cancer: the next generation. Cell 144 (5):646–674. doi:10.1016/j.cell.2011.02.013
2. Avis I, Hong SH, Martinez A, Moody T, Choi YH, Trepel J, Das R, Jett M, Mulshine JL (2001) Five-lipoxygenase inhibitors can mediate apoptosis in human breast cancer cell lines through complex eicosanoid interactions. FASEB J 15(11):2007–2009
3. Chen X, Sood S, Yang CS, Li N, Sun Z (2006) Five-lipoxygenase pathway of arachidonic acid metabolism in carcinogenesis and cancer chemoprevention. Curr Cancer Drug Targets 6 (7):613–622
4. Hong SH, Avis I, Vos MD, Martinez A, Treston AM, Mulshine JL (1999) Relationship of arachidonic acid metabolizing enzyme expression in epithelial cancer cell lines to the growth effect of selective biochemical inhibitors. Cancer Res 59(9):2223–2228
5. Hennig R, Ding XZ, Tong WG, Schneider MB, Standop J, Friess H, Buchler MW, Pour PM, Adrian TE (2002) 5-Lipoxygenase and leukotriene B(4) receptor are expressed in human pancreatic cancers but not in pancreatic ducts in normal tissue. Am J Pathol 161(2):421–428
6. Romano M, Catalano A, Nutini M, D'Urbano E, Crescenzi C, Claria J, Libner R, Davi G, Procopio A (2001) 5-lipoxygenase regulates malignant mesothelial cell survival: involvement of vascular endothelial growth factor. FASEB J 15(13):2326–2336. doi:10.1096/fj.01-0150com
7. Krieg P, Furstenberger G (2014) The role of lipoxygenases in epidermis. Biochim Biophys Acta 1841(3):390–400. doi:10.1016/j.bbalip.2013.08.005
8. Matsuyama M, Yoshimura R, Mitsuhashi M, Tsuchida K, Takemoto Y, Kawahito Y, Sano H, Nakatani T (2005) 5-Lipoxygenase inhibitors attenuate growth of human renal cell carcinoma and induce apoptosis through arachidonic acid pathway. Oncol Rep 14(1):73–79
9. Matsuyama M, Yoshimura R, Tsuchida K, Takemoto Y, Segawa Y, Shinnka T, Kawahito Y, Sano H, Nakatani T (2004) Lipoxygenase inhibitors prevent urological cancer cell growth. Int J Mol Med 13(5):665–668
10. van Rossum GS, Bijvelt JJ, van den Bosch H, Verkleij AJ, Boonstra J (2002) Cytosolic phospholipase A2 and lipoxygenase are involved in cell cycle progression in neuroblastoma cells. Cell Mol Life Sci 59(1):181–188
11. Catalano A, Procopio A (2005) New aspects on the role of lipoxygenases in cancer progression. Histol Histopathol 20(3):969–975

12. Ding XZ, Tong WG, Adrian TE (2003) Multiple signal pathways are involved in the mitogenic effect of 5(S)-HETE in human pancreatic cancer. Oncology 65(4):285–294
13. Tong WG, Ding XZ, Talamonti MS, Bell RH, Adrian TE (2005) LTB4 stimulates growth of human pancreatic cancer cells via MAPK and PI-3 kinase pathways. Biochem Biophys Res Commun 335(3):949–956
14. Sveinbjornsson B, Rasmuson A, Baryawno N, Wan M, Pettersen I, Ponthan F, Orrego A, Haeggstrom JZ, Johnsen JI, Kogner P (2008) Expression of enzymes and receptors of the leukotriene pathway in human neuroblastoma promotes tumor survival and provides a target for therapy. FASEB J 22(10):3525–3536
15. Ding XZ, Iversen P, Cluck MW, Knezetic JA, Adrian TE (1999) Lipoxygenase inhibitors abolish proliferation of human pancreatic cancer cells. Biochem Biophys Res Commun 261 (1):218–223
16. Ghosh J, Myers CE (1998) Inhibition of arachidonate 5-lipoxygenase triggers massive apoptosis in human prostate cancer cells. Proc Natl Acad Sci U S A 95(22):13182–13187
17. Wong BC, Wang WP, Cho CH, Fan XM, Lin MC, Kung HF, Lam SK (2001) 12-Lipoxygenase inhibition induced apoptosis in human gastric cancer cells. Carcinogenesis 22(9):1349–1354
18. Anderson KM, Seed T, Meng J, Ou D, Alrefai WA, Harris JE (1998) Five-lipoxygenase inhibitors reduce Panc-1 survival: the mode of cell death and synergism of MK886 with gamma linolenic acid. Anticancer Res 18(2A):791–800
19. Ding XZ, Kuszynski CA, El-Metwally TH, Adrian TE (1999) Lipoxygenase inhibition induced apoptosis, morphological changes, and carbonic anhydrase expression in human pancreatic cancer cells. Biochem Biophys Res Commun 266(2):392–399. doi:10.1006/bbrc.1999.1824
20. Hennig R, Ding XZ, Tong WG, Schneider MB, Standop J, Friess H, Buchler MW, Pour PM, Adrian TE (2002) 5-Lipoxygenase and leukotriene B(4) receptor are expressed in human pancreatic cancers but not in pancreatic ducts in normal tissue. Am J Pathol 161(2):421–428
21. Zhou GX, Ding XL, Huang JF, Zhang H, Wu SB (2007) Suppression of 5-lipoxygenase gene is involved in triptolide-induced apoptosis in pancreatic tumor cell lines. Biochim Biophys Acta 1770(7):1021–1027. doi:10.1016/j.bbagen.2007.03.002
22. Ding XZ, Talamonti MS, Bell RH Jr, Adrian TE (2005) A novel anti-pancreatic cancer agent, LY293111. Anticancer Drugs 16(5):467–473, 00001813-200506000-00001 [pii]
23. Tong WG, Ding XZ, Talamonti MS, Bell RH, Adrian TE (2007) Leukotriene B4 receptor antagonist LY293111 induces S-phase cell cycle arrest and apoptosis in human pancreatic cancer cells. Anticancer Drugs 18(5):535–541. doi:10.1097/01.cad.0000231477.22901.8a, 00001813-200706000-00003 [pii]
24. Hennig R, Grippo P, Ding XZ, Rao SM, Buchler MW, Friess H, Talamonti MS, Bell RH, Adrian TE (2005) 5-Lipoxygenase, a marker for early pancreatic intraepithelial neoplastic lesions. Cancer Res 65(14):6011–6016
25. Wenger FA, Kilian M, Achucarro P, Heinicken D, Schimke I, Guski H, Jacobi CA, Muller JM (2002) Effects of Celebrex and Zyflo on BOP-induced pancreatic cancer in Syrian hamsters. Pancreatology 2(1):54–60
26. Wenger FA, Kilian M, Bisevac M, Khodadayan C, von Seebach M, Schimke I, Guski H, Muller JM (2002) Effects of Celebrex and Zyflo on liver metastasis and lipidperoxidation in pancreatic cancer in Syrian hamsters. Clin Exp Metastasis 19(8):681–687
27. Baetz T, Eisenhauer E, Siu L, MacLean M, Doppler K, Walsh W, Fisher B, Khan AZ, de Alwis DP, Weitzman A, Brail LH, Moore M (2007) A phase I study of oral LY293111 given daily in combination with irinotecan in patients with solid tumours. Invest New Drugs 25 (3):217–225. doi:10.1007/s10637-006-9021-8
28. Saif MW, Oettle H, Vervenne WL, Thomas JP, Spitzer G, Visseren-Grul C, Enas N, Richards DA (2009) Randomized double-blind phase II trial comparing gemcitabine plus LY293111 versus gemcitabine plus placebo in advanced adenocarcinoma of the pancreas. Cancer J 15 (4):339–343. doi:10.1097/PPO.0b013e3181b36264, 00130404-200908000-00012 [pii]

29. Janne PA, Paz-Ares L, Oh Y, Eschbach C, Hirsh V, Enas N, Brail L, von Pawel J (2014) Randomized, double-blind, phase II trial comparing gemcitabine-cisplatin plus the LTB4 antagonist LY293111 versus gemcitabine-cisplatin plus placebo in first-line non-small-cell lung cancer. J Thorac Oncol 9(1):126–131. doi:10.1097/JTO.0000000000000037
30. Ding XZ, Tong WG, Adrian TE (2001) 12-lipoxygenase metabolite 12(S)-HETE stimulates human pancreatic cancer cell proliferation via protein tyrosine phosphorylation and ERK activation. Int J Cancer 94(5):630–636
31. Tong WG, Ding XZ, Witt RC, Adrian TE (2002) Lipoxygenase inhibitors attenuate growth of human pancreatic cancer xenografts and induce apoptosis through the mitochondrial pathway. Mol Cancer Ther 1(11):929–935
32. Hennig R, Kehl T, Noor S, Ding XZ, Rao SM, Bergmann F, Furstenberger G, Buchler MW, Friess H, Krieg P, Adrian TE (2007) 15-lipoxygenase-1 production is lost in pancreatic cancer and overexpression of the gene inhibits tumor cell growth. Neoplasia 9(11):917–926
33. Kort WJ, Bijma AM, van Dam JJ, van der Ham AC, Hekking JM, van der Ingh HF, Meijer WS, van Wilgenburg MG, Zijlstra FJ (1992) Eicosanoids in breast cancer patients before and after mastectomy. Prostaglandins Leukot Essent Fatty Acids 45(4):319–327
34. Natarajan R, Esworthy R, Bai W, Gu JL, Wilczynski S, Nadler J (1997) Increased 12-lipoxygenase expression in breast cancer tissues and cells. Regulation by epidermal growth factor. J Clin Endocrinol Metab 82(6):1790–1798. doi:10.1210/jcem.82.6.3990
35. Jiang WG, Douglas-Jones A, Mansel RE (2003) Levels of expression of lipoxygenases and cyclooxygenase-2 in human breast cancer. Prostaglandins Leukot Essent Fatty Acids 69(4):275–281, S0952327803001108 [pii]
36. Jiang WG, Watkins G, Douglas-Jones A, Mansel RE (2006) Reduction of isoforms of 15-lipoxygenase (15-LOX)-1 and 15-LOX-2 in human breast cancer. Prostaglandins Leukot Essent Fatty Acids 74(4):235–245. doi:10.1016/j.plefa.2006.01.009
37. Jiang WG, Douglas-Jones AG, Mansel RE (2006) Aberrant expression of 5-lipoxygenase-activating protein (5-LOXAP) has prognostic and survival significance in patients with breast cancer. Prostaglandins Leukot Essent Fatty Acids 74(2):125–134. doi:10.1016/j.plefa.2005.10.005, S0952-3278(05)00176-6 [pii]
38. Wang J, John EM, Ingles SA (2008) 5-lipoxygenase and 5-lipoxygenase-activating protein gene polymorphisms, dietary linoleic acid, and risk for breast cancer. Cancer Epidemiol Biomarkers Prev 17(10):2748–2754. doi:10.1158/1055-9965.epi-08-0439
39. Wang J, John EM, Horn-Ross PL, Ingles SA (2008) Dietary fat, cooking fat, and breast cancer risk in a multiethnic population. Nutr Cancer 60(4):492–504. doi:10.1080/01635580801956485
40. O'Flaherty JT, Rogers LC, Paumi CM, Hantgan RR, Thomas LR, Clay CE, High K, Chen YQ, Willingham MC, Smitherman PK, Kute TE, Rao A, Cramer SD, Morrow CS (2005) 5-Oxo-ETE analogs and the proliferation of cancer cells. Biochim Biophys Acta 1736 (3):228–236. doi:10.1016/j.bbalip.2005.08.009
41. Tong WG, Ding XZ, Adrian TE (2002) The mechanisms of lipoxygenase inhibitor-induced apoptosis in human breast cancer cells. Biochem Biophys Res Commun 296(4):942–948
42. O'Flaherty JT, Wooten RE, Samuel MP, Thomas MJ, Levine EA, Case LD, Akman SA, Edwards IJ (2013) Fatty acid metabolites in rapidly proliferating breast cancer. PLoS One 8 (5), e63076. doi:10.1371/journal.pone.0063076
43. Soberman RJ, Harper TW, Betteridge D, Lewis RA, Austen KF (1985) Characterization and separation of the arachidonic acid 5-lipoxygenase and linoleic acid omega-6 lipoxygenase (arachidonic acid 15-lipoxygenase) of human polymorphonuclear leukocytes. J Biol Chem 260(7):4508–4515
44. Reddy N, Everhart A, Eling T, Glasgow W (1997) Characterization of a 15-lipoxygenase in human breast carcinoma BT-20 cells: stimulation of 13-HODE formation by TGF alpha/EGF. Biochem Biophys Res Commun 231(1):111–116. doi:10.1006/bbrc.1997.6048

45. Hammamieh R, Sumaida D, Zhang X, Das R, Jett M (2007) Control of the growth of human breast cancer cells in culture by manipulation of arachidonate metabolism. BMC Cancer 7:138. doi:10.1186/1471-2407-7-138
46. Rao CV (2007) Regulation of COX and LOX by curcumin. Adv Exp Med Biol 595:213–226. doi:10.1007/978-0-387-46401-5_9
47. Neichi T, Koshihara Y, Murota S (1983) Inhibitory effect of esculetin on 5-lipoxygenase and leukotriene biosynthesis. Biochim Biophys Acta 753(1):130–132
48. Kitagawa H, Noguchi M (1994) Comparative effects of piroxicam and esculetin on incidence, proliferation, and cell kinetics of mammary carcinomas induced by 7,12-dimethylbenz[a] anthracene in rats on high- and low-fat diets. Oncology 51(5):401–410
49. Matsunaga K, Yoshimi N, Yamada Y, Shimizu M, Kawabata K, Ozawa Y, Hara A, Mori H (1998) Inhibitory effects of nabumetone, a cyclooxygenase-2 inhibitor, and esculetin, a lipoxygenase inhibitor, on N-methyl-N-nitrosourea-induced mammary carcinogenesis in rats. Jpn J Cancer Res 89(5):496–501
50. Inano H, Onoda M, Inafuku N, Kubota M, Kamada Y, Osawa T, Kobayashi H, Wakabayashi K (1999) Chemoprevention by curcumin during the promotion stage of tumorigenesis of mammary gland in rats irradiated with gamma-rays. Carcinogenesis 20(6):1011–1018
51. Inano H, Onoda M, Inafuku N, Kubota M, Kamada Y, Osawa T, Kobayashi H, Wakabayashi K (2000) Potent preventive action of curcumin on radiation-induced initiation of mammary tumorigenesis in rats. Carcinogenesis 21(10):1835–1841
52. Reddy NP, Aparoy P, Reddy TC, Achari C, Sridhar PR, Reddanna P (2010) Design, synthesis, and biological evaluation of prenylated chalcones as 5-LOX inhibitors. Bioorg Med Chem 18(16):5807–5815. doi:10.1016/j.bmc.2010.06.107
53. Singh AK, Singh R, Naz F, Chauhan SS, Dinda A, Shukla AA, Gill K, Kapoor V, Dey S (2012) Structure based design and synthesis of peptide inhibitor of human LOX-12: in vitro and in vivo analysis of a novel therapeutic agent for breast cancer. PLoS One 7(2), e32521. doi:10.1371/journal.pone.0032521
54. Armstrong B, Doll R (1975) Environmental factors and cancer incidence and mortality in different countries, with special reference to dietary practices. Int J Cancer 15(4):617–631
55. Fleshner N, Bagnell PS, Klotz L, Venkateswaran V (2004) Dietary fat and prostate cancer. J Urol 171(2 Pt 2):S19–S24. doi:10.1097/01.ju.0000107838.33623.19
56. Ghosh J, Myers CE (1997) Arachidonic acid stimulates prostate cancer cell growth: critical role of 5-lipoxygenase. Biochem Biophys Res Commun 235(2):418–423. doi:10.1006/bbrc. 1997.6799
57. Sarveswaran S, Thamilselvan V, Brodie C, Ghosh J (2011) Inhibition of 5-lipoxygenase triggers apoptosis in prostate cancer cells via down-regulation of protein kinase C-epsilon. Biochim Biophys Acta 1813(12):2108–2117. doi:10.1016/j.bbamcr.2011.07.015
58. Crawford ED, Eisenberger MA, McLeod DG, Spaulding JT, Benson R, Dorr FA, Blumenstein BA, Davis MA, Goodman PJ (1989) A controlled trial of leuprolide with and without flutamide in prostatic carcinoma. N Engl J Med 321(7):419–424. doi:10.1056/nejm198908173210702
59. Gupta S, Srivastava M, Ahmad N, Sakamoto K, Bostwick DG, Mukhtar H (2001) Lipoxygenase-5 is overexpressed in prostate adenocarcinoma. Cancer 91(4):737–743
60. Bishayee K, Khuda-Bukhsh AR (2013) 5-lipoxygenase antagonist therapy: a new approach towards targeted cancer chemotherapy. Acta Biochim Biophys Sin 45(9):709–719. doi:10.1093/abbs/gmt064
61. Meng Z, Cao R, Yang Z, Liu T, Wang Y, Wang X (2013) Inhibitor of 5-lipoxygenase, zileuton, suppresses prostate cancer metastasis by upregulating E-cadherin and paxillin. Urology 82(6):1452. doi:10.1016/j.urology.2013.08.060, e1457–1414
62. Frixen UH, Behrens J, Sachs M, Eberle G, Voss B, Warda A, Lochner D, Birchmeier W (1991) E-cadherin-mediated cell-cell adhesion prevents invasiveness of human carcinoma cells. J Cell Biol 113(1):173–185

63. Singhai R, Patil VW, Jaiswal SR, Patil SD, Tayade MB, Patil AV (2011) E-Cadherin as a diagnostic biomarker in breast cancer. N Am J Med Sci 3(5):227–233. doi:10.4297/najms. 2011.3227
64. Sen A, De Castro I, Defranco DB, Deng FM, Melamed J, Kapur P, Raj GV, Rossi R, Hammes SR (2012) Paxillin mediates extranuclear and intranuclear signaling in prostate cancer proliferation. J Clin Invest 122(7):2469–2481. doi:10.1172/jci62044
65. Sarveswaran S, Myers CE, Ghosh J (2010) MK591, a leukotriene biosynthesis inhibitor, induces apoptosis in prostate cancer cells: synergistic action with LY294002, an inhibitor of phosphatidylinositol 3′-kinase. Cancer Lett 291(2):167–176. doi:10.1016/j.canlet.2009.10. 008
66. Huang B, Cao K, Li X, Guo S, Mao X, Wang Z, Zhuang J, Pan J, Mo C, Chen J, Qiu S (2011) The expression and role of protein kinase C (PKC) epsilon in clear cell renal cell carcinoma. J Exp Clin Cancer Res 30:88. doi:10.1186/1756-9966-30-88
67. Ferlay J, Steliarova-Foucher E, Lortet-Tieulent J, Rosso S, Coebergh JW, Comber H, Forman D, Bray F (2013) Cancer incidence and mortality patterns in Europe: estimates for 40 countries in 2012. Eur J Cancer 49(6):1374–1403. doi:10.1016/j.ejca.2012.12.027
68. Hussey HJ, Tisdale MJ (1996) Inhibition of tumour growth by lipoxygenase inhibitors. Br J Cancer 74:683–687
69. Hussey HJ, Bibby MC, Tisdale MJ (1996) Novel anti-tumour activity of 2,3,5-trimethyl-6-(3-pyridylmethyl)-1,4- benzoquinone (CV-6504) against established murine adenocarcinomas (MAC). Br J Cancer 73:1187–1192
70. Öhd JF, Nielsen CK, Campbell J, Landberg G, Löfberg H, Sjölander A (2003) Expression of the leukotriene D4 receptor CysLT1, COX-2, and other cell survival factors in colorectal adenocarcinomas. Gastroenterology 124:57–70. doi:10.1053/gast.2003.50011
71. Soumaoro LT, Iida S, Uetake H, Ishiguro M, Takagi Y, Higuchi T, Yasuno M, Enomoto M, Sugihara K (2006) Expression of 5-lipoxygenase in human colorectal cancer. World J Gastroenterol 12:6355–6360
72. Paruchuri S, Mezhybovska M, Juhas M, Sjölander A (2006) Endogenous production of leukotriene D4 mediates autocrine survival and proliferation via CysLT1 receptor signalling in intestinal epithelial cells. Oncogene 25:6660–6665. doi:10.1038/sj.onc.1209666
73. Nielsen CK, Campbell JIA, Ohd JF, Mörgelin M, Riesbeck K, Landberg G, Sjölander A (2005) A novel localization of the G-protein-coupled CysLT1 receptor in the nucleus of colorectal adenocarcinoma cells. Cancer Res 65:732–742
74. Ihara A, Wada K, Yoneda M, Fujisawa N, Takahashi H, Nakajima A (2007) Blockade of leukotriene B4 signaling pathway induces apoptosis and suppresses cell proliferation in colon cancer. J Pharmacol Sci 103:24–32
75. Melstrom LG, Bentrem DJ, Salabat MR, Kennedy TJ, Ding X-Z, Strouch M, Rao SM, Witt RC, Ternent CA, Talamonti MS, Bell RH, Adrian TA (2008) Overexpression of 5-lipoxygenase in colon polyps and cancer and the effect of 5-LOX inhibitors in vitro and in a murine model. Clin Cancer Res 14:6525–6530. doi:10.1158/1078-0432.CCR-07-4631
76. Cheon EC, Khazaie K, Khan MW, Strouch MJ, Krantz SB, Phillips J, Blatner NR, Hix LM, Zhang M, Dennis KL, Salabat MR, Heiferman M, Grippo PJ, Munshi HG, Gounaris E, Bentrem DJ (2011) Mast cell 5-lipoxygenase activity promotes intestinal polyposis in APC 468 mice. Cancer Res 71:1627–1636. doi:10.1158/0008-5472.CAN-10-1923
77. Cope GF, Wyatt JI, Pinder IF, Lee PN, Heatley RV, Kelleher J (1991) Alcohol consumption in patients with colorectal adenomatous polyps. Gut 32:70–72
78. Ye Y-N, Liu ES-L, Shin VY, Wu WK-K, Cho C-H (2004) Contributory role of 5-lipoxygenase and its association with angiogenesis in the promotion of inflammation-associated colonic tumorigenesis by cigarette smoking. Toxicology 203:179–188. doi:10. 1016/j.tox.2004.06.004
79. Nishida N, Yano H, Nishida T, Kamura T, Kojiro M (2006) Angiogenesis in cancer. Vasc Health Risk Manag 2(3):213–219

80. Ye YN, Wu WK, Shin VY, Bruce IC, Wong BC, Cho CH (2005) Dual inhibition of 5-LOX and COX-2 suppresses colon cancer formation promoted by cigarette smoke. Carcinogenesis 26(4):827–834. doi:10.1093/carcin/bgi012
81. Arber N, Eagle CJ, Spicak J, Rácz I, Dite P, Hajer J, Zavoral M, Lechuga MJ, Gerletti P, Tang J, Rosenstein RB, Macdonald K, Bhadra P, Fowler R, Wittes J, Zauber AG, Solomon SD, Levin B, Investigators PT (2006) Celecoxib for the prevention of colorectal adenomatous polyps. N Engl J Med 355:885–895. doi:10.1056/NEJMoa061652
82. Cimen I, Tuncay S, Banerjee S (2009) 15-Lipoxygenase-1 expression suppresses the invasive properties of colorectal carcinoma cell lines HCT-116 and HT-29. Cancer Sci 100(12):2283–2291. doi:10.1111/j.1349-7006.2009.01313.x
83. Zuo X, Peng Z, Wu Y, Moussalli MJ, Yang XL, Wang Y, Parker-Thornburg J, Morris JS, Broaddus RR, Fischer SM, Shureiqi I (2012) Effects of gut-targeted 15-LOX-1 transgene expression on colonic tumorigenesis in mice. J Natl Cancer Inst 104(9):709–716. doi:10.1093/jnci/djs187
84. Wu Y, Mao F, Zuo X, Moussalli MJ, Elias E, Xu W, Shureiqi I (2014) 15-LOX-1 suppression of hypoxia-induced metastatic phenotype and HIF-1alpha expression in human colon cancer cells. Cancer Med 3(3):472–484. doi:10.1002/cam4.222
85. de Carvalho DD, Sadok A, Bourgarel-Rey V, Gattacceca F, Penel C, Lehmann M, Kovacic H (2008) Nox1 downstream of 12-lipoxygenase controls cell proliferation but not cell spreading of colon cancer cells. Int J Cancer 122(8):1757–1764. doi:10.1002/ijc.23300
86. Rao CV, Rivenson A, Simi B, Reddy BS (1995) Chemoprevention of colon carcinogenesis by dietary curcumin, a naturally occurring plant phenolic compound. Cancer Res 55:259–266
87. Hong J, Bose M, Ju J, Ryu J-H, Chen X, Sang S, Lee M-J, Yang CS (2004) Modulation of arachidonic acid metabolism by curcumin and related beta-diketone derivatives: effects on cytosolic phospholipase A(2), cyclooxygenases and 5-lipoxygenase. Carcinogenesis 25:1671–1679. doi:10.1093/carcin/bgh165
88. Rao CV, Desai D, Rivenson A, Simi B, Amin S, Reddy BS (1995) Chemoprevention of colon carcinogenesis by phenylethyl-3-methylcaffeate. Cancer Res 55:2310–2315
89. Mohammed A, Janakiram NB, Li Q, Choi C-I, Zhang Y, Steele VE, Rao CV (2011) Chemoprevention of colon and small intestinal tumorigenesis in APCMin/+ mice by licofelone, a novel dual 5-LOX/COX inhibitor: potential implications for human colon cancer prevention. Cancer Prev Res 4:2015–2026. doi:10.1158/1940-6207.CAPR-11-0233
90. Ghatak S, Vyas A, Misra S, O'Brien P, Zambre A, Fresco VM, Markwald RR, Swamy KV, Afrasiabi Z, Choudhury A, Khetmalas M, Padhye S (2014) Novel di-tertiary-butyl phenylhydrazones as dual cyclooxygenase-2/5-lipoxygenase inhibitors: synthesis, COX/LOX inhibition, molecular modeling, and insights into their cytotoxicities. Bioorg Med Chem Lett 24:317–324. doi:10.1016/j.bmcl.2013.11.015
91. Misra S, Ghatak S, Patil N, Dandawate P, Ambike V, Adsule S, Unni D, Venkateswara Swamy K, Padhye S (2013) Novel dual cyclooxygenase and lipoxygenase inhibitors targeting hyaluronan–CD44v6 pathway and inducing cytotoxicity in colon cancer cells. Bioorg Med Chem 21:2551–2559. doi:10.1016/j.bmc.2013.02.033
92. Tsukada T, Nakashima K, Shirakawa S (1986) Arachidonate 5-lipoxygenase inhibitors show potent antiproliferative effects on human leukemia cell lines. Biochem Biophys Res Commun 140(3):832–836
93. Anderson KM, Seed T, Plate JM, Jajeh A, Meng J, Harris JE (1995) Selective inhibitors of 5-lipoxygenase reduce CML blast cell proliferation and induce limited differentiation and apoptosis. Leuk Res 19(11):789–801
94. Gillis RC, Daley BJ, Enderson BL, Kestler DP, Karlstad MD (2007) Regulation of apoptosis in eicosapentaenoic acid-treated HL-60 cells. J Surg Res 137(1):141–150. doi:10.1016/j.jss.2006.08.012
95. Feltenmark S, Runarsson G, Larsson P, Jakobsson PJ, Bjorkholm M, Claesson HE (1995) Diverse expression of cytosolic phospholipase A2, 5-lipoxygenase and prostaglandin H synthase 2 in acute pre-B-lymphocytic leukaemia cells. Br J Haematol 90(3):585–594

96. Vincent C, Fiancette R, Donnard M, Bordessoule D, Turlure P, Trimoreau F, Denizot Y (2008) 5-LOX, 12-LOX and 15-LOX in immature forms of human leukemic blasts. Leuk Res 32(11):1756–1762. doi:10.1016/j.leukres.2008.05.005
97. Graham SM, Vass JK, Holyoake TL, Graham GJ (2007) Transcriptional analysis of quiescent and proliferating CD34+ human hemopoietic cells from normal and chronic myeloid leukemia sources. Stem Cells 25(12):3111–3120. doi:10.1634/stemcells.2007-0250
98. Radich JP, Dai H, Mao M, Oehler V, Schelter J, Druker B, Sawyers C, Shah N, Stock W, Willman CL, Friend S, Linsley PS (2006) Gene expression changes associated with progression and response in chronic myeloid leukemia. Proc Natl Acad Sci U S A 103(8):2794–2799. doi:10.1073/pnas.0510423103
99. Valk PJ, Verhaak RG, Beijen MA, Erpelinck CA, Barjesteh van Waalwijk van Doorn-Khosrovani S, Boer JM, Beverloo HB, Moorhouse MJ, van der Spek PJ, Lowenberg B, Delwel R (2004) Prognostically useful gene-expression profiles in acute myeloid leukemia. N Engl J Med 350(16):1617–1628. doi:10.1056/NEJMoa040465
100. Guriec N, Le Jossic-Corcos C, Simon B, Ianotto JC, Tempescul A, Dreano Y, Salaun JP, Berthou C, Corcos L (2014) The arachidonic acid-LTB4-BLT2 pathway enhances human B-CLL aggressiveness. Biochim Biophys Acta 1842(11):2096–2105. doi:10.1016/j.bbadis.2014.07.016
101. Rosenwald A, Alizadeh AA, Widhopf G, Simon R, Davis RE, Yu X, Yang L, Pickeral OK, Rassenti LZ, Powell J, Botstein D, Byrd JC, Grever MR, Cheson BD, Chiorazzi N, Wilson WH, Kipps TJ, Brown PO, Staudt LM (2001) Relation of gene expression phenotype to immunoglobulin mutation genotype in B cell chronic lymphocytic leukemia. J Exp Med 194 (11):1639–1647
102. Runarsson G, Liu A, Mahshid Y, Feltenmark S, Pettersson A, Klein E, Bjorkholm M, Claesson HE (2005) Leukotriene B4 plays a pivotal role in CD40-dependent activation of chronic B lymphocytic leukemia cells. Blood 105(3):1274–1279. doi:10.1182/blood-2004-07-2546
103. Stratowa C, Loffler G, Lichter P, Stilgenbauer S, Haberl P, Schweifer N, Dohner H, Wilgenbus KK (2001) CDNA microarray gene expression analysis of B-cell chronic lymphocytic leukemia proposes potential new prognostic markers involved in lymphocyte trafficking. Int J Cancer 91(4):474–480
104. Boyd RS, Jukes-Jones R, Walewska R, Brown D, Dyer MJ, Cain K (2009) Protein profiling of plasma membranes defines aberrant signaling pathways in mantle cell lymphoma. Mol Cell Proteomics 8(7):1501–1515. doi:10.1074/mcp.M800515-MCP200
105. Mahshid Y, Lisy MR, Wang X, Spanbroek R, Flygare J, Christensson B, Bjorkholm M, Sander B, Habenicht AJ, Claesson HE (2009) High expression of 5-lipoxygenase in normal and malignant mantle zone B lymphocytes. BMC Immunol 10:2. doi:10.1186/1471-2172-10-2
106. Runarsson G, Feltenmark S, Forsell PK, Sjoberg J, Bjorkholm M, Claesson HE (2007) The expression of cytosolic phospholipase A2 and biosynthesis of leukotriene B4 in acute myeloid leukemia cells. Eur J Haematol 79(6):468–476. doi:10.1111/j.1600-0609.2007.00967.x, EJH967 [pii]
107. Stenke L, Sjolinder M, Miale TD, Lindgren JA (1998) Novel enzymatic abnormalities in AML and CML in blast crisis: elevated leucocyte leukotriene C4 synthase activity paralleled by deficient leukotriene biosynthesis from endogenous substrate. Br J Haematol 101 (4):728–736
108. Steinhilber D, Fischer AS, Metzner J, Steinbrink SD, Roos J, Ruthardt M, Maier TJ (2010) 5-lipoxygenase: underappreciated role of a pro-inflammatory enzyme in tumorigenesis. Front Pharmacol 1:143. doi:10.3389/fphar.2010.00143
109. Chen Y, Hu Y, Zhang H, Peng C, Li S (2009) Loss of the Alox5 gene impairs leukemia stem cells and prevents chronic myeloid leukemia. Nat Genet 41(7):783–792
110. Lucas CM, Harris RJ, Giannoudis A, McDonald E, Clark RE (2014) Low leukotriene B4 receptor 1 (LTB4R1) leads to ALOX5 down-regulation at diagnosis of chronic myeloid leukemia. Haematologica. doi:10.3324/haematol.2013.101972

111. DeKelver RC, Lewin B, Lam K, Komeno Y, Yan M, Rundle C, Lo MC, Zhang DE (2013) Cooperation between RUNX1-ETO9a and novel transcriptional partner KLF6 in upregulation of Alox5 in acute myeloid leukemia. PLoS Genet 9(10), e1003765. doi:10.1371/journal.pgen.1003765
112. Lo MC, Peterson LF, Yan M, Cong X, Jin F, Shia WJ, Matsuura S, Ahn EY, Komeno Y, Ly M, Ommen HB, Chen IM, Hokland P, Willman CL, Ren B, Zhang DE (2012) Combined gene expression and DNA occupancy profiling identifies potential therapeutic targets of t (8;21) AML. Blood 120(7):1473–1484. doi:10.1182/blood-2011-12-395335
113. Matsumoto N, Kubo A, Liu H, Akita K, Laub F, Ramirez F, Keller G, Friedman SL (2006) Developmental regulation of yolk sac hematopoiesis by Kruppel-like factor 6. Blood 107 (4):1357–1365. doi:10.1182/blood-2005-05-1916
114. Zhao JL, Austen KF, Lam BK (2000) Cell-specific transcription of leukotriene C(4) synthase involves a Kruppel-like transcription factor and Sp1. J Biol Chem 275(12):8903–8910
115. Roos J, Oancea C, Heinssmann M, Khan D, Held H, Kahnt AS, Capelo R, la Buscato E, Proschak E, Puccetti E, Steinhilber D, Fleming I, Maier TJ, Ruthardt M (2014) 5-Lipoxygenase is a candidate target for therapeutic management of stem cell-like cells in acute myeloid leukemia. Cancer Res 74(18):5244–5255. doi:10.1158/0008-5472.CAN-13-3012
116. Alison MR, Lin WR, Lim SM, Nicholson LJ (2012) Cancer stem cells: in the line of fire. Cancer Treat Rev 38(6):589–598. doi:10.1016/j.ctrv.2012.03.003
117. Debeb BG, Lacerda L, Xu W, Larson R, Solley T, Atkinson R, Sulman EP, Ueno NT, Krishnamurthy S, Reuben JM, Buchholz TA, Woodward WA (2012) Histone deacetylase inhibitors stimulate dedifferentiation of human breast cancer cells through WNT/beta-catenin signaling. Stem Cells 30(11):2366–2377. doi:10.1002/stem.1219
118. Roarty K, Rosen JM (2010) Wnt and mammary stem cells: hormones cannot fly wingless. Curr Opin Pharmacol 10(6):643–649. doi:10.1016/j.coph.2010.07.004
119. Vermeulen L, De Sousa EMF, van der Heijden M, Cameron K, de Jong JH, Borovski T, Tuynman JB, Todaro M, Merz C, Rodermond H, Sprick MR, Kemper K, Richel DJ, Stassi G, Medema JP (2010) Wnt activity defines colon cancer stem cells and is regulated by the microenvironment. Nat Cell Biol 12(5):468–476. doi:10.1038/ncb2048
120. Heidel FH, Bullinger L, Feng Z, Wang Z, Neff TA, Stein L, Kalaitzidis D, Lane SW, Armstrong SA (2012) Genetic and pharmacologic inhibition of beta-catenin targets imatinib-resistant leukemia stem cells in CML. Cell Stem Cell 10(4):412–424. doi:10.1016/j.stem.2012.02.017
121. Wang Y, Krivtsov AV, Sinha AU, North TE, Goessling W, Feng Z, Zon LI, Armstrong SA (2010) The Wnt/beta-catenin pathway is required for the development of leukemia stem cells in AML. Science 327(5973):1650–1653. doi:10.1126/science.1186624
122. Zhao C, Blum J, Chen A, Kwon HY, Jung SH, Cook JM, Lagoo A, Reya T (2007) Loss of beta-catenin impairs the renewal of normal and CML stem cells in vivo. Cancer Cell 12 (6):528–541. doi:10.1016/j.ccr.2007.11.003
123. Chen Y, Peng C, Abraham SA, Shan Y, Guo Z, Desouza N, Cheloni G, Li D, Holyoake TL, Li S (2014) Arachidonate 15-lipoxygenase is required for chronic myeloid leukemia stem cell survival. J Clin Invest 124(9):3847–3862. doi:10.1172/JCI66129
124. Ohgami RS, Ma L, Ren L, Weinberg OK, Seetharam M, Gotlib JR, Arber DA (2012) DNA methylation analysis of ALOX12 and GSTM1 in acute myeloid leukaemia identifies prognostically significant groups. Br J Haematol 159(2):182–190. doi:10.1111/bjh.12029
125. Stenke L, Edenius C, Samuelsson J, Lindgren JA (1991) Deficient lipoxin synthesis: a novel platelet dysfunction in myeloproliferative disorders with special reference to blastic crisis of chronic myelogenous leukemia. Blood 78(11):2989–2995
126. Stenke L, Lauren L, Reizenstein P, Lindgren JA (1987) Leukotriene production by fresh human bone marrow cells: evidence of altered lipoxygenase activity in chronic myelocytic leukemia. Exp Hematol 15(2):203–207

127. Desplat V, Ivanovic Z, Dupuis F, Faucher JL, Denizot Y, Praloran V (2000) Effects of lipoxygenase metabolites of arachidonic acid on the growth of human blood CD34(+) progenitors. Blood Cells Mol Dis 26(5):427–436. doi:10.1006/bcmd.2000.0321
128. Mahipal SV, Subhashini J, Reddy MC, Reddy MM, Anilkumar K, Roy KR, Reddy GV, Reddanna P (2007) Effect of 15-lipoxygenase metabolites, 15-(S)-HPETE and 15-(S)-HETE on chronic myelogenous leukemia cell line K-562: reactive oxygen species (ROS) mediate caspase-dependent apoptosis. Biochem Pharmacol 74(2):202–214. doi:10.1016/j.bcp.2007.04.005
129. Middleton MK, Zukas AM, Rubinstein T, Jacob M, Zhu P, Zhao L, Blair I, Pure E (2006) Identification of 12/15-lipoxygenase as a suppressor of myeloproliferative disease. J Exp Med 203(11):2529–2540. doi:10.1084/jem.20061444
130. Burchert A, Cai D, Hofbauer LC, Samuelsson MK, Slater EP, Duyster J, Ritter M, Hochhaus A, Muller R, Eilers M, Schmidt M, Neubauer A (2004) Interferon consensus sequence binding protein (ICSBP; IRF-8) antagonizes BCR/ABL and down-regulates bcl-2. Blood 103(9):3480–3489. doi:10.1182/blood-2003-08-2970
131. Hao SX, Ren R (2000) Expression of interferon consensus sequence binding protein (ICSBP) is downregulated in Bcr-Abl-induced murine chronic myelogenous leukemia-like disease, and forced coexpression of ICSBP inhibits Bcr-Abl-induced myeloproliferative disorder. Mol Cell Biol 20(4):1149–1161
132. Schmidt M, Nagel S, Proba J, Thiede C, Ritter M, Waring JF, Rosenbauer F, Huhn D, Wittig B, Horak I, Neubauer A (1998) Lack of interferon consensus sequence binding protein (ICSBP) transcripts in human myeloid leukemias. Blood 91(1):22–29
133. Pelletier SD, Hong DS, Hu Y, Liu Y, Li S (2004) Lack of the adhesion molecules P-selectin and intercellular adhesion molecule-1 accelerate the development of BCR/ABL-induced chronic myeloid leukemia-like myeloproliferative disease in mice. Blood 104(7):2163–2171. doi:10.1182/blood-2003-09-3033
134. Sullivan C, Chen Y, Shan Y, Hu Y, Peng C, Zhang H, Kong L, Li S (2011) Functional ramifications for the loss of P-selectin expression on hematopoietic and leukemic stem cells. PLoS One 6(10), e26246. doi:10.1371/journal.pone.0026246
135. Hu Y, Chen Y, Douglas L, Li S (2009) beta-Catenin is essential for survival of leukemic stem cells insensitive to kinase inhibition in mice with BCR-ABL-induced chronic myeloid leukemia. Leukemia 23(1):109–116. doi:10.1038/leu.2008.262
136. Peng C, Chen Y, Yang Z, Zhang H, Osterby L, Rosmarin AG, Li S (2010) PTEN is a tumor suppressor in CML stem cells and BCR-ABL-induced leukemias in mice. Blood 115(3):626–635. doi:10.1182/blood-2009-06-228130
137. Fischer AS, Metzner J, Steinbrink SD, Ulrich S, Angioni C, Geisslinger G, Steinhilber D, Maier TJ (2010) 5-Lipoxygenase inhibitors induce potent anti-proliferative and cytotoxic effects in human tumour cells independently of suppression of 5-lipoxygenase activity. Br J Pharmacol 161(4):936–949. doi:10.1111/j.1476-5381.2010.00915.x
138. Datta K, Biswal SS, Kehrer JP (1999) The 5-lipoxygenase-activating protein (FLAP) inhibitor, MK886, induces apoptosis independently of FLAP. Biochem J 340(Pt 2):371–375
139. Sabirsh A, Bristulf J, Karlsson U, Owman C, Haeggstrom JZ (2005) Non-specific effects of leukotriene synthesis inhibitors on HeLa cell physiology. Prostaglandins Leukot Essent Fatty Acids 73(6):431–440. doi:10.1016/j.plefa.2005.08.004
140. Tavolari S, Bonafe M, Marini M, Ferreri C, Bartolini G, Brighenti E, Manara S, Tomasi V, Laufer S, Guarnieri T (2008) Licofelone, a dual COX/5-LOX inhibitor, induces apoptosis in HCA-7 colon cancer cells through the mitochondrial pathway independently from its ability to affect the arachidonic acid cascade. Carcinogenesis 29(2):371–380. doi:10.1093/carcin/bgm265, bgm265 [pii]
141. Hoque A, Lippman SM, Wu TT, Xu Y, Liang ZD, Swisher S, Zhang H, Cao L, Ajani JA, Xu XC (2005) Increased 5-lipoxygenase expression and induction of apoptosis by its inhibitors in esophageal cancer: a potential target for prevention. Carcinogenesis 26(4):785–791

142. Schwartz GK, Weitzman A, O'Reilly E, Brail L, de Alwis DP, Cleverly A, Barile-Thiem B, Vinciguerra V, Budman DR (2005) Phase I and pharmacokinetic study of LY293111, an orally bioavailable LTB4 receptor antagonist, in patients with advanced solid tumors. J Clin Oncol 23(23):5365–5373. doi:10.1200/JCO.2005.02.766, JCO.2005.02.766 [pii]
143. Marder P, Sawyer JS, Froelich LL, Mann LL, Spaethe SM (1995) Blockade of human neutrophil activation by 2-[2-propyl-3-[3-[2-ethyl-4-(4-fluorophenyl)-5- hydroxyphenoxy] propoxy]phenoxy]benzoic acid (LY293111), a novel leukotriene B4 receptor antagonist. Biochem Pharmacol 49(11):1683–1690, 0006-2952(95)00078-E [pii]
144. Adrian TE, Hennig R, Friess H, Ding X (2008) The role of PPARgamma receptors and LEUKOTRIENE B(4) receptors in mediating the effects of LY293111 in pancreatic cancer. PPAR Res 2008:827096. doi:10.1155/2008/827096
145. Lepley RA, Fitzpatrick FA (1994) 5-Lipoxygenase contains a functional Src homology 3-binding motif that interacts with the Src homology 3 domain of Grb2 and cytoskeletal proteins. J Biol Chem 269(39):24163–24168

The Physiology and Pathophysiology of Lipoxygenases in the Skin

Peter Krieg and Gerhard Fürstenberger

Abstract The skin is the primary barrier between the external environment and the internal milieu of the host protecting the body from physical and chemical insults and injury and preventing the loss of water. An active lipid metabolism and fatty acid-derived oxylipins are crucially involved in the structural integrity and functionality of the skin. Among them are lipoxygenases (LOX)-derived autacoids generated by an abundant and diverse cutaneous LOX metabolism. LOX products fulfill substantial functions in epithelial tissue homeostasis, inflammation as the general skin response to external damage, wound healing, and disease-related processes including numerous inflammatory skin conditions and the development of skin cancer. Recent results point to a critical role of a distinct LOX pathway in the development and maintenance of the epidermal barrier. This review focuses on the activities and mechanisms of actions of individual LOX-derived oxylipins, and the dysregulation of the corresponding LOX enzymes in diseased skin.

Abbreviations

ARCI	Autosomal recessive congenital ichthyoses
CE	Cornified cell envelope
CLE	Corneocyte lipid envelope
DC	Dendritic cell(s)
DMBA	7,12-dimethylbenz[a]anthrazene
e12-LOX	Epidermis-type 12-lipoxygenase
EFA	Essential fatty acid(s)
eLOX-3	Epidermis-type lipoxygenase-3
EOS	Esterified ω-hydroxyacylsphingosine(s)
(F)FA	(Free) fatty acid(s)
H(P)ETE	Hydro(pero)xyeicosatetraenoic acid
H(P)ODE	Hydro(pero)xyoctadecadienoic acid
HxA (B)	hepoxilin A (B)

P. Krieg (✉) • G. Fürstenberger
Molecular Diagnostics of Oncogenic Infections, German Cancer Research Center, Im Neuenheimer Feld 280, 69120 Heidelberg, Germany
e-mail: p.krieg@dkfz.de; g.fuerstenberger@gmx.de

l12-LOX	Leukocyte-type 12-lipoxygenase
LOX	Lipoxygenases(s)
LT	Leukotriene(s)
OS	ω-hydroxyacylsphingosine(s)
p12-LOX	Platelet-type 12-lipoxygenase
PCR	Polymerase chain reaction
PMN	Polymorphonuclear neutrophils
PPAR	Peroxisome Proliferator-activated receptor
PUFA	Polyunsaturated fatty acids
RT	Reverse Transcriptase
TEWL	Transepidermal water loss
TPA	12-O-tetradecanoyl phorbol-13-acetate

1 Introduction: Structural and Metabolic Functions of LOX in Skin

Shortly after the first discovery of lipoxygenases (LOX) in animal tissues in 1974 [1] LOX activities and LOX products have been identified in human and murine skin and epidermal cells [2] as well. Since then a plethora of findings showed that mammalian skin is characterized by an active and diverse LOX metabolism that plays crucial roles both in epithelial tissue homeostasis and in disease-related processes. LOX products identified in skin include a huge variety of oxidized free polyunsaturated fatty acids (PUFA) that have been associated with the modulation of epithelial proliferation and differentiation, inflammatory processes, and with several pathological skin conditions including cancer development. The discovery of epidermis-type LOX isoforms from skin has paved the way to new insights concerning the role of LOX metabolism in terms of structural integrity and functionality of skin. Recent studies revealed that a LOX-catalyzed oxygenation of the linoleate moiety of distinct ceramide species is critically involved in building a crucial structure of the epidermal barrier [3, 4].

2 Structure and Function of the Skin

The skin is the largest mammalian organ performing various crucial functions, including sensation of temperature, touch and pain, thermoregulation, and, most importantly, to serve as a protective barrier against mechanical, chemical, and biological insults and as a water-impermeable barrier that prevents excessive loss of body fluids, a function that is of particular importance for the survival of all terrestrial vertebrates immediately after birth.

The skin is composed of three distinct layers, the epidermis, the dermis and the hypodermis (Fig. 1). While the deepest section, the hypodermis, comprises adipose tissue and blood vessels, the dermis represents a fibrous network composed of

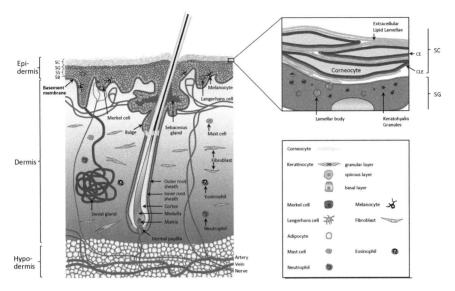

Fig. 1 Structure and cellular components of the mammalian skin. The mammalian skin is composed of hypodermis, dermis and the multi-layered epidermis. The structural components of the epidermal barrier function located in the stratum corneum are depicted within the inset

collagen and elastic fibers containing hair follicles, sweat glands, smaller blood vessels and sensory nerves, as well as fibroblasts and various immune cells including mast cells that play an important role in triggering the skin's inflammatory response to invading microorganisms, allergens, and physical injury.

Separated from the dermis by the basement membrane the multi-layered epidermis is composed mainly of keratinocytes which in the course of a terminal differentiation process mature and move from the proliferative cell type of the stratum basale through the spinous and granular layer consisting of post-mitotic keratinocytes and finally associate as flat dead corneocytes within the uppermost layer, the stratum corneum from where they eventually are shed by the process of desquamation. The epidermis furthermore contains specialized epidermal cells including pigment-producing melanocytes and Merkel cells, which are believed to be involved in mechanosensation and skin homeostasis, as well as antigen-presenting Langerhans cells constituting the frontline defense of the skin's immune system.

The epidermal barrier function mainly resides in the stratum corneum whose outermost layers are organised in a way that is often described as "bricks and mortar model" (Fig. 1, inset). The bricks are the anucleated corneocytes packed with keratin filaments and endowed with an inner rigid protein envelope, the cornified cell envelope (CE), and an outer corneocyte lipid envelope (CLE), a lipid monolayer of ω-OH ceramides and ω-OH fatty acids (FA) covalently bound to the CE. The corneocytes are embedded in an intercellular lipid matrix (the "mortar") forming tightly packed lipid bilayers with an equimolar composition of ceramides, cholesterol and free fatty acids (FFA). The formation of these lipid structures depends on the synthesis, processing and extrusion of distinct lipid species, consisting of delicately orchestrated processes in which a particular LOX-catalysed step has been recently shown to play a pivotal role [3, 4].

3 LOX and LOX Metabolites in Human and Mouse Skin: Expression, Activity and Physiological Function

Six different functional LOX isoforms have been identified in humans and seven in mice.

All of them have been shown to be expressed in skin by polymerase chain reaction (PCR) of mRNA, immunohistochemistry, enzyme activity and determination of products. They include the "classical" LOX isoforms originally isolated from distinct blood cell types as there are: 5-LOX, 12/15-LOX [i. e. human reticulocyte 15-LOX-1 and mouse leukocyte-type 12-LOX (l12-LOX)] and platelet-type 12-LOX (p12-LOX), as well as the skin-derived epidermis-type LOX comprising mouse epidermis-type 12-LOX (e12-LOX), human 15-LOX-2 and its mouse ortholog 8-LOX, 12R-LOX and epidermis-type LOX-3 (eLOX-3). The cellular sources for the distinct LOX isoforms in skin, their preferred substrates and products are summarized in Table 1.

3.1 5-Lipoxygenase (5-LOX, ALOX5)

5-LOX catalyzes the rate-limiting step of the leukotriene (LT) and lipoxin (LX) synthesis which requires the activity of various other enzymes including 5-Lipoxygenase-Activating Protein, LTA4 hydrolase, and LTC4 synthase as well as 12-LOX for LX formation [5].

The formation of LT in skin has first been detected in 1985 in exudates from human allergic skin [6] and thereafter in lesional skin of patients with atopic dermatitis [7] and psoriasis [8] as well as in inflamed periodontal tissues [9]. Their cellular origin, however, is still under debate [10]. While an earlier study reported the expression of 5-LOX in differentiating human HaCaT keratinocytes and normal foreskin epidermal cells [11], more recent data support the view that in epidermis in vivo 5-LOX activity is restricted to Langerhans cells [12] and other immune cells. In lesional skin of patients with systemic sclerosis 5-LOX was found to be expressed in inflammatory cells resembling macrophages and in addition to that in skin fibroblasts [13]. Apart from immune cells and fibroblasts, sebocytes may be another source of 5-LOX activity in skin as indicated by mRNA and protein expression of 5-LOX and LTA4 hydrolase in sebaceous glands in vivo and in SZ95 sebocytes in vitro [14]. In mouse skin, the expression of

Table 1 Expression and products of lipoxygenases in skin

Name (Abbreviation, Gene name)	Expression in skin	Substrate → Primary product	Secondary products
5-Lipoxygenase (5-LOX, *ALOX5*)	Langerhans cells, inflammatory cells Skin fibroblasts Sebocytes	AA → 5-HPETE, LTA4 (EPA→)18R-HEPE → **Resolvins**	**LTB4, LTC4 LXA4 Resolvins**
12/15-lipoxygenases (12/15-LOX)			
15-lipoxygenase-1 (human) (15-LOX-1, *ALOX15*)	Inflammatory cells? Keratinocytes in vitro	AA → 15-HPETE > 12-HPETE LA → 13-HPODE DHA → 17H-DHA	**12-HETE, 15-HETE, LXA4, 13-HODE, Resolvins, Protectins, Maresins**
Leukocyte-type 12-lipoxygenase (mouse) (l-12-LOX, *Alox15*)	Inflammatory cells?	AA → 12-HPETE > 15-HPETE LA → 13-HPODE DHA → 17H-DHA	**12-HETE, 15-HETE, LXA4 13-HODE, Resolvins, Protectins, Maresins**
Epidermis-type 12-lipoxygenase (mouse) (e12-LOX, *Aloxe*)	Differentiated layers of epidermis Hair follicle Sebaceous gland	AA → 12-HPETE > 15-HPETE LA → 13-HPODE	**12-HETE, 15-HETE, LXA4, 13-HODE, Resolvins, Protectins, Maresins**
Platelet-type 12-lipoxygenase (p12-LOX, *ALOX12*)	Human: basal layers of epidermis Mouse: stratum granulosum Hair follicle	AA → 12-HPETE	**12-HETE**
Human 15-lipoxygenase-2 (15-LOX-2, *ALOX15B*)	Basal layers of epidermis Hair follicle Sebaceous gland Eccrine, apocrine glands	AA → 15-HPETE	**15-HETE**
8-lipoxygenase (mouse) (8-LOX, *Alox8* alias *Alox15b*)	Differentiated layers of epidermis Hair follicle Sebaceous gland Eccrine, apocrine glands	AA → 8-HPETE LA → 9-HPODE 5-HPETE → **LTA4**	**8-HETE 9-HODE**

(continued)

Table 1 (continued)

Name (Abbreviation, Gene name)	Expression in skin	Substrate → Primary product	Secondary products
12R-lipoxygenase (12R-LOX, *ALOX12B*)	Differentiated layers of epidermis Hair follicle	AA → 12R-HPETE EOS → **9R-HPODE-EOS**	**12R-HETE** ω-OH-OS + 9R-HODE
Epidermis-type lipoxygenase 3 (eLOX-3, *ALOXE3*)	Differentiated layers of epidermis Hair follicle	12-HPETE → **Hepoxilins** 9R-HPODE-EOS → **9R (10R)-Epoxy-13-HOME-EOS**	ω-OH-OS + 9R(10R)-Epoxy-13-HOME 9,10,13-TriHOME

AA Arachidonic acid, *HEPE* Hydroxyeicosapentaenic acid, *LA* Linoleic acid, *EPA* Eicosapentaenoic acid, *HODE* Hydroxyoctadecadienoic acid, *LT* Leukotriene, *EOS* esterified ω-hydroxyacylsphingosine, *HOME* Hydroxyoctadecenoic acid, *TriHOME* Tri-Hydroxyoctadecenoic acid, *DHA* Docosahexaenoic acid, *HPETE* Hydroperoxyeicosatetraenoic acid, *LX* Lipoxin, *HETE* Hydroxyeicosatetrenoic acid, *HPODE* Hydroperoxyoctadecadienoic acid, *ω-OH-OS* ω-hydroxyacylsphingosine

5-LOX mRNA detected by reverse transcriptase (RT) PCR has not been attributed to a specific cellular source [15].

Apparently, the epidermis itself cannot form LTB4 directly from arachidonic acid but relies on a transcellular mechanism. LTA4 is secreted by migrating neutrophils and it has been shown that human keratinocytes and human epidermis are able to transform neutrophil-derived LTA4 into LTB4 by means of LTA4 hydrolase identified in keratinocytes [16, 17]. Transcellular LT synthesis may therefore be an important mechanism by which human epidermis can contribute to LTB4 formation in inflammatory skin diseases.

5-LOX metabolism evidently does not play a prominent role in normal epidermal homeostasis since 5-LOX-deficient mice do not show an obvious skin phenotype [18]. Notwithstanding this observation, 5-LOX deficiency has been shown to impede activation-dependent migration of dendritic cells (DC) to skin-draining lymph nodes but had no effect on cell growth or DC maturation [19]. LTB4 produced by skin sebocytes has been shown to be involved in the regulation of sebum synthesis [14, 20]. Moreover, LT were found to exhibit different effects on keratinocytes. According to this, LTB4 has been reported to stimulate proliferation and migration of human keratinocytes and to accelerate wound healing in the skin of mice mediated via the specific LTB4 receptor BLT2 on these cells [21–23]. Topical application of LTB4 to skin in vivo increases keratinocyte proliferation indirectly following the formation of microabscesses [22]. In human melanocytes, LTB4 and LTC4 have been reported to induce pigmentation and growth, respectively [24].

3.2 12/15-Lipoxygenases: Human 15-LOX-1 (ALOX15), Mouse Leukocyte-Type 12-LOX (Alox15), Mouse Epidermis-Type 12-LOX (Aloxe)

12/15-LOX are characterized by a dual positional specificity of oxygenation. While human 15-LOX-1 converts arachidonic acid predominantly to 15-HPETE, the mouse l12-LOX mainly produces 12-HPETE. Secondary oxidized lipids of the 12/15-LOX pathway comprise various proresolving mediators such as lipoxilins and the ω-3 PUFA derived resolvins, protectins, and maresins (for review see [25]). 12/15-LOX also efficiently metabolize linoleic acid to 13-HPODE and are furthermore able to oxygenate esterified PUFA such as arachidonyl-phosphatidylcholine and -ethanolamine as well as cholesterol esters (for review see [26]).

Expression of 12/15-LOX has been shown preferentially in cells of hematopoietic origin: human 15-LOX-1 in immature red blood cells and eosinophils, whereas the highest expression of mouse l12-LOX is found in peritoneal macrophages. In mammalian skin, 15-LOX is consistently reported as one of the main LOX activities. Of note, 15-lipoxygenation has also been assigned to other LOX-isoenzymes: in humans to a second 15-LOX isoenzyme, 15-LOX-2 which exclusively forms 15-HPETE [27], and in mice to e12-LOX which also functions as a dual 12/15-LOX similar to the l12-LOX [28, 29]. Furthermore, 15-HETE can also be formed via the peroxidase activity of cyclooxygenase.

Increased levels of 15-HETE were first found in psoriatic scales [30], and subsequently 15-LOX activity was identified in rat epidermis [31], cultivated human epidermal cells [32], psoriatic scales [33] and in freshly isolated human basal cells [34]. Since epidermal 15-LOX-2 was detected only recently it still remains to be elucidated in detail which isoenzyme accounts for which 15-LOX activities in human skin.

Expression of 15-LOX-1 mRNA in keratinocytes as detected by RT-PCR has been reported so far only in human epidermal and oral keratinocytes in vitro [35, 36], whereas in vivo evidence is lacking. In psoriatic skin, 15-LOX-1 was found by immunohistochemical staining to be localised in mast cells [37] indicating that 15-LOX-1 expression in skin may be restricted to immune cells. Similarly, expression of mouse l12-LOX was not detectable in normal mouse skin and 12-O-tetradecanoyl phorbol-13-acetate (TPA)-induced mRNA expression has been suggested to result from invading inflammatory cells [38]. UV irradiation of human keratinocytes upregulated the expression of 15-LOX, although the specific 15-LOX isoform was not determined [35]. In mouse keratinocytes 13-HODE has been shown to counteract the effects of 12-HETE on epidermal proliferation and the inhibition of terminal differentiation by down regulation of p12-LOX expression [35, 39]. In addition, both 13-HODE and 15-HETE application to human keratinocytes inhibited proliferation and increased UVB-induced apoptosis [35]. Induction of apoptosis is mediated through binding of 13-HODE to the peroxisome proliferator-activated receptor (PPAR) δ and downregulation of its activity and expression as identified in colorectal cancer cells [40]. Moreover,

9-HODE-effected inhibition of proliferation of human keratinocytes has been shown to be mediated via the stress-inducible G protein-coupled receptor G2A under oxidative conditions [41].

The second 12/15-LOX in mice, e12-LOX, in contrast, is abundantly expressed throughout the neonatal mouse epidermis [15] and in differentiated layers of adult epidermis, in hair follicles, conjunctiva of the eyelid and sebaceous glands [42]. Worthy of note is that the corresponding human gene has been found to be nonfunctional [43].

3.3 Platelet-Type 12-LOX (p12-LOX, ALOX12)

12-LOX activity was the first documented evidence of a LOX in animals initially discovered in platelets [44] and thereafter in several other tissues including human epidermal cells [45]. p12-LOX preferentially uses arachidonic acid as substrate to form almost exclusively 12-HPETE while linoleic acid is converted much less efficiently to 13-HPODE [28, 46]. p12-LOX has been reported to be highly active in human and murine skin epidermis. In human skin, it is selectively expressed in the basal layer where it was found to be overexpressed in psoriatic skin [47]. p12-LOX expression was also demonstrated in human epidermal and oral keratinocytes [36,48]. UV-irradiation suppressed p12-LOX expression but up-regulated 15-LOX-1 expression in human keratinocytes [35]. In mouse skin, p12-LOX expression was found in basal and suprabasal epidermal layers of neonatal skin [15] but restricted to the stratum granulosum and hair follicles in the adult skin [49]. During epidermal development expression starts at embryonic day 13.5 and reaches maximum expression around E14.5 and E15.5 when commitment to terminal differentiation occurs [50].

Consistent with expression of p12-LOX in the basal compartment of human skin is the proliferation stimulating activity of 12-HETE in cultured human keratinocytes and in guinea pig skin in vivo [51, 52]. In addition, 12-HETE has been reported to stimulate proliferation and to inhibit terminal differentiation of mouse keratinocytes [53]. On the other hand, in the mouse tail p-12-LOX expression was restricted to the stratum granulosum [47]. In addition, the transcription factor p63 isoform δNp63α, being present in postnatal epidermis and involved in the regulation of late stages of keratinocyte differentiation and barrier formation, was shown to induce the expression of p12-LOX in differentiating keratinocytes exclusively [50]. Accordingly, p12-LOX deficient mice exhibited a slightly increased epidermal water loss indicating a minor role of p12-LOX in epidermal barrier function [49]. Thus p12-LOX and its product 12-HETE exhibit both proliferating and differentiating effects in mouse skin which apparently depend on its expression in either basal or suprabasal compartments of skin epidermis.

3.4 Human 15-Lipoxygenase-2 (15-LOX-2, ALOX15B) and Mouse 8-Lipoxygenase (8-LOX, Alox8 alias Alox15b)

Human 15-LOX-2 and mouse 8-LOX represent orthologous members of the LOX family but display different positional specificities and tissue distributions. 15-LOX-2 converts arachidonic acid almost exclusively to 15-HPETE, and linoleic acid is less well metabolized to 13-HPODE [27]. 8-LOX which is solely found in mice metabolizes arachidonic acid exclusively into 8-HPETE, whereas linoleic acid is much less efficiently transformed into 9-HPODE [54–56]. 8-LOX is also able to oxygenate α-linolenic acid and docosahexaenoic acid and to exhibit LTA4 synthase activity by converting 5-HPETE to LTA4 [57].

In skin, both LOX exhibit a similar expression pattern in cutaneous adnexa and hair follicles and are observed most abundantly in sebaceous glands and eccrine and apocrine glands. Their expression in the interfollicular epidermis, however, is different [15, 58]. 15-LOX-2 was found to be constitutively expressed in the basal layers of normal human skin epidermis while in psoriatic lesions 15-LOX-2 expression was reported to be increased in all living layers of the epidermis [59]. 8-LOX activity, in contrast, could not be detected in normal mouse epidermis but was induced in the upper differentiating layers upon treatment of the skin with TPA [60].

Biological functions of the orthologous 8-LOX and 15-LOX-2 products in keratinocytes and other epithelia are related to modulation of differentiation and suppression of growth. Thus, 8-HETE was reported to stimulate differentiation as indicated by increased keratin-1 expression in primary mouse keratinocytes along a PPARα-mediated pathway [61] while 15-HETE exhibited a differentiation-modulating activity in prostate epithelial cells and suppression of proliferation of prostatic, colorectal cancer cells through activation of PPARγ [62, 63]. The co-expression of 15-LOX-2 and PPARγ in keratinocytes points to a critical role of the latter in 15-HETE-induced growth inhibition [64]. The localization of 15-LOX-2 and 8-LOX in secretory cells of cutaneous adnexa, sebaceous glands and prostatic secretory cells suggests a general role in regulating secretory differentiation or some aspect of the secretory process. Actually, 15-HETE, the product of 15-LOX-2, has been suggested to contribute to secretory differentiation in sebaceous glands and prostate via activation of PPARγ that is expressed in human sebocytes and prostate tissue [58]. Inducible expression of both 15-LOX-2 and 8-LOX in the mouse papilloma cell line 308 caused suppression of cell growth associated with inhibition of DNA synthesis along a p38 mitogen-activated protein kinase dependent pathway [65].

3.5 12R-Lipoxygenase (12R-LOX, ALOX12B) and Epidermis-Type Lipoxygenase 3 (eLOX-3, ALOXE3)

The last discovered members of the mammalian LOX family, 12R-LOX and eLOX-3 represent two functionally closely linked isoenzymes with unconventional enzymatic and structural features that play a pivotal role in skin physiology.

The identification of the unusual arachidonic acid metabolite 12R-HETE in psoriatic scales indicated the presence of a 12R-oxygenase in skin which initially was suspected to belong to the cytochrome P450 family [66]. In 1998, Brash and coworkers provided mechanistic evidence for a LOX-catalyzed biosynthesis of 12R-HETE in psoriatic skin and cloned the cDNA for the human 12R-LOX [67] which turned out to be the structural homologue of a previously uncharacterized LOX cloned from mouse epidermis [56]. 12R-LOX represents the only mammalian LOX that forms R enantiomeric products. Human 12R-LOX oxygenates arachidonic acid with low catalytic activity exclusively to 12R-HPETE and few other PUFA at the ω-9 position. The mouse enzyme, in contrast, does not accept free arachidonic acid but only the corresponding fatty acid methyl ester [67, 151]. Recent studies showed that a linoleic acid ester, O-linoleoyl-ω-hydroxylacyl-sphingosine (EOS *alias* ceramide 1), is a natural substrate for both, mouse and human 12R-LOX [68].

eLOX-3 was cloned from TPA-treated mouse epidermis and subsequently from human keratinocytes [69, 70]. Although the gene was clearly identified as a member of the mammalian LOX family based on its primary sequence, the expressed protein failed to exhibit any catalytic activity when incubated under standard conditions with prototypical LOX substrates, and thus was denoted epidermis-type lipoxygenase-3 (eLOX-3, *Aloxe3*) [69, 70]. Eventually, eLOX-3 was found to exhibit only a latent dioxygenase activity [71] and to function as a hydroperoxide isomerase converting hydroperoxides, the primary products of LOX reactions, to the corresponding epoxyalcohol derivatives [72]. Human and mouse eLOX-3 transform a variety of LOX-derived hydroperoxides with somewhat different substrate specificities [73]. By converting 12-HPETE to hepoxilins eLOX-3 may represent the genuine hepoxilin synthase in vivo as evidenced by eLOX-3 knockout studies and recent reports on nociception in rat spinal cord [74, 75]. Relevant for its physiological role in skin, eLOX-3 was also shown to convert the 12R-LOX generated hydroperoxide of ceramide 1 to the corresponding epoxyalcohol- and ketoester of ceramide 1 [68].

12R-LOX and eLOX-3 exhibit a similar tissue expression pattern in skin and few other epithelia, with the highest expression found in the differentiated layers of the epidermis and in hair follicles indicating functional linkage and a role in epidermal differentiation [67, 69, 70, 76]. Immunofluorescence analyses showed 12R-LOX and eLOX-3 proteins colocalized at the surface of keratinocytes in the stratum granulosum of mouse skin [77], and in situ hybridization showed induction of

12R-LOX expression in embryonic mouse skin on embryonic day 15.5 at the onset of skin development [78].

4 Structural Function of LOX in Skin: Role of 12R-LOX/ eLOX-3 in Epidermal Barrier Formation and Maintenance

A major breakthrough in understanding a pivotal physiological function of LOX metabolism in normal skin was provided by a genetic study showing that mutations in *ALOX12B* and *ALOXE3*, the genes encoding 12R-LOX and eLOX-3, are linked to the development of autosomal recessive congenital ichthyosis (ARCI) [79].

ARCI refers to a clinically and genetically heterogeneous group of cornification disorders including harlequin ichthyosis, lamellar ichthyosis, and congenital ichthyosiform erythroderma [80]. The main symptoms are scaling of the skin and variable degrees of erythema. It is generally acknowledged that the ARCI phenotype results from a physical compensation for the defective permeability barrier that underlies all ichthyosis disorders [81]. To date, nine different genes have been identified to be associated with the ARCI phenotype. The majority of the corresponding proteins have been shown to be involved in the processing, transport or assembly of lipid components of the stratum corneum which constitute essential elements of the permeability barrier of the skin [82]. Mutations in *ALOX12B* and *ALOXE3* have been found to be the second most common cause of ARCI being observed in about 10 % of the cases. At present 40 mutations have been identified in *ALOX12B* and 13 in *ALOXE3* [79, 83–90]. About a quarter of them are nonsense mutations resulting in a premature stop codon while the majority represents missense mutations. Studies on the functional impact of the observed mutations showed that although most of them apparently did not interfere with substrate binding or catalysis directly all mutations led to complete loss of the catalytic activity of the LOX enzymes [84, 85, 91], thus supporting the concept that the loss of function of 12R-LOX or eLOX-3 is fundamental to the pathogenesis of the LOX-dependent form of ARCI.

The essential role of 12R-LOX and eLOX-3 in barrier function was further evidenced by data from mouse knockout studies that also provided insights into the mechanism of LOX action in barrier formation. Constitutive ablation of either 12R-LOX or eLOX-3 resulted in perinatal lethality which was shown to be due to excessive transepidermal water loss (TEWL) [75, 77]. The mature LOX-deficient mouse skin, as analyzed in skin grafts and in the conditional knockout model with inducible gene inactivation in the epidermis, developed an ichthyosiform phenotype that closely resembled the disease phenotype observed in ichthyosis patients [92]. Studies on the molecular mechanisms involved in the disruption of the barrier function showed that LOX deficiency in mouse skin was associated with ultrastructural anomalies of varying degrees in the upper granular layer and the stratum corneum, impaired processing of profilaggrin to filaggrin monomers, and disordered composition of ester-bound ceramides. As decisive defect compromising barrier function in LOX-deficient skin the loss of protein-bound ω-hydroxyceramides could be proven. These ceramides (-ω-hydroxyacylsphingosines, OS) are covalently bound to proteins of the CE and constitute the main components of the CLE surrounding the corneocyte. The CLE is deemed indispensable for skin barrier function representing a necessary scaffold for the lamellar bilayer organization of the intercellular lipids. The covalent linkage of OS is brought about by transglutaminase-catalyzed transesterification of the precursor metabolites—esterified ω-hydroxyacylsphingosines (EOS), which are acylated in ω-position with linoleic acid. While initially a function of the 12R-LOX/eLOX-3 enzyme sequence in eicosanoid signaling had been hypothesized [93], recent work showed that these EOS ceramides are natural substrates for the 12R-LOX/LOX-3 pathway and are oxygenated by the consecutive actions of the two enzymes to 9R-hydroperoxide- and corresponding 9R,10R-epoxy-13R-hydroxy-derivatives [68]. This LOX-mediated modification was suggested to be a prerequisite for the subsequent hydrolysis and release of the linoleate moiety and consequently for the successful linkage of ceramides to CE proteins. Whether the oxidized linoleate cleavage products may have additionally signaling functions during barrier formation or other aspects of skin physiology remains to be resolved [3, 4].

While the pivotal role of 12R-LOX and eLOX-3 in barrier function is documented by a great number of genetic and biochemical evidence there is also one report suggesting the involvement of another LOX isoform in this aspect of skin physiology. As otherwise exhibiting a normal skin phenotype p12-LOX-deficient mice were shown to exhibit slightly increased transepidermal water loss (TEWL) indicating a minor role of p12-LOX in epidermal barrier function which still remains to be defined [49].

5 Role of LOX in Skin Inflammation, Injury and Wound Healing

Although LOX have been suspected of playing a role in cutaneous homeostasis, under normal physiological skin conditions most LOX products are generally regarded as mediators of inflammation and as molecules involved in pathophysiological processes. Skin inflammation is an early host defense reaction against irritation and injury. It includes the recruitment of polymorphonuclear neutrophils (PMN) as first effector cells at the inflamed tissue site, followed by macrophages at the transition of the inflammatory phase to the proliferative phase of resolution of inflammation and wound healing. Resident mast cells and T-lymphocytes are thought to be involved in early and late phases of wound healing. Generally, leukocytes and other immune cells contribute by releasing growth factors involved in the initiation and resolution of inflammation [94, 95]. Sequential profiles of early and late LOX-derived arachidonic acid products together with COX- and P450-derived eicosanoids are observed in response to pro-inflammatory stimuli and injury (e.g. TPA-induced inflammation in mouse skin [96, 97] and sunburn response in human skin [96, 98] as well as during resolution of inflammation and wound healing [99]. Among them, LTB4 exhibits a potent pro-inflammatory activity. Topically applied to human skin LTB4 induced inflammation accompanied by infiltration of neutrophils and mononuclear cells. In fact, after binding to specific G-protein coupled receptors LTB4 has been shown to promote PMN adhesion to endothelial cells, diapedesis, chemotaxis and chemokinesis and PMN-dependent plasma leakage pointing to a central role of LTB4 as a regulator of cell-trafficking [100]. Cysteinyl LT promote vascular permeability and smooth muscle contraction [101]. Furthermore, inhibitors of 5-LOX or LTB4 receptor antagonists have been shown to curtail inflammation in different models [102]. Moreover, 12-HETE was reported to induce inflammatory effects in skin encompassing formation of erythema and accumulation of neutrophils and macrophages [103]. Amongst the two 12-HETE epimers 12R-HETE was more potent than 12S-HETE in this respect [104]. LOX-derived eicosanoids not only initiate, amplify and perpetuate inflammation but they act also as key regulators of resolution of inflammatory responses [105]. During an inflammatory process the profile of pro-inflammatory eicosanoids including LT and 12-HETE is shifted to anti-inflammatory and pro-resolving activities by agents such as LXA4, resolvins and protectins with 12/15-LOX being key enzymes in their biosynthesis in macrophages [25]. In humans LXA4 has been shown to curtail pro-inflammatory neutrophil recruitment and promote wound-healing by macrophage recruitment through specific receptor signaling. In addition, LXA4 increased macrophage viability via an anti-apoptotic activity thus providing a beneficial effect on the resolution of inflammation [106]. Common activities of these lipids encompass the confinement of the activation of PMN and lymphocytes and of the generation of inflammatory mediators as well as the stimulation of PMN phagocytosis by macrophages [105]. Accordingly, in a keratinocyte wound healing assay resolvin D1 has been shown to increase closure rates [107]. Interestingly, 5-LOX knockout mice exhibited faster wound healing compared to wild-type mice. This is caused by a reduced inflammatory response due to the lack of pro-inflammatory LT thus allowing an earlier initiation of the resolution phase. Furthermore upregulation of heme oxygenase-1, which is negatively modulated by the pro-inflammatory LTB4 and LTC4 but induced by LXA4, resulted in an alleviation of inflammation [108, 109]. On the other hand, the activity of 12/15-LOX turned out to be Janus-faced in inflammation because both pro-inflammatory mediators as well as pro-resolving factors can be produced along this pathway [110].

6 Role of LOX in Inflammatory Skin Diseases

Dysregulation of eicosanoid metabolism in particular that of LOX is critically involved in initiation and maintenance of inflammatory skin diseases including psoriasis, atopic dermatitis and systemic sclerosis.

6.1 Psoriasis

Psoriasis is a hyperinflammatory and hyperproliferative skin disease which can be influenced by genetic and environmental factors including stress, infection and medications. It is an immune-related disease which involves a complex interrelationship between hyperproliferative keratinocytes and several immune cells including T cells, neutrophils, dendritic cells and macrophages [111]. The clinical appearance is characterized by sharply defined symmetric plaques of various sizes.

Disease-specific alterations of LOX-catalyzed arachidonic acid metabolism have been observed in psoriasis patients [100]. Thus, large amounts of 12S-HETE, the product of p12-LOX overexpressed in the basal compartment of psoriatic skin, is a prominent feature of the disease [47]. The growth factor IGF-II which is elevated in psoriatic skin contributes to the upregulation of p12-LOX mRNA and protein through the extracellular-signal-regulated kinases (ERK), the mitogen activated protein kinase (MAPK) and the phosphatidyl 3-kinase (PI3K) pathways [112]. In addition, significant amounts of 12R-HETE, the product of 12R-LOX being strongly expressed in differentiating cell layers, have been extracted from psoriatic plaques [33, 67]. Contrariwise, 15-LOX-2 expression has been found to be downregulated in psoriatic skin along with a decrease of 15-HETE generation [113] suggesting that the balance between p12-LOX and 15-LOX-2 is crucial for the etiology of psoriasis. UV radiation treatment of psoriasis exhibits beneficial effects and this correlates with the upregulation of 15-LOX-1 expression and 15-HETE generation [35]. 15-HETE is known to suppress the generation of the pro-inflammatory 12-HETE and LTB_4 by downregulating the expression of the corresponding LOX [112]. On the other hand, several observations point to a pathological function of LT in psoriatic skin. Thus, biologically active concentrations of LTB4 have been extracted from psoriatic skin lesions and LTB4 was reported to provoke intraepidermal microabscesses upon topical application to human skin, resembling the characteristic alterations seen in early phases of psoriasis [100]. In addition, pharmacologic inhibition of 5-LOX has been shown to have some beneficial effects in the treatment of psoriasis [114].

6.2 Atopic Dermatitis

Atopic dermatitis is a chronic inflammatory skin disease. Patients experience dry skin and are prone to inflammatory rashes and itches. Key events in atopic dermatitis are epidermal barrier dysfunction and ineffective cellular immunity [115]. The major criterion is itch [116].

It has long been known that LT, in particular LTB4, are markedly increased in lesional skin of patients with atopic dermatitis [117, 118]. In addition, LTB4 was identified as an endogenous itch mediator in skin of mice [119] and was shown to mediate the effect of the itch eliciting lipid sphingosylphosphorylcholine in atopic skin [120]. Furthermore, LTB4 has been reported to be involved in the pruritus response in allergic skin reactions [121]. The main source of LT generation in atopic dermatitis are infiltrating immune cells including eosinophils, basophils and mast cells [122]. Cysteinyl-LT have been shown to promote and maintain inflammation in atopic skin due to their effects on chemotaxis, vasodilation and edema. In agreement with this are the beneficial effects of 5-LOX inhibition and LT receptor antagonists in the treatment of atopic dermatitis [123, 124]. Thus, this data support the view that LT are strongly associated with atopic dermatitis and may be an appropriate target for therapeutic interventions.

6.3 Systemic Sclerosis

Systemic sclerosis is a chronic disease that is clinically characterized by thickening and progressive fibrosis of the skin and internal organs. Fibrosis is thought to be the final result of inflammation and vascular injury [125]. Products of 5-LOX have been involved in diverse inflammatory skin diseases [99] and in systemic sclerosis in skin [126]. In addition, overexpression of 5-LOX was observed in the skin of patients, predominantly in cells of the perivascular inflammatory infiltrate and in skin fibroblasts [13]. In this scenario, LT are reported to be involved in the regulation of the inflammatory response and blood vessel function and to play a crucial role in connective tissue remodeling [126]. LTB4 stimulates fibroblast migration [127] and cysteinyl LT induce collagen synthesis and stimulate fibroblast differentiation into myofibroblasts [128]. In addition, LTB4 may augment itching since inhibition of LTB4 production suppresses it [120, 129]. A significant exacerbation of experimental fibrosis has been observed in 12/15-LOX null mice pointing to an antifibrotic action of 12/15-LOX and its metabolites including LX [130]. Given a significant role of LT in the pathogenesis of systemic sclerosis anti-leukotriene therapies using LTB4 receptor antagonists and/or LX analogs are under development [126].

7 LOX and Skin Cancer

Multiple evidence suggests that LOX-derived products play a prominent role in skin cancer development and experimental skin cancers (for reviews see [131–134]). This evidence is based on the observation of aberrant LOX expression and overshooting product formation as well as the cancer preventive activity of specific and non-specific LOX inhibitors.

Some evidence for a role of 5-LOX metabolism in experimental mouse skin carcinogenesis comes from studies using 5-LOX inhibitors. Thus, orally applied TMK866, a 5-LOX inhibitor with cross reactivity towards 8-LOX and COX, was shown to reduce tumor formation induced by 7,12-dimethylbenz[a]anthrazene (DMBA)/TPA or benzo[a]pyrene alone [135]. Topical application of a lower dose of the same compound prevented TPA-induced LT formation and edema but did not impair the TPA-induced epidermal hyperproliferation and tumor formation [136]. The inhibitory activity observed upon oral application of a high dose of MK866 used in the former experiment might be due to the concomitant inhibition of 8-LOX and COX in addition to 5-LOX. In fact, combined inhibitors of 5-LOX and COX-2 were reported to synergistically reduce tumor growth of human squamous carcinoma cells inoculated in nude mice [137] and oral carcinogenesis in the hamster cheek pouch [138, 139].

p12-LOX expression is transiently induced by the tumor promoter TPA and strongly increased in benign and malignant skin tumors generated by the two-stage carcinogenesis approach. In these tumors, p12-LOX activity and 12-HETE production are highly elevated as well [38]. In line with this is the observation that p12-LOX deficient mice are less sensitive to tumor induction by this protocol [140]. In addition, increased expression of p12-LOX is widely found in human cancer tissues that develop in the colon, prostate, stomach and skin as compared to their normal counterparts where p12-LOX expression is absent or at a basal level. This data suggest that aberrant expression and activity of p12-LOX may contribute to the development and progression of various types of cancers [132]. Effects of 12-HETE related to cancer development are up-regulation of tumor cell adhesion molecules, stimulation of angiogenesis, increase of tumor cell spreading and inhibition of apoptosis [141]. In JB6 epidermal cells, the expression of p12-LOX and the activity of 12-HETE were shown to increase cloning efficiency leading to accelerated transformation of these cells [142]. Significant increases of p12-LOX from normal skin melanocytes and benign nevi through dysplastic to malignant metastatic melanoma have been observed. In murine melanoma lines 12-HETE is crucial for the invasion potential upon implantation in vivo. High concentrations in metastatic murine melanoma cell lines point to an important function of 12-HETE in the metastatic cascade including survival, cellular adhesion, invasion and motility and angiogenesis as well [141, 143, 144].

In contrast to p12-LOX, which was shown to be overexpressed in mouse skin tumors, e12-LOX was found to be down regulated during tumor formation suggesting an anti-tumorigenic effect of the latter. Overexpression of e12-LOX under the control of keratin 6 promoter in transgenic mice showed depending on the expression level opposite effects on the tumor response in skin [145]. Moderate transgenic e12-LOX expression resulted in decreased tumor formation and was paralleled by an up-regulation of l12-LOX and an accumulation of the linoleic acid metabolite 13-HODE while high transgene expression increased tumor response which was accompanied by an accumulation of the arachidonic acid derivative 12-HETE as the predominant LOX product.

While 5-LOX and 12-LOX predominantly exert pro-tumorigenic effects, 8-LOX and 15-LOX-2 have been rather associated with anti-tumorigenic activities. 8-LOX was shown to be constitutively expressed, and product levels were strongly elevated in benign highly differentiated papillomas obtained by the initiation-promotion protocol of mouse skin carcinogenesis [146]. Evidence from cell culture experiments points to a potent differentiation stimulating activity of 8-LOX and its metabolite 8-HETE. Forced expression of 8-LOX in the papilloma cell line MT1/2 brought about a more differentiated phenotype as indicated by induction of keratin 1 expression. Furthermore, overexpression of 8-LOX in the murine carcinoma cell line CH72 inhibited cell proliferation in vitro and in in vivo xenografts [147]. Similarly, transgenic mice overexpressing 8-LOX under the control of the loricrin promoter showed a more differentiated epidermal phenotype. Moreover, a reduced papilloma multiplicity was observed in initiation-promotion experiments with these 8-LOX transgenic mice suggesting a tumor suppressing function of the transgene due to the pro-differentiating and growth-inhibitory activity observed in papilloma and carcinoma cell lines ectopically expressing 8-LOX [61]. Consistent with this data is the loss of 8-LOX expression and activity upon conversion of papillomas to squamous cell carcinomas in mouse skin [146]. However, the roles of 8-HETE and 12-HETE in skin carcinogenesis in mice appear to be Janus-faced, in that extraordinarily high levels of these metabolites in papillomas were shown to induce chromosomal alterations which may contribute to the genetic instability observed in malignant progression of papillomas to squamous cell carcinomas [146, 148].

As has been observed for 8-LOX, the expression of the human 15-LOX-2 homolog has also been shown to be lost in the course of malignant progression of oral keratinocytes [149] and in prostate neoplasia [150] suggesting that the enzyme may act as a tumor suppressor. In agreement with this, the constitutive basal expression of 15-LOX-2 in human skin was lost in invasive malignant cells of basal cell carcinomas as well [58]. In addition, the strong uniform staining of 15-LOX-2 in secretory cells of skin sebaceous glands and sebaceous adenoma was significantly reduced in skin sebaceous carcinomas [58].

As a resume, antipodal functions of distinct LOX are observed in epithelial carcinogenesis in that 5-LOX and p12-LOX exhibit procarcinogenic activities while 15-LOX-2/8-LOX tend to suppress carcinogenesis.

In summary, the mammalian skin represents a tissue with a highly abundant and diverse LOX metabolism. The various LOX isoenzymes are key enzymes in the biosynthesis of a variety of oxylipins which act as signaling molecules involved in the modulation of keratinocyte proliferation and differentiation, in acute and

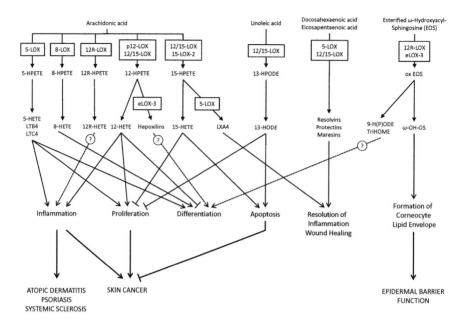

Fig. 2 Effects of lipoxygenases and their metabolites on physiology and pathophysiology of the skin. Arachidonic acid-derived leukotrienes and 12-HETE generated by 5-LOX, p12-LOX and 12/15-LOX have pro-inflammatory and proliferation-stimulating effects and are associated with inflammatory skin diseases like atopic dermatitis, psoriasis and systemic sclerosis, and skin cancer. 15-HETE and linoleic acid-derived 13-HODE, in contrast, have opposing effects by inhibiting proliferation and stimulating apoptosis. The 15-LOX/5-LOX arachidonic acid-derived product lipoxin A4 as well as the docosahexaenoic acid- and eicosapentaenoic acid-derived resolvins, protectins and maresins are potent anti-inflammatory autacoids involved in the active resolution of inflammation. 12R-LOX/eLOX-3-catalysed oxygenation of the linoleate moiety of ceramide EOS represents an essential step in the formation of the corneocyte lipid envelope, an indispensable structure of the epidermal barrier function

chronic inflammatory processes, in epithelial regeneration and cancer development in the skin. Recent results point to a pivotal role of a distinct LOX pathway in the development and maintenance of the epidermal barrier. The diverse effects of LOX and their metabolites on physiology and pathophysiology of the skin are summarized in Fig. 2.

References

1. Hamberg M, Svensson J, Samuelsson B (1974) Prostaglandin endoperoxides. A new concept concerning the mode of action and release of prostaglandins. Proc Natl Acad Sci USA 71:3824–3828
2. Hammarström S, Hamberg M, Samuelsson B et al (1975) Increased concentrations of nonesterified arachidonic acid, 12L-hydroxy-5,8,14-eicosatetraenoic acid, prostaglandin

E2, and prostaglandin F2alpha in epidermis of psoriasis. Proc Nal Acad Sci USA 72:5130–5134
3. Krieg P, Fürstenberger G (2014) The role of lipoxygenases in epidermis. Biochim Biophys Acta 1841:390–400
4. Munoz-Garcia A, Thomas CP, Keeney DS et al (2014) The importance of the lipoxygenase-hepoxilin pathway in the mammalian epidermal barrier. Biochim Biophys Acta 1841:401–408
5. Haeggstrom JZ, Funk CD (2011) Lipoxygenase and leukotriene pathways: biochemistry, biology, and roles in disease. Chem Rev 111:5866–5898
6. Talbot SF, Atkins PC, Goetzl EJ et al (1985) Accumulation of leukotriene C4 and histamine in human allergic skin reactions. J Clin Invest 76:650–656
7. Fogh K, Herlin T, Kragballe K (1989) Eicosanoids in skin of patients with atopic dermatitis: prostaglandin E2 and leukotriene B4 are present in biologically active concentrations. J Allergy Clin Immunol 83:450–455
8. Fogh K, Kiil J, Herlin T et al (1987) Heterogeneous distribution of lipoxygenase products in psoriatic skin lesions. Arch Dermatol Res 279:504–511
9. Eberhard J, Jepsen S, Albers HK et al (2000) Quantitation of arachidonic acid metabolites in small tissue biopsies by reversed-phase high-performance liquid chromatography. Anal Biochem 280:258–263
10. Luo M, Lee S, Brock TG (2003) Leukotriene synthesis by epithelial cells. Histol Histopathol 18:587–595
11. Janssen-Timmen U, Vickers P, Wittig U et al (1995) Expression of 5-lipoxygenase in differentiating human skin keratinocytes. Proc Natl Acad Sci USA 92:6966–6970
12. Spanbroek R, Stark HJ, Janssen Timmen U et al (1998) 5-Lipoxygenase expression in Langerhans cells of normal human epidermis. Proc Natl Acad Sci USA 95:663–668
13. Kowal-Bielecka O, Distler O, Neidhart M et al (2001) Evidence of 5-lipoxygenase overexpression in the skin of patients with systemic sclerosis: a newly identified pathway to skin inflammation in systemic sclerosis. Arthritis Rheum 44:1865–1875
14. Alestas T, Ganceviciene R, Fimmel S et al (2006) Enzymes involved in the biosynthesis of leukotriene B4 and prostaglandin E2 are active in sebaceous glands. J Mol Med (Berl) 84:75–87
15. Heidt M, Fürstenberger G, Vogel S et al (2000) Diversity of mouse lipoxygenases: identification of a subfamily of epidermal isozymes exhibiting a differentiation-dependent mRNA expression pattern. Lipids 35:701–707
16. Iversen L, Kragballe K, Ziboh VA (1997) Significance of leukotriene-A4 hydrolase in the pathogenesis of psoriasis. Skin Pharmacol 10:169–177
17. Iversen L, Kristensen P, Nissen JB et al (1995) Purification and characterization of leukotriene A4 hydrolase from human epidermis. FEBS Lett 358:316–322
18. Funk CD, Chen XS, Johnson EN et al (2002) Lipoxygenase genes and their targeted disruption. Prostaglandins Other Lipid Mediat 68–69:303–312
19. Doepping S, Funk CD, Habenicht AJ et al (2007) Selective 5-lipoxygenase expression in Langerhans cells and impaired dendritic cell migration in 5-LO-deficient mice reveal leukotriene action in skin. J Invest Dermatol 127:1692–1700
20. Ottaviani M, Camera E, Picardo M (2010) Lipid mediators in acne. Mediators Inflamm 2010:858176
21. Iizuka Y, Yokomizo T, Terawaki K et al (2005) Characterization of a mouse second leukotriene B4 receptor, mBLT2: BLT2-dependent ERK activation and cell migration of primary mouse keratinocytes. J Biol Chem 280:24816–24823
22. Kragballe K, Voorhees JJ (1985) Arachidonic acid in psoriasis. Pathogenic role and pharmacological regulation. Acta Derm Venereol Suppl Stockh 120:12–17
23. Reusch MK, Wastek GJ (1989) Human keratinocytes in vitro have receptors for leukotriene B4. Acta Derm Venereol 69:429–431

24. Morelli JG, Hake SS, Murphy RC et al (1992) Leukotriene B4-induced human melanocyte pigmentation and leukotriene C4-induced human melanocyte growth are inhibited by different isoquinolinesulfonamides. J Invest Dermatol 98:55–58
25. Serhan CN, Petasis NA (2011) Resolvins and protectins in inflammation resolution. Chem Rev 111:5922–5943
26. Kühn H, O'donnell VB (2006) Inflammation and immune regulation by 12/15-lipoxygenases. Prog Lipid Res 45:334–356
27. Brash AR, Boeglin WE, Chang MS (1997) Discovery of a second 15S-lipoxygenase in humans. Proc Natl Acad Sci USA 94:6148–6152
28. Bürger F, Krieg P, Marks F et al (2000) Positional- and stereo-selectivity of fatty acid oxygenation catalysed by mouse (12S)-lipoxygenase isoenzymes. Biochem J 348:329–335
29. Mcdonnell M, Davis W Jr, Li H et al (2001) Characterization of the murine epidermal 12/15-lipoxygenase. Prostaglandins Other Lipid Mediat 63:93–107
30. Camp RDR, Mallet AI, Woollard PM et al (1983) The identification of hydroxy fatty acids in psoriatic skin. Prostaglandins 26:431–447
31. Nugteren DH, Kivits GA (1987) Conversion of linoleic acid and arachidonic acid by skin epidermal lipoxygenases. Biochim Biophys Acta 921:135–141
32. Burrall BA, Cheung M, Chiu A et al (1988) Enzymatic properties of the 15-lipoxygenase of human cultured keratinocytes. J Invest Dermatol 91:294–297
33. Baer AN, Costello PB, Green FA (1991) Stereospecificity of the products of the fatty acid oxygenases derived from psoriatic scales. J Lipid Res 32:341–347
34. Henneicke-Von Zepelin HH, Schroder JM, Smid P et al (1991) Metabolism of arachidonic acid by human epidermal cells depends upon maturational stage. J Invest Dermatol 97:291–297
35. Yoo H, Jeon B, Jeon MS et al (2008) Reciprocal regulation of 12- and 15-lipoxygenases by UV-irradiation in human keratinocytes. FEBS Lett 582:3249–3253
36. Zhao H, Richards-Smith B, Baer AN et al (1995) Lipoxygenase mRNA in cultured human epidermal and oral keratinocytes. J Lipid Res 36:2444–2449
37. Gulliksson M, Brunnstrom A, Johannesson M et al (2007) Expression of 15-lipoxygenase type-1 in human mast cells. Biochim Biophys Acta 1771:1156–1165
38. Krieg P, Kinzig A, Ress-Loschke M et al (1995) 12-Lipoxygenase isoenzymes in mouse skin tumor development. Mol Carcinog 14:118–129
39. Fischer SM, Hagerman RA, Li-Stiles E et al (1996) Arachidonate has protumor-promoting action that is inhibited by linoleate in mouse skin carcinogenesis. J Nutr 126:1099S–1104S
40. Shureiqi I, Jiang W, Zuo X et al (2003) The 15-lipoxygenase-1 product 13-S-hydroxyoctadecadienoic acid down-regulates PPAR-delta to induce apoptosis in colorectal cancer cells. Proc Natl Acad Sci USA 100:9968–9973
41. Hattori T, Obinata H, Ogawa A et al (2008) G2A plays proinflammatory roles in human keratinocytes under oxidative stress as a receptor for 9-hydroxyoctadecadienoic acid. J Invest Dermatol 128:1123–1133
42. Funk CD, Keeney DS, Oliw EH et al (1996) Functional expression and cellular localization of a mouse epidermal lipoxygenase. J Biol Chem 271:23338–23344
43. Sun D, Elsea SH, Patel PI et al (1998) Cloning of a human "epidermal-type" 12-lipoxygenase-related gene and chromosomal localization to 17p13. Cytogenet Cell Genet 81:79–82
44. Hamberg M, Samuelsson B (1974) Prostaglandin endoperoxides. Novel transformations of arachidonic acid in human platelets. Proc Natl Acad Sci USA 71:3400–3404
45. Holtzman MJ, Turk J, Pentland A (1989) A regiospecific monooxygenase with novel stereopreference is the major pathway for arachidonic acid oxygenation in isolated epidermal cells. J Clin Invest 84:1446–1453
46. Funk CD, Furci L, Fitzgerald GA (1990) Molecular cloning, primary structure, and expression of the human platelet/erythroleukemia cell 12-lipoxygenase. Proc Natl Acad Sci USA 87:5638–5642

47. Hussain H, Shornick LP, Shannon VR et al (1994) Epidermis contains platelet-type 12-lipoxygenase that is overexpressed in germinal layer keratinocytes in psoriasis. Am J Physiol 266:C243–C253
48. Takahashi Y, Reddy GR, Ueda N et al (1993) Arachidonate 12-lipoxygenase of platelet-type in human epidermal cells. J Biol Chem 268:16443–16448
49. Johnson EN, Nanney LB, Virmani J et al (1999) Basal transepidermal water loss is increased in platetel-type 12-lipoxygenase deficient mice. J Invest Dermatol 112:861–865
50. Kim S, Choi IF, Quante JR et al (2009) p63 directly induces expression of Alox12, a regulator of epidermal barrier formation. Exp Dermatol 18:1016–1021
51. Chan CC, Duhamel L, Ford-Hutchison A (1985) Leukotriene B4 and 12-hydroxyeicosatetraenoic acid stimulate epidermal proliferation in vivo in the guinea pig. J Invest Dermatol 85:333–334
52. Kragballe K, Fallon JD (1986) Increased aggregation and arachidonic acid transformation by psoriatic platelets: evidence that platelet-derived 12-hydroxy-eicosatetraenoic acid increases keratinocyte DNA synthesis in vitro. Arch Dermatol Res 278:449–453
53. Hagerman RA, Fischer SM, Locniskar MF (1997) Effect of 12-O-tetradecanoylphorbol-13-acetate on inhibition of expression of keratin 1 mRNA in mouse keratinocytes mimicked by 12(S)-hydroxyeicosatetraenoic acid. Mol Carcinog 19:157–164
54. Fürstenberger G, Marks F, Krieg P (2002) Arachidonate 8(S)-lipoxygenase. Prostaglandins Other Lipid Mediat 68–69:235–243
55. Jisaka M, Kim RB, Boeglin WE et al (1997) Molecular cloning and functional expression of a phorbol ester-inducible 8S-lipoxygenase from mouse skin. J Biol Chem 272:24410–24416
56. Krieg P, Kinzig A, Heidt M et al (1998) cDNA cloning of a 8-lipoxygenase and a novel epidermis-type lipoxygenase from phorbol ester-treated mouse skin. Biochim Biophys Acta 1391:7–12
57. Qiao N, Takahashi Y, Takamatsu H et al (1999) Leukotriene A synthase activity of purified mouse skin arachidonate 8-lipoxygenase expressed in escherichia coli. Biochim Biophys Acta 1438:131–139
58. Shappell SB, Keeney DS, Zhang J et al (2001) 15-Lipoxygenase-2 expression in benign and neoplastic sebaceous glands and other cutaneous adnexa. J Invest Dermatol 117:36–43
59. Setsu N, Matsuura H, Hirakawa S et al (2006) Interferon-gamma-induced 15-lipoxygenase-2 expression in normal human epidermal keratinocytes and a pathogenic link to psoriasis vulgaris. Eur J Dermatol 16:141–145
60. Fürstenberger G, Hagedorn H, Jacobi T et al (1991) Characterization of an 8-lipoxygenase activity induced by the phorbol ester tumor promoter 12-O-tetradecanoylphorbol-13-acetate in mouse skin in vivo. J Biol Chem 266:15738–15745
61. Muga SJ, Thuillier P, Pavone A et al (2000) 8S-lipoxygenase products activate peroxisome proliferator- activated receptor alpha and induce differentiation in murine keratinocytes. Cell Growth Differ 11:447–454
62. Bhatia B, Tang S, Yang P et al (2005) Cell-autonomous induction of functional tumor suppressor 15-lipoxygenase 2 (15-LOX2) contributes to replicative senescence of human prostate progenitor cells. Oncogene 24:3583–3595
63. Tang DG, Bhatia B, Tang S et al (2007) 15-lipoxygenase 2 (15-LOX2) is a functional tumor suppressor that regulates human prostate epithelial cell differentiation, senescence, and growth (size). Prostaglandins Other Lipid Mediat 82:135–146
64. Flores AM, Li L, Mchugh NG et al (2005) Enzyme association with PPARgamma: evidence of a new role for 15-lipoxygenase type 2. Chem Biol Interact 151:121–132
65. Schweiger D, Fürstenberger G, Krieg P (2007) Inducible expression of 15-lipoxygenase-2 and 8-lipoxygenase inhibits cell growth via common signaling pathways. J Lipid Res 48:553–564
66. Woollard PM (1986) Stereochemical difference between 12-hydroxy-5,8,10,14-eicosatetraenoic acid in platelets and psoriatic lesions. Biochem Biophys Res Commun 136:169–176

67. Boeglin WE, Kim RB, Brash AR (1998) A 12R-lipoxygenase in human skin: mechanistic evidence, molecular cloning, and expression. Proc Natl Acad Sci USA 95:6744–6749
68. Zheng Y, Yin H, Boeglin WE et al (2011) Lipoxygenases mediate the effect of essential fatty acid in skin barrier formation: a proposed role in releasing omega-hydroxyceramide for construction of the corneocyte lipid envelope. J Biol Chem 286:24046–24056
69. Kinzig A, Heidt M, Fürstenberger G et al (1999) cDNA cloning, genomic structure, and chromosomal localization of a novel murine epidermis-type lipoxygenase. Genomics 58:158–164
70. Krieg P, Marks F, Fürstenberger G (2001) A gene cluster encoding human epidermis-type lipoxygenases at chromosome 17p13.1: cloning, physical mapping, and expression. Genomics 73:323–330
71. Zheng Y, Brash AR (2010) Dioxygenase activity of epidermal lipoxygenase-3 unveiled: typical and atypical features of its catalytic activity with natural and synthetic polyunsaturated fatty acids. J Biol Chem 285:39866–39875
72. Yu Z, Schneider C, Boeglin WE et al (2003) The lipoxygenase gene ALOXE3 implicated in skin differentiation encodes a hydroperoxide isomerase. Proc Natl Acad Sci USA 100:9162–9167
73. Yu Z, Schneider C, Boeglin WE et al (2006) Human and mouse eLOX3 have distinct substrate specificities: implications for their linkage with lipoxygenases in skin. Arch Biochem Biophys 455:188–196
74. Gregus AM, Dumlao DS, Wei SC et al (2013) Systematic analysis of rat 12/15-lipoxygenase enzymes reveals critical role for spinal eLOX3 hepoxilin synthase activity in inflammatory hyperalgesia. FASEB J 27:1939–1949
75. Krieg P, Rosenberger S, De Juanes S et al (2013) Aloxe3 knockout mice reveal a function of epidermal lipoxygenase-3 as hepoxilin synthase and its pivotal role in barrier formation. J Invest Dermatol 133:172–180
76. Schneider C, Keeney DS, Boeglin WE et al (2001) Detection and cellular localization of 12R-lipoxygenase in human tonsils. Arch Biochem Biophys 386:268–274
77. Epp N, Fürstenberger G, Müller K et al (2007) 12R-lipoxygenase deficiency disrupts epidermal barrier function. J Cell Biol 177:173–182
78. Sun D, Mcdonnell M, Chen XS et al (1998) Human 12(R)-lipoxygenase and the mouse ortholog: molecular cloning, expression, and gene chromosomal assignment. J Biol Chem 50:33540–33547
79. Jobard F, Lefevre C, Karaduman A et al (2002) Lipoxygenase-3 (ALOXE3) and 12(R)-lipoxygenase (ALOX12B) are mutated in non-bullous congenital ichthyosiform erythroderma (NCIE) linked to chromosome 17p13.1. Hum Mol Genet 11:107–113
80. Oji V, Tadini G, Akiyama M et al (2010) Revised nomenclature and classification of inherited ichthyoses: results of the First Ichthyosis Consensus Conference in Soreze 2009. J Am Acad Dermatol 63:607–641
81. Elias PM, Williams ML, Feingold KR (2012) Abnormal barrier function in the pathogenesis of ichthyosis: therapeutic implications for lipid metabolic disorders. Clin Dermatol 30:311–322
82. Feingold KR, Elias PM (2014) Role of lipids in the formation and maintenance of the cutaneous permeability barrier. Biochim Biophys Acta 1841:280–294
83. Akiyama M, Sakai K, Yanagi T et al (2010) Partially disturbed lamellar granule secretion in mild congenital ichthyosiform erythroderma with ALOX12B mutations. Br J Dermatol 163:201–204
84. Eckl KM, De Juanes S, Kurtenbach J et al (2009) Molecular analysis of 250 patients with autosomal recessive congenital ichthyosis: evidence for mutation hotspots in ALOXE3 and allelic heterogeneity in ALOX12B. J Invest Dermatol 129:1421–1428
85. Eckl KM, Krieg P, Küster W et al (2005) Mutation spectrum and functional analysis of epidermis-type lipoxygenases in patients with autosomal recessive congenital ichthyosis. Hum Mutat 26:351–361

86. Harting M, Brunetti-Pierri N, Chan CS et al (2008) Self-healing collodion membrane and mild nonbullous congenital ichthyosiform erythroderma due to 2 novel mutations in the ALOX12B gene. Arch Dermatol 144:351–356
87. Israeli S, Goldberg I, Fuchs-Telem D et al (2013) Non-syndromic autosomal recessive congenital ichthyosis in the Israeli population. Clin Exp Dermatol 38(8):911–916
88. Lesueur F, Bouadjar B, Lefevre C et al (2007) Novel mutations in ALOX12B in patients with autosomal recessive congenital ichthyosis and evidence for genetic heterogeneity on chromosome 17p13. J Invest Dermatol 127:829–834
89. Rodriguez-Pazos L, Ginarte M, Fachal L et al (2011) Analysis of TGM1, ALOX12B, ALOXE3, NIPAL4 and CYP4F22 in autosomal recessive congenital ichthyosis from Galicia (NW Spain): evidence of founder effects. Br J Dermatol 165:906–911
90. Vahlquist A, Bygum A, Ganemo A et al (2010) Genotypic and clinical spectrum of self-improving collodion ichthyosis: ALOX12B, ALOXE3, and TGM1 mutations in Scandinavian patients. J Invest Dermatol 130:438–443
91. Yu Z, Schneider C, Boeglin WE et al (2005) Mutations associated with a congenital form of ichthyosis (NCIE) inactivate the epidermal lipoxygenases 12R-LOX and eLOX3. Biochim Biophys Acta 1686(3):238–247
92. De Juanes S, Epp N, Latzko S et al (2009) Development of an ichthyosiform phenotype in Alox12b-deficient mouse skin transplants. J Invest Dermatol 129:1429–1436
93. Brash AR, Yu Z, Boeglin WE et al (2007) The hepoxilin connection in the epidermis. FEBS J 274:3494–3502
94. Gronert K (2008) Lipid autacoids in inflammation and injury responses: a matter of privilege. Mol Interv 8:28–35
95. Oberyszyn TM (2007) Inflammation and wound healing. Front Biosci 12:2993–2999
96. Nicolaou A, Masoodi M, Gledhill K et al (2012) The eicosanoid response to high dose UVR exposure of individuals prone and resistant to sunburn. Photochem Photobiol Sci 11:371–380
97. Zhang G, Liu X, Wang C et al (2013) Resolution of PMA-induced skin inflammation involves interaction of IFN-gamma and ALOX15. Mediators Inflamm 2013:930124
98. Rhodes LE, Gledhill K, Masoodi M et al (2009) The sunburn response in human skin is characterized by sequential eicosanoid profiles that may mediate its early and late phases. FASEB J 23:3947–3956
99. Kendall AC, Nicolaou A (2013) Bioactive lipid mediators in skin inflammation and immunity. Prog Lipid Res 52:141–164
100. Fogh K, Kragballe K (2000) Eicosanoids in inflammatory skin diseases. Prostaglandins Other Lipid Mediat 63:43–54
101. Hui Y, Cheng Y, Smalera I et al (2004) Directed vascular expression of human cysteinyl leukotriene 2 receptor modulates endothelial permeability and systemic blood pressure. Circulation 110:3360–3366
102. Fretland DJ, Widomski DL, Zemaitis JM et al (1990) Inflammation of guinea pig dermis. Effects of leukotriene B4 receptor antagonist, SC-41930. Inflammation 14:727–739
103. Dowd PM, Black AK, Woollard PW et al (1987) Cutaneous responses to 12-hydroxy-5,8,10,14-eicosatetraenoic acid (12-HETE) and 5,12-dihydroxyeicosatetraenoic acid (leukotriene B4) in psoriasis and normal human skin. Arch Dermatol Res 279:427–434
104. Wollard PM, Cunnigham FM, Murphy GM et al (1989) A comparison of the proinflammatory effects of 12(R)- and 12(S)-hydroxy-5,8,10,14-eicosatetraenoic acid in human skin. Prostaglandins 38:465–471
105. Serhan CN, Chiang N, Van Dyke TE (2008) Resolving inflammation: dual anti-inflammatory and pro-resolution lipid mediators. Nat Rev Immunol 8:349–361
106. Prieto P, Cuenca J, Traves PG et al (2010) Lipoxin A4 impairment of apoptotic signaling in macrophages: implication of the PI3K/Akt and the ERK/Nrf-2 defense pathways. Cell Death Differ 17:1179–1188

107. Norling LV, Spite M, Yang R et al (2011) Cutting edge: humanized nano-proresolving medicines mimic inflammation-resolution and enhance wound healing. J Immunol 186:5543–5547
108. Brogliato AR, Moor AN, Kesl SL et al (2014) Critical role of 5-lipoxygenase and heme oxygenase-1 in wound healing. J Invest Dermatol 134:1436–1445
109. Hanselmann C, Mauch C, Werner S (2001) Haem oxygenase-1: a novel player in cutaneous wound repair and psoriasis? Biochem J 353:459–466
110. Uderhardt S, Herrmann M, Oskolkova OV et al (2012) 12/15-lipoxygenase orchestrates the clearance of apoptotic cells and maintains immunologic tolerance. Immunity 36:834–846
111. Lowes MA, Bowcock AM, Krueger JG (2007) Pathogenesis and therapy of psoriasis. Nature 445:866–873
112. Yoo H, Kim SJ, Kim Y et al (2007) Insulin-like growth factor-II regulates the 12-lipoxygenase gene expression and promotes cell proliferation in human keratinocytes via the extracellular regulatory kinase and phosphatidylinositol 3-kinase pathways. Int J Biochem Cell Biol 39:1248–1259
113. Gudjonsson JE, Ding J, Li X et al (2009) Global gene expression analysis reveals evidence for decreased lipid biosynthesis and increased innate immunity in uninvolved psoriatic skin. J Invest Dermatol 129:2795–2804
114. Black AK, Camp RD, Mallet AI et al (1990) Pharmacologic and clinical effects of lonapalene (RS 43179), a 5-lipoxygenase inhibitor, in psoriasis. J Invest Dermatol 95:50–54
115. Boguniewicz M, Leung DY (2011) Atopic dermatitis: a disease of altered skin barrier and immune dysregulation. Immunol Rev 242:233–246
116. Oyoshi MK, He R, Li Y et al (2012) Leukotriene B4-driven neutrophil recruitment to the skin is essential for allergic skin inflammation. Immunity 37:747–758
117. Hua Z, Fei H, Mingming X (2006) Evaluation and interference of serum and skin lesion levels of leukotrienes in patients with eczema. Prostaglandins Leukot Essent Fatty Acids 75:51–55
118. Ruzicka T, Simmet T, Peskar BA et al (1986) Skin levels of arachidonic acid-derived inflammatory mediators and histamine in atopic dermatitis and psoriasis. J Invest Dermatol 86:105–108
119. Andoh T, Kuraishi Y (1998) Intradermal leukotriene B4, but not prostaglandin E2, induces itch-associated responses in mice. Eur J Pharmacol 353:93–96
120. Andoh T, Saito A, Kuraishi Y (2009) Leukotriene B(4) mediates sphingosylphosphorylcholine-induced itch-associated responses in mouse skin. J Invest Dermatol 129:2854–2860
121. Tsuji F, Aono H, Tsuboi T et al (2010) Role of leukotriene B4 in 5-lipoxygenase metabolite- and allergy-induced itch-associated responses in mice. Biol Pharm Bull 33:1050–1053
122. Sampson AP, Thomas RU, Costello JF et al (1992) Enhanced leukotriene synthesis in leukocytes of atopic and asthmatic subjects. Br J Clin Pharmacol 33:423–430
123. Nettis E, D'erasmo M, Di Leo E et al (2010) The employment of leukotriene antagonists in cutaneous diseases belonging to allergological field. Mediators Inflamm 2010, 628171
124. Rubin P, Mollison KW (2007) Pharmacotherapy of diseases mediated by 5-lipoxygenase pathway eicosanoids. Prostaglandins Other Lipid Mediat 83:188–197
125. Krieg T, Takehara K (2009) Skin disease: a cardinal feature of systemic sclerosis. Rheumatology (Oxford) 48(Suppl 3):iii14–iii18
126. Chwiesko-Minarowska S, Kowal K, Bielecki M et al (2012) The role of leukotrienes in the pathogenesis of systemic sclerosis. Folia Histochem Cytobiol 50:180–185
127. Mensing H, Czarnetzki BM (1984) Leukotriene B4 induces in vitro fibroblast chemotaxis. J Invest Dermatol 82:9–12
128. Vannella KM, Mcmillan TR, Charbeneau RP et al (2007) Cysteinyl leukotrienes are autocrine and paracrine regulators of fibrocyte function. J Immunol 179:7883–7890
129. Andoh T, Haza S, Saito A et al (2011) Involvement of leukotriene B4 in spontaneous itch-related behaviour in NC mice with atopic dermatitis-like skin lesions. Exp Dermatol 20:894–898

130. Kronke G, Reich N, Scholtysek C et al (2012) The 12/15-lipoxygenase pathway counteracts fibroblast activation and experimental fibrosis. Ann Rheum Dis 71:1081–1087
131. Catalano A, Procopio A (2005) New aspects on the role of lipoxygenases in cancer progression. Histol Histopathol 20:969–975
132. Fürstenberger G, Krieg P, Müller-Decker K et al (2006) What are cyclooxygenases and lipoxygenases doing in the driver's seat of carcinogenesis? Int J Cancer 119:2247–2254
133. Schneider C, Pozzi A (2011) Cyclooxygenases and lipoxygenases in cancer. Cancer Metastasis Rev 30:277–294
134. Steele VE, Holmes CA, Hawk ET et al (1999) Lipoxygenase inhibitors as potential cancer chemopreventives. Cancer Epidemiol Biomarkers Prev 8:467–483
135. Jiang H, Yamamoto S, Kato R (1994) Inhibition of two-stage skin carcinogenesis as well as complete skin carcinogenesis by oral administration of TMK688, a potent lipoxygenase inhibitor. Carcinogenesis 15:807–812
136. Fürstenberger G, Csuk-Glanzer BI, Marks F et al (1994) Phorbol ester-induced leukotriene biosynthesis and tumor promotion in mouse epidermis. Carcinogenesis 15:2823–2827
137. Fegn L, Wang Z (2009) Topical chemoprevention of skin cancer in mice, using combined inhibitors of 5-lipoxygenase and cyclo-oxygenase-2. J Laryngol Otol 123:880–884
138. Li N, Sood S, Wang S et al (2005) Overexpression of 5-lipoxygenase and cyclooxygenase 2 in hamster and human oral cancer and chemopreventive effects of zileuton and celecoxib. Clin Cancer Res 11:2089–2096
139. Sun Z, Sood S, Li N et al (2006) Involvement of the 5-lipoxygenase/leukotriene A4 hydrolase pathway in 7,12-dimethylbenz[a]anthracene (DMBA)-induced oral carcinogenesis in hamster cheek pouch, and inhibition of carcinogenesis by its inhibitors. Carcinogenesis 27:1902–1908
140. Virmani J, Johnson EN, Klein-Szanto AJ et al (2001) Role of 'platelet-type' 12-lipoxygenase in skin carcinogenesis. Cancer Lett 162:161–165
141. Pidgeon GP, Lysaght J, Krishnamoorthy S et al (2007) Lipoxygenase metabolism: roles in tumor progression and survival. Cancer Metastasis Rev 26:503–524
142. Piao YS, Du YC, Oshima H et al (2008) Platelet-type 12-lipoxygenase accelerates tumor promotion of mouse epidermal cells through enhancement of cloning efficiency. Carcinogenesis 29:440–447
143. Raso E, Dome B, Somlai B et al (2004) Molecular identification, localization and function of platelet-type 12-lipoxygenase in human melanoma progression, under experimental and clinical conditions. Melanoma Res 14:245–250
144. Winer I, Normolle DP, Shureiqi I et al (2002) Expression of 12-lipoxygenase as a biomarker for melanoma carcinogenesis. Melanoma Res 12:429–434
145. Müller K, Siebert M, Heidt M et al (2002) Modulation of epidermal tumor development caused by targeted overexpression of epidermis-type 12S-lipoxygenase. Cancer Res 62:4610–4616
146. Bürger F, Krieg P, Kinzig A et al (1999) Constitutive expression of 8-lipoxygenase in papillomas and clastogenic effects of lipoxygenase-derived arachidonic acid metabolites in keratinocytes. Mol Carcinog 24:108–117
147. Kim E, Rundhaug JE, Benavides F et al (2005) An antitumorigenic role for murine 8S-lipoxygenase in skin carcinogenesis. Oncogene 24:1174–1187
148. Nair J, Fürstenberger G, Bürger F et al (2000) Promutagenic etheno-DNA adducts in multistage mouse skin carcinogenesis: correlation with lipoxygenase-catalyzed arachidonic acid metabolism. Chem Res Toxicol 13:703–709
149. Wang D, Chen S, Feng Y et al (2006) Reduced expression of 15-lipoxygenase 2 in human head and neck carcinomas. Tumour Biol 27:261–273
150. Shappell SB, Boeglin WE, Olson SJ et al (1999) 15-lipoxygenase-2 (15-LOX-2) is expressed in benign prostatic epithelium and reduced in prostate adenocarcinoma. Am J Pathol 155:235–245
151. Siebert M, Krieg P, Lehmann WD et al (2001) Enzymatic characterization of epidermis-derived 12-lipoxygenase isozymes. Biochem J 355:97–104

5-Oxo-ETE and Inflammation

William S. Powell and Joshua Rokach

Abstract 5-Oxo-6E,8Z,11Z,14Z-eicosatetraenoic acid (5-oxo-ETE) is a potent proinflammatory 5-lipoxygenase product. It is formed by oxidation of 5S-HETE by the highly selective dehydrogenase 5-HEDH, which is expressed in both inflammatory and structural cells. Its synthesis is regulated by the availability of the obligate cofactor $NADP^+$ and is favored by conditions such as oxidative stress and activation of the respiratory burst in phagocytic cells. The actions of 5-oxo-ETE are mediated by the G_i-coupled OXE receptor that is found in many species but not in rodents. This receptor is highly expressed on eosinophils and basophils, and is also found on neutrophils, monocytes/macrophages, and various cancer cell lines. Because of its potent actions on eosinophils 5-oxo-ETE may be an important proinflammatory mediator in asthma and other eosinophilic diseases. 5-Oxo-ETE may also be involved in cancer, as it promotes the survival and proliferation of a number of cancer cell lines and rescues these cells from apoptosis induced by inhibitors of the 5-LO pathway. OXE receptor antagonists, which are currently under development, may be useful therapeutic agents in asthma and other eosinophilic disorders and possibly also in cancer.

Abbreviations

12S-HETE	12S-Hydroxy-5Z,8Z,10E,14Z-eicosatetraenoic acid
5-HEDH	5-Hydroxyeicosanoid dehydrogenase
5-LO	5-Lipoxygenase
5-oxo-12S-HETE	5-Oxo,12S-hydroxy-6E,8Z,10E,14Z-eicosatetraenoic acid
5-oxo-15S-HETE	5-Oxo,15S-hydroxy-6E,8Z,11Z,13E-eicosatetraenoic acid

---

W.S. Powell (✉)
Meakins-Christie Laboratories, McGill University Health Centre Research Institute, Centre for Translational Biology, 1001 Decarie Blvd, Montreal, QC, Canada, H4A 3J1
e-mail: William.Powell@McGill.ca

J. Rokach
Department of Chemistry, Claude Pepper Institute, Florida Institute of Technology, Melbourne, FL 32901-6982, USA

5-oxo-20-HETE	5-Oxo-20-hydroxy-6E,8Z,11Z,14Z-eicosatetraenoic acid
5-oxo-EPE	5-Oxo-6E,8Z,11Z,14Z,17Z-eicosapentaenoic acid
5-oxo-ETE	5-Oxo-6E,8Z,11Z,14Z-eicosatetraenoic acid
5-oxo-ETrE	5-Oxo-6E,8Z,11Z-eicosatrienoic acid
5-oxo-ODE	5-Oxo-6E,8Z-octadecadienoic acid
5S,12S-diHETE	5S,12S-Dihydroxy-6E,8Z,10E,14Z-eicosatetraenoic acid
5S-HEPE	5S-Hydroxy-6E,8Z,11Z,14Z,17Z-eicosapentaenoic acid
5S-HETE	5S-Hydroxy-6E,8Z,11Z,14Z-eicosatetraenoic acid
5S-HETrE	5S-Hydroxy-6E,8Z,11Z-eicosatrienoic
5S-HODE	5S-Hydroxy-6E,8Z-octadecadienoic acid
5S-HpETE	5S-Hydroperoxy-6E,8Z,11Z,14Z-eicosatetraenoic acid
AA	Arachidonic acid
DAG	Diacyl glycerol
EPA	5Z,8Z,11Z,14Z,17Z-Eicosapentaenoic acid
FLAP	5-Lipoxygenase activating protein
FOG_7	5-Oxo-7-glutathionyl-8,11,14-eicosatrienoic acid
GM-CSF	Granulocyte-macrophage colony stimulating factor
HCA	Hydroxyl-carboxylic acid receptor
LT	Leukotriene
MAP kinases	Mitogen-activated protein kinases
MMP-9	Matrix metalloproteinase-9
PAF	Platelet-activating factor
PI3K	Phosphoinositide 3-kinase
PKC	Protein kinase C
PLC	Phospholipase C
PUFA	Polyunsaturated fatty acid
TNF-α	Tumor necrosis factor-α
uPAR	Urokinase-type plasminogen activator receptor

1 Introduction

The 5-lipoxygenase (5-LO) pathway was discovered in 1976 with the identification of 5S-hydroxy-6,8,11,14-eicosatetraenoic acid (5S-HETE) as a product of arachidonic acid (AA) metabolism by rabbit neutrophils [1]. It was subsequently shown that 5S-hydroperoxy-6,8,11,14-eicosatetraenoic acid (5S-HpETE) the hydroperoxy precursor of 5S-HETE, is transformed to the unstable epoxide leukotriene (LT) A_4, which gives rise to the proinflammatory leukotrienes LTB_4 [2] and LTC_4 [3] (Fig. 1). The 5-LO pathway was subsequently found to also be involved in the formation of the anti-inflammatory lipoxins [4].

LTA_4 is highly unstable and is rapidly hydrolyzed nonenzymatically to two biologically inactive 6-trans isomers of LTB_4 [2]. While investigating the metabolism of LTB_4 by an ω-hydroxylase enzyme in neutrophils we found that these

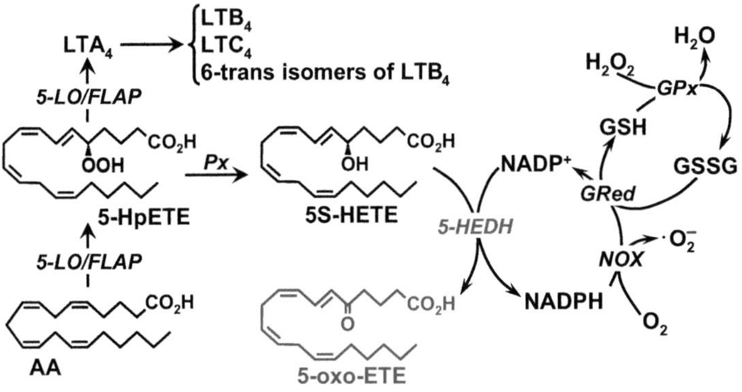

Fig. 1 Biosynthesis of 5-oxo-ETE. 5-Oxo-ETE is formed by oxidation of the 5-LO product 5S-HETE by 5-HEDH in the presence of $NADP^+$. For appreciable 5-oxo-ETE synthesis to occur, intracellular $NADP^+$ levels must be raised, as occurs during activation of NADPH oxidase (NOX) or as a result of oxidative stress through the GSH redox cycle. *Px* peroxidase, *GRed* glutathione reductase, *GPx* glutathione peroxidase

6-trans isomers are converted enzymatically to dihydro metabolites [5]. LTB_4 was not a substrate for this pathway, which appeared to be initiated by oxidation of the 5-hydroxyl group. The high degree of selectivity of this pathway for 6-trans isomers of LTB_4 over (6-cis) LTB_4 was surprising, and suggested that it might be of some significance. In searching for a more biologically relevant substrate we found that 5S-HETE, which also has a 6-trans double bond, is the preferred substrate and is oxidized to 5-oxo-6,8,11,14-eicosatetraenoic acid (5-oxo-ETE) by 5-hydroxyeicosanoid dehydrogenase (5-HEDH) [6]. The discovery of this pathway clarified earlier findings showing that 5S-HETE is a weak but selective activator of neutrophils that acts by a mechanism independent of the receptors for other lipid mediators [7]. Examination of the effects of 5-oxo-ETE on neutrophils revealed that it activates these cells with a potency about 100 times greater than that of 5S-HETE [8, 9]. As with 5S-HETE, these effects are independent of the receptors for other lipid mediators. Furthermore, there was complete cross-desensitization between 5S-HETE and 5-oxo-ETE, indicating that they act through the same receptor.

2 Formation of 5-Oxo-ETE

2.1 Substrate Selectivity of 5-HEDH

5-HEDH is a microsomal enzyme that oxidizes 5S-HETE to 5-oxo-ETE in the presence of $NADP^+$ [6]. It is highly selective for a 5S-hydroxyl group and displays little or no activity against 5R-HETE or positional isomers of 5S-HETE (8- 9-, 11-, 12-, or 15- HETE). Extensive studies with a series of synthetic substrates revealed that the 5S-hydroxyl group must be followed by at least one trans double bond but that additional double bonds are not a requirement [10]. A free carboxyl group as well as an extended hydrophobic terminal portion of the substrate are also required for appreciable activity. 5-Oxo fatty acids containing less than 16 carbons, an ω-hydroxyl group, or an esterified carboxyl group are all poor substrates for 5-HEDH.

In addition to AA 5-oxo-PUFA can be formed by 5-HEDH from a number of naturally occurring PUFA, including the C_{18} PUFA sebaleic acid, a major fatty acid in human sebaceous glands, the C_{20} PUFA Mead acid, which accumulates as a result of dietary essential fatty acid deficiency, and the ω3-PUFA eicosapentaenoic acid (EPA). 5S-Hydroxy-6E,8Z-octadecadienoic acid (5S-HODE; derived from sebaleic acid) is converted to 5-oxo-6E,8Z-octadecadienoic acid (5-oxo-ODE) [11], 5S-hydroxy-6E,8Z,11Z-eicosatrienoic acid (5S-HETrE; derived from Mead acid) is converted to 5-oxo-6E,8Z,11Z-eicosatrienoic acid (5-oxo-ETrE) [10] and 5S-hydroxy-6E,8Z,11Z,14Z,17Z-eicosapentaenoic acid (5S-HEPE; derived from EPA) is converted to 5-oxo-6E,8Z,11Z,14Z,17Z-eicosatrienoic acid (5-oxo-EPE) [12].

DiHETEs containing the requisite 5S-hydroxy-$\Delta^{6\text{-trans}}$ moiety are also substrates for 5-HEDH although they are metabolized somewhat more slowly than 5S-HETE [6]. The two 6-trans isomers of LTB_4 formed nonenzymatically from LTA_4 are converted to the corresponding 5-oxo-12-hydroxy-6E,8E,10E,14Z-eicosatetraenoic acid isomers, whereas 5S,12S-dihydroxy-6E,8Z,10E,14Z-eicosatetraenoic acid (5S,12S-diHETE) is converted to 5-oxo-12S-HETE [13]. The eosinophil-derived diHETE 5S,15S-dihydroxy-6E,8Z,11Z,13E-eicosatetraenoic acid is converted to 5-oxo-15S-HETE [6].

2.2 Properties of 5-HEDH

In addition to its selectivity for 5S-HETE, 5-HEDH is also highly selective for the cofactor $NADP^+$, for which it displays a 10,000-fold preference over NAD^+ [14]. The K_m for $NADP^+$ (~140 nM) is even lower than that for 5-oxo-ETE (~600 nM). However, the formation of 5-oxo-ETE is complicated by a strong inhibitory effect of NADPH (IC_{50} ~ 220 nM) and it is the ratio of $NADP^+$ to NADPH rather than the absolute concentration of $NADP^+$ that is critical in determining enzyme activity. Since the intracellular concentration of NADPH is normally much higher than that of $NADP^+$, the synthesis of 5-oxo-ETE is suppressed in resting intact cells as discussed below. For example, the ratio of $NADP^+$ to NADPH is about 0.08 in both resting neutrophils [15] and U937 cells (compared to a ratio of up to 3 in stimulated cells) [14] and this strongly limits 5-HEDH activity. The formation of 5-oxo-ETE by 5-HEDH appears to follow a ping-pong mechanism whereby one substrate (presumably $NADP^+$) is bound, transformed, and released by the enzyme followed by the binding and transformation of the second substrate (presumably 5S-HETE).

5-HEDH also catalyzes the reverse reaction, converting 5-oxo-ETE exclusively to the S enantiomer of 5-HETE in the presence of NADPH [6]. The pH optima for the two reactions are quite different: pH 10 for the formation of 5-oxo-ETE and pH 6 for the formation of 5S-HETE [14]. The high pH optimum for the oxidation reaction is similar to those observed for other dehydrogenases such as 15-hydroxyprostaglandin dehydrogenase [16]. At neutral pH, although the K_m for the eicosanoid substrate for the reduction reaction is somewhat lower, the oxidation reaction is favored because the V_{max} is nearly 10 times greater.

2.3 Distribution of 5-HEDH

Although first discovered in neutrophils 5-HEDH is found in a wide range of human cell types, including many types of inflammatory cells: eosinophils [17], monocytes [18], monocyte-derived dendritic cells [19], B lymphocytes [20], and platelets [13]. It is also present in structural cells, including airway epithelial and smooth muscle cells [21], vascular smooth muscle cells [22], and keratinocytes [11]. A variety of tumor cell lines derived from colon, lung, breast, prostate and myeloid cells also express 5-HEDH activity [21, 23, 24].

2.4 Transcellular Biosynthesis of 5-Oxo-ETE

The distribution of 5-HEDH is considerably wider than that of 5-LO/FLAP, which are largely restricted to inflammatory cells. This is also true for other enzymes in the 5-LO pathway, including LTA_4 hydrolase and LTC_4 synthase [25]. Although cells expressing these downstream enzymes in the absence of 5-LO cannot on their own synthesize LTs or 5-oxo-ETE, they can utilize substrates (LTA_4 or 5S-HETE) released from neighboring inflammatory cells. Both LTB_4 and LTC_4 have been shown to be synthesized by this process of transcellular biosynthesis [25]. We have shown that PC3 prostate tumor cells can synthesize substantial amounts of 5-oxo-ETE by transcellular biosynthesis [24]. In coincubations with human neutrophils over 75 % of the 5-oxo-ETE produced following stimulation could be attributed to PC3 cells. Thus structural cells expressing 5-HEDH could be an important source of 5-oxo-ETE in the presence of infiltrating inflammatory cells as occurs in inflammation and allergy.

2.5 Regulation of Cellular Synthesis of 5-Oxo-ETE

The synthesis of 5-oxo-ETE is regulated by the availability of the two 5-HEDH substrates 5S-HETE and $NADP^+$. To provide sufficient levels of intracellular $NADP^+$ to support 5-oxo-ETE synthesis cells must be exposed to an environment that promotes the oxidation of NADPH. In phagocytic cells such as neutrophils, eosinophils and monocytes, $NADP^+$ levels are increased during the respiratory burst due to activation of NADPH oxidase, which converts molecular oxygen to superoxide using NADPH as a cofactor (Fig. 1). Oxidative stress also results in oxidation of NADPH to $NADP^+$ through the GSH redox cycle [26]. For example, H_2O_2 and other hydroperoxides are reduced by glutathione peroxidase, which is accompanied by oxidation of GSH to GSSG. The GSSG is then recycled back to GSH by the NADPH-selective enzyme glutathione reductase, resulting in a very rapid increase in intracellular $NADP^+$. Cell death is also an important regulator of the synthesis of 5-oxo-ETE in both inflammatory cells [15] and tumor cells [24]. When cultured for 24 h neutrophils undergo apoptosis accompanied by a dramatic increase in the ratio of intracellular $NADP^+$ to NADPH, resulting in a profound increase in their ability to synthesize 5-oxo-ETE [15]. Similarly, exposure of tumor cells to cytotoxic agents dramatically increases 5-oxo-ETE synthesis by these cells [24].

2.6 Other Pathways for the Synthesis of 5-Oxo-ETE

Oxidation of 5S-HETE by 5-HEDH is the major pathway for the formation of 5-oxo-ETE, at least in human cells. However, oxo-fatty acids derived from polyunsaturated fatty acids (PUFA) can also be formed by the non-enzymatic dehydration of the corresponding hydroperoxy-PUFA, a process that is accelerated in the presence of heme-containing compounds [27]. Thus 5-oxo-ETE can be formed directly from 5-HpETE generated during lipid peroxidation. For example, appreciable amounts of 5-oxo-ETE esterified to membrane phospholipids were detected after exposure of red blood cell ghosts to t-butyl hydroperoxide [28]. Esterified 5-oxo-ETE was also detected after oxidation of the plasmalogen 1-O-hexadec-19-enyl-arachidonoyl glycerophosphocholine in the presence of Cu^{++} and H_2O_2 [29]. 5-Oxo-ETE is formed by a similar mechanism from 5-HpETE rather than from 5-HETE by murine macrophages, which do not contain detectable levels of 5-HEDH [30]. This activity was promoted by a cytosolic protein, but was not prevented by heating or treatment with trypsin, suggesting that the active component was not the protein itself, but possibly a heme compound bound to the protein. 5-Oxo-ETE was also reported to be formed by a murine mast cell line (MC-9 cells), probably by a similar mechanism [31].

5-Oxo-ETE can also be formed enzymatically from 5S-HpETE by cytochrome P450 enzymes, although this is unlikely to be a major biosynthetic pathway for this compound. Human CYP2S1 was reported to convert 5S-HpETE, as well as 12S-HpETE and 15S-HpETE, to their corresponding oxo derivatives in the absence of NADPH [32]. These reactions can also be catalyzed by other cytochrome P450 enzymes, including CYP1A1, 1A2, 1B1, and 3A4, but at slower rates.

3 Metabolism of 5-Oxo-ETE

The formation of 5-oxo-ETE by 5-HEDH is a reversible reaction as 5-oxo-ETE can be reduced in the presence of NADPH specifically to 5S-HETE [6]. However, there are considerable differences between the forward and reverse reaction [14]. The rate of formation of 5-oxo-ETE is favored by alkaline pH whereas the formation of 5S-HETE is favored by acid pH. Although both reactions have quite low K_m values, the V_{max} for the oxidation reaction is nearly 10 times higher than that for reduction of the keto group, strongly favoring the forward reaction.

5-Oxo-ETE is metabolized by a number of pathways in common with AA and other eicosanoids, including oxidation to 5-oxo-HETEs by both 12-LO and 15-LO as well as by ω-oxidation (Fig. 2). 15-LO in human eosinophils converts 5-oxo-ETE to 5-oxo-15S-HETE [33]. This product can also be formed by the oxidation of 5S,15S-diHETE by 5-HEDH [6]. Similarly, platelet 12-LO oxidizes 5-oxo-ETE to 5-oxo-12S-HETE [13]. 5-Oxo-ETE is also metabolized by cytochrome P450 enzymes to ω-oxidation products. Human neutrophils, which highly express LTB$_4$ 20-hydroxylase (CYP4F3), biologically inactivate 5-oxo-ETE by converting

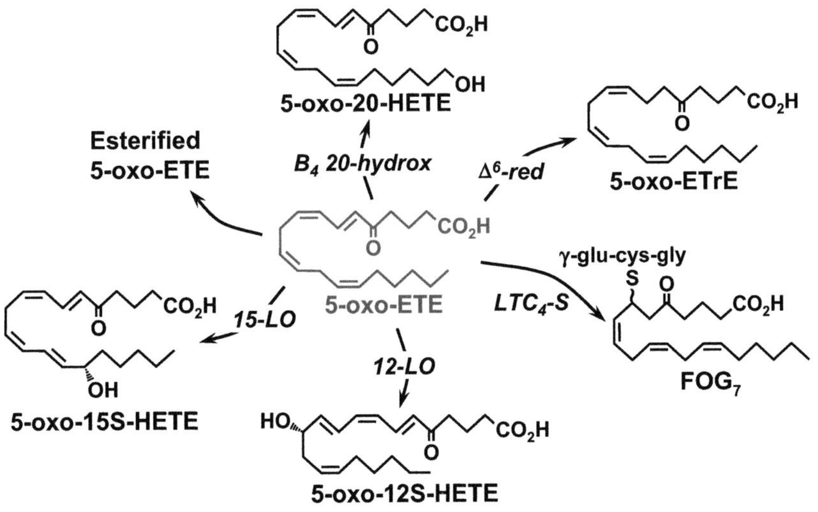

Fig. 2 Metabolism of 5-oxo-ETE. 5-Oxo-ETE is metabolized by LTB$_4$ 20-hydroxylase, eicosanoid Δ^6-reductase (Δ^6-red), LTC$_4$ synthase (LTC$_4$-S), and 12- and 15- lipoxygenases, as well as by incorporation into lipids

it to 5-oxo-20-HETE, presumably due to the action of this enzyme [34]. In contrast, human monocytes, which do not express appreciable levels of LTB_4 20-hydroxylase, do not metabolize 5-oxo-ETE to significant amounts of this product [18]. Murine macrophages also metabolize 5-oxo-ETE by ω-oxidation, but in this case the products are 5-oxo-18-HETE and 5-oxo-19-HETE rather than 5-oxo-20-HETE [35].

Neutrophils contain a Δ^6-reductase that biologically inactivates 5-oxo-ETE by converting it to its 6,7-dihydro metabolite 5-oxo-8,10,14-eicosatrienoic acid (5-oxo-ETrE) [36]. Interestingly this is a calcium/calmodulin-dependent enzyme, in contrast to prostaglandin Δ^{13}-reductase, which is also present in neutrophils and converts 15-oxo-ETE to its dihydro metabolite 15-oxo-5,8,11-eicosatrienoic acid (15-oxo-ETrE). The eicosanoid Δ^6-reductase also converts 5-oxo-15S-HETE to its dihydro metabolite 5-oxo-8,11,13-eicosatrienoic acid [36]. A similar enzyme may exist in murine macrophages, since 6,7-dihydro derivatives of ω-oxidation products of 5-oxo-ETE were identified after incubation of 5-oxo-ETE with these cells [35], but it is not known whether the formation of these products is dependent on calcium and calmodulin.

Because of its oxo-diene system 5-oxo-ETE is susceptible to Michael addition reactions and is a substrate for various glutathione transferases. Both rat liver and human placental glutathione transferases convert 5-oxo-ETE to a biologically inactive glutathione conjugate [37]. In the presence of murine macrophages 5-oxo-ETE is converted to a 7-glutathionyl conjugate (5-oxo-7-glutathionyl-8,11,14-eicosatrienoic acid) termed FOG_7 which promotes granulocyte migration, presumably by a mechanism distinct from that of 5-oxo-ETE because of the substantial structural differences [15, 84]. This reaction appears to be catalyzed by LTC_4 synthase, as FOG_7 is also formed from 5-oxo-ETE by the recombinant enzyme [37].

It has long been known that 5S-HETE is rapidly incorporated into cellular lipids, principally triglycerides, in neutrophils as well as other cell types [38]. 5-Oxo-ETE can also be esterified in neutrophils, both when incubated with intact cells as well as with isolated membrane fractions [39]. As with 5-HETE, the major site for esterified 5-oxo-ETE is the triglyceride fraction.

4 Effects of 5-Oxo-ETE on Leukocytes

4.1 *Effects on Granulocytes*

5-Oxo-ETE induces an array of responses in both neutrophils and eosinophils and a more limited range of responses in basophils (Table 1). In general eosinophils tend to respond more strongly to 5-oxo-ETE than neutrophils [40, 41], possibly due to their higher expression of the OXE receptor [42] or to differences in downstream signaling.

Table 1 Effects of 5-oxo-ETE on granulocytes and monocytes

Cell type	Ca^{++}	F-actin	CD11b	CD69	L-sel	Chtx	Degran	$O_2^{-\bullet}$	AA rel	GM-CSF
Eosinophils	+++	+++	+++	+++	+++	+++	+[b]	+++		
Neutrophils	+++	+++	+++			+++	+[b]	+/−[b]	+	
Basophils	+	++[a]	+			+++	−			
Monocytes	−	++				++			+	++

5-Oxo-ETE elicits a variety of responses in inflammatory cells, including calcium mobilization, actin polymerization (F-actin), surface expression of CD11b and CD69, shedding of L-selectin (L-sel), chemotaxis (Chtx), superoxide formation ($O_2^{-\bullet}$), and the release of AA (AA rel) and GM-CSF
[a] Shape change was measured rather than actin polymerization
[b] These responses are strongly enhanced following treatment with certain cytokines or inflammatory mediators

5-Oxo-ETE very rapidly stimulates mobilization of intracellular calcium and actin polymerization (i.e. formation of F-actin) in neutrophils [8, 9, 43–45] and eosinophils [41, 46] with both reaching near peak levels by 5 s. Calcium levels drop fairly rapidly thereafter, whereas F-actin levels decline more slowly, reaching close to baseline levels by 5 min. In contrast, 5-oxo-ETE elicits a relatively modest calcium response in basophils [47] but does induce a rapid shape change response [47], presumably linked to actin polymerization. It also induces L-selectin shedding in both eosinophils [46] and neutrophils (Powell, unpublished). Surface expression of CD11b [46] and the activation marker CD69 [48] are also increased in response to 5-oxo-ETE in eosinophils, and in neutrophils the surface expression of both CD11b and CD11c is increased [45]. On the other hand, 5-oxo-ETE has only a very modest effect on the surface expression of CD11b and the activation marker CD203c in basophils [49]. In addition, it has been reported to induce the release of AA from both neutrophils [44] and monocytes [50].

In contrast to its potent stimulatory effects on the above granulocyte responses 5-oxo-ETE induces only a relatively modest degranulation response in unprimed neutrophils [44] and eosinophils [40], and does not elicit this response in basophils [51]. However, priming with TNF-α [9] or GM-CSF [44] or addition of low concentrations of PAF or ATP [52] strongly enhance 5-oxo-ETE-induced neutrophil degranulation. GM-CSF also markedly increases eosinophil degranulation in response to 5-oxo-ETE and synergistic interactions have been observed between 5-oxo-ETE and other proinflammatory mediators in eliciting this response [40]. Although it strongly induces superoxide production in eosinophils [41], 5-oxo-ETE has either a relatively weak effect [43] or no effect at all [44] on this response in unprimed neutrophils, whereas in GM-CSF-primed neutrophils is has a strong stimulatory effect [44].

In addition to inducing the above relatively rapid responses in granulocytes 5-oxo-ETE is a potent chemoattractant for neutrophils [8], eosinophils [17, 33], and basophils [47, 51]. Moreover, synergistic interaction exist between 5-oxo-ETE and other eosinophil chemoattractants including eotaxin, RANTES [53], and PAF

[17]. Apart from its metabolites 5-oxo-15S-HETE and Fog_7, 5-oxo-ETE is the only 5-LO product with appreciable chemotactic activity for human eosinophils. In contrast to its potent chemoattractant effects for guinea pig eosinophils [54], LTB_4 displays little chemotactic activity for human eosinophils [17, 54, 55].

4.2 Effects of 5-Oxo-ETE on Monocytes

5-Oxo-ETE induces a more limited range of response in monocytes compared to granulocytes. It is a monocyte chemoattractant and interacts synergistically with monocyte chemotactic proteins 1 and 3 in inducing this response. It has also been reported to induce actin polymerization but not calcium mobilization in these cells [50]. In contrast, another study found that 5-oxo-ETE does not elicit a shape change response in monocytes but does stimulate calcium mobilization [47]. In our studies we have never observed calcium mobilization in response to 5-oxo-ETE in populations of monocytes gated out using flow cytometry (Powell, unpublished data) but we did find that it is a potent activator of GM-CSF secretion from these cells [56].

5 5-Oxo-ETE Receptor

The first evidence for a selective receptor for 5-oxo-ETE came from studies in neutrophils showing that this substance is a potent activator of calcium mobilization (EC_{50} 2 nM) in these cells [8, 9]. This response was subject to homologous desensitization following pretreatment with a low concentration of 5-oxo-ETE. It could also be blocked by pretreatment with high concentrations of 5S-HETE, which appears to activate neutrophils by the same mechanism as 5-oxo-ETE but is about 100 times less potent. In contrast, pretreatment with 5-oxo-ETE had no effect on the responses to other neutrophil agonists, including LTB_4, platelet-activating factor (PAF), or IL-8 [57]. Similarly, pretreatment with LTB_4 does not inhibit 5-oxo-ETE-induced calcium mobilization in neutrophils [8]. Activation of neutrophils by 5-oxo-ETE can be blocked by pretreatment with pertussis toxin, indicating that its receptor is a member of the G protein-coupled receptor (GPCR) family [34, 43, 44]. Binding studies with 5-oxo-ETE were complicated by its incorporation into cellular lipids following incubation with either intact cells or membrane fractions as noted above. However, O'Flaherty was able to demonstrate selective 5-oxo-ETE binding sites (K_d, 4 nM) in neutrophils following treatment with the acetyl CoA transferase inhibitor triacsin C, which blocked 5-oxo-ETE esterification [39].

5.1 Cloning of the 5-Oxo-ETE Receptor

The receptor for 5-oxo-ETE was cloned independently by three groups and shown to be a member of the GPCR family. From a database search Hosoi and coworkers identified an intronless gene on chromosome 2p21 that was predicted to code for a candidate GPCR containing 423 amino acids, which they tentatively named TG1019 [58]. To find the ligand for this orphan receptor they fused TG1019 to $G\alpha_i$, $G\alpha_s$, or $G\alpha_q$ and screened a library in search of agonists that stimulated GTPγS binding. This led to the identification of the most potent ligand for TG1019 as 5-oxo-ETE (5-oxo-ETE>>5S-HpETE>>5S-HETE=5R-HETE). It is possible that the activity of 5S-HpETE found in this and other studies could be due at least in part to its facile dehydration to 5-oxo-ETE. Takeda and coworkers also used a GTPγS binding assay to screen a library of potential agonists and identified 5-oxo-ETE as the preferred ligand for an orphan GPCR named hGPCR48 that they identified and expressed following a database search for intronless GPCRs [59]. Jones and coworkers used a different approach in which they coexpressed the orphan GPCR R527 and $G\alpha_{16}$ in HEK cells and used a calcium mobilization assay to screen a library to search for its ligand, which they identified as 5-oxo-ETE [60]. All three proteins were identical, except that R527 lacked the first 39 amino acids.

The 5-oxo-ETE receptor has been named the OXE receptor by the IUPHAR Nomenclature Subcommittee for Leukotriene Receptors [61] and the corresponding gene has been designated *OXER1*. Compared to other GPCRs the sequence of the OXE receptor most closely resembles those of the hydroxyl-carboxylic acid

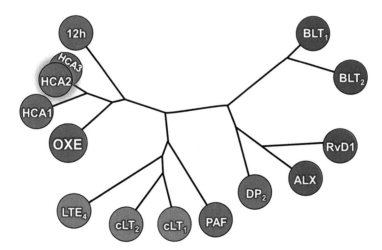

Fig. 3 Dendrogram showing the relationship of the OXE receptor to other related receptors. This was constructed using the facilities at www.phylogeny.fr [107]. The receptors for 12-HETE (12h) [63], LXA$_4$ (ALX) [61], resolvin D1 (RvD1) [108] and LTE$_4$ [109] are also known as GPR31, FPR2, GPR32 and GPR99 (IUPAC designation OXGR1), respectively. Receptors for products of the 5-LO pathway are shown in *color*

receptors HCA1 HCA2, and HCA3, with which it has about 30 % identity (Fig. 3). The endogenous ligands for these receptors are small carboxylic acids, including lactate, hydroxybutyrate, and 3-hydroxyoctanoic acid [62]. HCA2 and HCA3 also bind nicotinic acid. Among eicosanoid receptors, the OXE receptor most closely resembles GPR31 (also about 30 % identity), which has recently been reported to be a high affinity receptor for 12S-HETE [63], and also bears some similarity to the cysLT$_1$ and cysLT$_2$ receptors.

5.2 Expression of the OXE Receptor

In human tissues OXE receptor mRNA is highly expressed in peripheral leukocytes spleen, lung, liver and kidney [58]. Among leukocytes it is most highly expressed in eosinophils [42] and basophils [47, 51], with lower expression levels in neutrophils, monocytes [47], and macrophages [42]. It is also expressed in a number of cancer cell lines [64, 65] and was recently identified in an adrenocortical cell line [66].

Although orthologs of the human OXE receptor occur in a large number of mammalian species as diverse as bats whales, manatees, elephants, and monkeys, as well as in some species of fish, they are not present in rodents. Interestingly, zebrafish possess a 5-oxo-ETE receptor (GPR81-like protein 4) that, among human proteins, resembles most closely the OXE receptor (34 % identity) and is also similar to the HCA 1, 2, and 3 receptors (about 30 % identity). 5-Oxo-ETE has been shown to activate zebrafish leukocytes via this receptor and appears to play an important role in host defence in this species, inducing the accumulation of leukocytes at sites of tissue injury [67].

The unavailability of mouse animal models to investigate the pathophysiological role of 5-oxo-ETE and the OXE receptor has markedly impeded progress in this area. However we have recently developed potent synthetic OXE receptor antagonists [57] that should be useful in addressing this issue as discussed in Sect. 7.

5.3 OXE Receptor Signaling Mechanisms

The OXE receptor is coupled to $G_{i/o}$ proteins as the actions of 5-oxo-ETE can be blocked by pertussis toxin [34, 43, 44]. However, in contrast to many other chemoattractant receptors, it is not coupled to G_q [58, 68]. Phospholipase C (PLC) is very rapidly activated following activation of the OXE receptor, resulting in the formation of inositol trisphosphate and diacyl glycerol (DAG) and consequent calcium mobilization and activation of DAG-sensitive protein kinase C (PKC) isoforms [69]. Some of the actions of 5-oxo-ETE on eosinophils appear to be mediated by PKC-δ [70], whereas PKC-ε has been reported to mediate the proliferative effects of 5-oxo-ETE on prostate cancer cells [71]. There is also evidence for the involvement of PKC-ζ in 5-oxo-ETE-induced eosinophil migration [70], but this is unlikely to involve PLC, as this PKC isoform is not activated by calcium or DAG. Interaction of 5-oxo-ETE with the OXE receptor in neutrophils and OXE-transfected CHO cells also activates phosphoinositide 3-kinase (PI3K), resulting in the formation of phosphatidylinositol (3,4,5)-trisphosphate and phosphorylation of Akt [43, 69]. 5-Oxo-ETE-induced cell migration was blocked by both PKC and PI3K inhibitors, suggesting the involvement of both of these pathways in this response [69]. MAP kinases are also involved in mediating cellular responses to 5-oxo-ETE in neutrophils [44], eosinophils [40, 70, 72] and transfected CHO cells [69]. ERK-1, ERK-1 and p38 are phosphorylated in response to 5-oxo-ETE and p38 inhibitors have been shown to block 5-oxo-ETE-induced eosinophil migration through Matrigel [70]. 5-Oxo-ETE also promotes the phosphorylation of cPLA$_2$ in neutrophils, resulting in the release of AA [44], which is likely to be mediated by activation of MAP kinases.

Although 5-oxo-ETE inhibits Gα$_i$ protein-dependent inhibition of adenylyl cyclase [58, 73] this does not appear to play a significant role in mediating its effects on cellular responses [73, 74]. It was found that the biased OXE receptor antagonist Gue1654 does not inhibit Gα$_i$-mediated signaling (i.e. activation of adenylyl cyclase) but rather blocks βγ-mediated signalling [73] and recruitment of β-arrestin [74]. The fact that Gue1654 inhibits the stimulatory effects of 5-oxo-ETE on calcium mobilization, actin polymerization, cell migration, adhesion to endothelial cells, and the respiratory burst in granulocytes, indicates that the OXE receptor-induced activation of these cells is mediated by Gβγ and possibly β-arrestin rather than by Gα$_i$.

6 Structure-Activity Relationships for Activation of the OXE Receptor

6.1 Structural Requirements for Activation of the OXE Receptor

The structure-activity relationship for activation of the OXE receptor has been investigated in considerable detail in neutrophils. As noted above cross-desensitization studies indicated that this receptor is selective for 5-oxo-ETE over other structurally distinct lipid mediators. Furthermore, minor structural changes resulting from metabolism or other structural modifications of 5-oxo-ETE dramatically reduce potency. Both reduction of the keto group (5S-HETE) [8] and addition of a hydroxyl group at C_{20} (5-oxo-20-HETE) [34] reduce potency by about 100-fold. Methylation of the carboxyl group of 5-oxo-ETE reduces potency by about 20-fold, indicating the importance of a free carboxyl group for biological activity [34]. The Δ^6-trans double bond is critical for biological activity, as its reduction (5-oxo-8Z,11Z,14Z-ETrE) dramatically reduces potency by about

1000-fold [36]. The Δ^8 double bond is also very important, as 5-oxo-6E-eicosenoic acid has negligible activity, whereas the remaining Δ^{11} and Δ^{14} double bonds are much less so [75]. The hydrophobic portion of 5-oxo-ETE (C_6 to C_{20}) is also critical for its biological activity, as activity is dramatically reduced by shortening the length of the molecule to 16 carbons and abolished by shortening it to 14 carbons. It can be concluded that the C_1 to C_{10} portion of 5-oxo-ETE is critical for its biological activity and is very sensitive to even minor changes and that a hydrophobic region at the other end of the molecule is also required.

6.2 Biological Activities of 5-Oxo Fatty Acids Derived from Endogenous PUFA

In addition to AA-derived 5-oxo-ETE, 5-oxo fatty acid metabolites formed from other PUFA also have substantial biological activities. 5-Oxo-PUFA derived from sebaleic acid (5-oxo-ODE) [11] and Mead acid (5-oxo-ETrE) [75] are nearly as potent as 5-oxo-ETE in activating eosinophils and neutrophils, whereas the EPA metabolite 5-oxo-EPE is about 5–10 times less potent [12, 75]. 5-Oxo-ODE could be produced in sebaceous glands where its substrate sebaleic acid is located, and could play a role in dermal inflammation, whereas 5-oxo-ETrE could be important in cases of essential fatty acid deficiency due to the accumulation of its precursor Mead acid. It should be noted that 5-oxo-ETrE would be the major granulocyte chemoattractant produced from Mead acid, since this substrate is not converted to significant amounts of LTB_3 due to inhibition of LTA_4 hydrolase by LTA_3 [76], nor to the eosinophil chemoattractant PGD_2, as Mead acid lacks the Δ^{14} double bond required for prostaglandin formation.

Lipoxygenase-derived metabolites of 5-oxo-ETE are also biologically active. 5-Oxo-15S-HETE formed by the soybean lipoxygenase-catalyzed oxidation of AA, was shown to be a chemoattractant for human eosinophils [55]. Although this compound has been reported to be equipotent with 5-oxo-ETE in activating neutrophils [43] and eosinophils [33], we [8, 17, 34] and others [52] have found it to be less potent. We found that the authentic compound (5-oxo-15S-hydroxy-6E,8Z,11Z,13E-eicosatetraenoic acid), prepared by total chemical synthesis, is about one-tenth as potent as 5-oxo-ETE in stimulating calcium mobilization in human neutrophils [77]. In contrast, 5-oxo-12S-HETE, the 12-LO metabolite of 5-oxo-ETE, has no effect on intracellular calcium levels in neutrophils, but rather acts as an antagonist, blocking 5-oxo-ETE-induced calcium mobilization (see Sect. 7). Thus the addition of a hydroxyl group to 5-oxo-ETE has dramatically different effects depending on its location, with the 12-hydroxy derivative being an antagonist, the 15-hydroxy compound a moderately potent agonist, and the 20-hydroxy compound a very weak agonist.

7 OXE Receptor Antagonists

OXE receptor antagonists would be very useful in determining the pathophysiological role of 5-oxo-ETE, especially considering the unavailability of rodent models to address this question. Such compounds could also prove to be useful therapeutic agents, given the potent proinflammatory effects of 5-oxo-ETE on eosinophils and other cells. The first OXE receptor antagonist to be identified was 5-oxo-12S-HETE (Fig. 4), which is formed by transcellular biosynthesis in coincubations of neutrophils with platelets [13]. This compound completely blocks 5-oxo-ETE-induced calcium mobilization in human neutrophils with an IC_{50} of about 500 nM. However, it is not suitable for development as a selective antagonist because it is relatively unstable and would likely be rapidly metabolized in vivo.

We used a structure-based approach to design a series of indoles with potent OXE receptor antagonist activity. In structure-activity studies (see Sect. 6.1) we initially identified three regions of 5-oxo-ETE that are essential for agonist activity: (1) the initial 5-carbon 5-oxo-valeric acid portion of the molecule (2) the adjacent 6-trans-8-cis diene system, and (3) the terminal hydrophobic region of the molecule [75, 78]. In addition, the chain length needed to be at least 18 carbons. Indole was used as a backbone to fix the conformation of substituents and to mimic the diene portion of 5-oxo-ETE. We found that indoles containing adjacent 5-oxo-valeric acid and hexyl moieties had OXE receptor antagonist activity [79]. Further refinement of the structures of these compounds resulted in the identification of compounds *230* and *264* as potent OXE receptor antagonists [57]. These compounds contain a chiral carbon in the 3-position of the 5-oxo-valeric acid side chain (Fig. 4). Only the S-enantiomers possess potent antagonist activity, having IC_{50} values of between 5 and 10 nM in blocking 5-oxo-ETE-induced calcium mobilization, with the R-enantiomers being 700-fold less potent [57, 80]. Both *230* and *264* are also potent inhibitors of 5-oxo-ETE-induced actin polymerization and chemotaxis in human neutrophils and eosinophils [57].

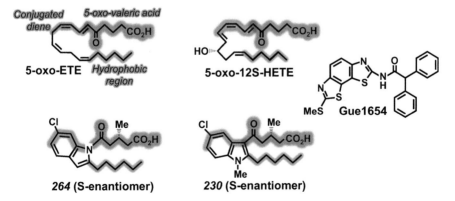

Fig. 4 OXE receptor antagonists. The highlighted regions are important for both agonist and antagonist activity

Another OXE receptor antagonist (Gue1654) has recently been identified using a different approach involving both virtual and biological screening of compound libraries [73] (Fig. 4). This compound blocks Gβγ-mediated activation of neutrophils and eosinophils [73, 74] (see Sect. 5.3) but is considerably less potent than *230* and *264* [57].

8 Potential Roles of 5-Oxo-ETE in Disease

8.1 Asthma and Other Eosinophilic Diseases

The eosinophil appears to be a prime target of 5-oxo-ETE suggesting the involvement of this substance in eosinophilic diseases such as asthma. In addition to its direct effect on eosinophil migration, 5-oxo-ETE induces the production of superoxide and the release of enzymes that can cause cell damage, especially in the presence of cytokines such as GM-CSF. Furthermore, 5-oxo-ETE can promote eosinophil survival by inducing the release of GM-CSF from monocytes (and perhaps macrophages). We found that 5-oxo-ETE does not directly prolong the survival of these cells when added to pure eosinophils, but dramatically increases their survival in the presence of small numbers of monocytes [56]. This effect could be blocked by an antibody to GM-CSF. The potent chemoattractant effects of 5-oxo-ETE on basophils [47, 51] could also contribute to its potential role as a proinflammatory mediator in asthma, as these cells may also play a role in this disease [81].

We have shown that 5-oxo-ETE can elicit the tissue accumulation of eosinophils in humans in vivo [82]. Intradermal injection of this substance induced the infiltration of eosinophils along with smaller numbers of neutrophils into the skin. Eosinophil (but not neutrophil) accumulation was considerably enhanced in asthmatic subjects compared to healthy controls. This effect may be due not only to the direct chemotactic effect of 5-oxo-ETE on eosinophils but also to the release of proteolytic enzymes that promote tissue infiltration. 5-Oxo-ETE has been reported to be a powerful inducer of the migration of eosinophils across endothelial cell monolayers [83] and Matrigel artificial basement membranes [84]. This effect was dependent on increased expression of MMP-9 and the urokinase-type plasminogen activator receptor (uPAR) [70, 84, 85], resulting in the degradation of the matrix, permitting the passage of eosinophils.

Definitive studies implicating 5-oxo-ETE in asthma and allergy have been hampered by the absence of an ortholog of the OXE receptor in rodents. In spite of this we found that intratracheal administration of 5-oxo-ETE to rats induced pulmonary infiltration of eosinophils [86, 87], raising the possibility that in this species it may act by a different receptor. We have recently explored the possibility that the cat may be a useful animal model, since it has an OXE receptor ortholog, the sequence of which is 75 % identical to that of the human receptor [88]. 5-Oxo-

ETE is a very potent activator of actin polymerization in feline eosinophils (EC_{50}, 700 pM). These cells also possess the DP_2 receptor and respond to PGD_2, but it is about 20 times less potent than 5-oxo-ETE. Furthermore, feline leukocytes obtained from blood and after bronchoalveolar lavage synthesize 5-oxo-ETE and 5-oxo-ETE is present in bronchoalveolar lavage fluid in physiologically relevant concentrations (~400 pM). Unfortunately, the OXE receptor antagonists (i.e. *230* and *264*) that we developed are not very potent in blocking the feline OXE receptor, whereas Gue1654 had no detectable effect on 5-oxo-ETE-induced feline leukocyte responses [88].

5-Oxo-ETE is obviously not the only chemoattractant that could play a role in asthma. However the synergistic interactions that occur between 5-oxo-ETE and other eosinophil chemoattractants and cytokines such as GM-CSF may be highly relevant and raise the possibility that in the absence of 5-oxo-ETE, eosinophilic infiltration could be markedly reduced. Furthermore, the contributions of 5-oxo-ETE and other chemoattractants may be time and location-dependent. For example, PGD_2, acting through the DP_2 receptor, might be more relevant for the acute response, as it is released following the interaction of allergen with mast cells and rapidly appears in airway fluid after allergen challenge [89]. The synthesis of 5-oxo-ETE could occur later in this process, as it requires high levels of intracellular $NADP^+$, as would occur following oxidative stress and activation of inflammatory cells. Therefore it is possible that 5-oxo-ETE could be involved at a later stage and could potentially play a role in chronic asthma.

8.2 Cancer

Ghosh and Myers reported that AA stimulates the proliferation prostate cancer-derived PC3 and LNCaP cells [90, 91]. This effect appeared to be mediated by its conversion to a 5-LO product as it was blocked by various 5-LO or FLAP inhibitors but not by inhibitors of cyclooxygenase, 12-LO, or cytochrome P450. Furthermore, treatment of these cells with a FLAP inhibitor impeded their in vitro migration through Matrigel as well as colony formation on soft-agar [92] and induced apoptosis [90, 92], suggesting that a 5-LO product may be required for the growth and metastatic potential of these cells. These results are consistent with other studies showing that blockade of the 5-LO pathway induces apoptosis in tumor cells derived from a variety of other tissues [93–95] as well as with in vivo studies using animal models [96]. Because the concentrations of 5-LO/FLAP inhibitors used in the above studies were much higher than those required for inhibition of 5-LO product formation in inflammatory cells, there has been some concern that their anti-tumor properties may be due, at least in part, to off-target effects [97, 98]. However, similar antiproliferative effects have been observed after down-regulation of 5-LO in prostate cancer cells using siRNA [99] and in malignant pleural mesothelial cells using an antisense oligonucleotide [100].

There is evidence that 5-oxo-ETE is the 5-LO product that is responsible for the proliferative effect of this pathway on tumor cells. 5-Oxo-ETE and its precursor 5S-HETE were the only eicosanoids that could mimic the proliferative effect of AA on prostate cancer cells and reverse the proapoptotic effect of 5-LO/FLAP inhibitors [90]. Furthermore 5-oxo-ETE inhibited selenium-induced apoptosis in prostate tumor cells and stimulated the proliferation of a variety of tumor cells from other sources [64]. The OXE receptor is expressed in tumor cells derived from the prostate [65] as well as from various other tissues [64]. Moreover, the viability of PC3 prostate tumor cells was reduced following knockdown of this receptor using siRNA [65], further supporting a role for endogenous 5-oxo-ETE as an important regulator of tumor cell proliferation. Prostate cancer cells have been reported to synthesize both the 5-oxo-ETE precursor 5-HETE [90, 101, 102] as well as 5-oxo-ETE itself [99]. Furthermore, human malignant pleural mesothelial cells but not normal mesothelial cells were found to express 5-LO and to produce 5S-HETE [100]. We detected high 5-HEDH activity in a variety of tumor cell lines and found that PC3 prostate tumor cells can synthesize 5-oxo-ETE by transcellular biosynthesis from neutrophil-derived 5-HETE in coincubation experiments, especially when the cells are subjected to oxidative stress [24]. Thus inflammatory cells, which are known to be present within tumors, could also contribute to the formation of 5-oxo-ETE. In addition to promoting tumor cell proliferation, 5-oxo-ETE could contribute to the infiltration of inflammatory cells into the tumor, and could possibly account for the eosinophil chemoattractant activity that has been reported to be released from dying cells within tumors [103]. This would be consistent with the stimulatory effect of cell death on 5-oxo-ETE synthesis as discussed is Sect. 2.5, and raises the possibility that the effectiveness of cytotoxic agents might be compromised due to cell death-induced synthesis of 5-oxo-ETE.

The proliferative effect of 5-oxo-ETE on prostate cancer cells appears to be initiated by the activation of PLC-β, resulting in the release of DAG and activation of PKC-ε [71, 99] and c-Myc. Induction of prostate tumor cell apoptosis by the FLAP antagonist MK591 was associated with a decline in c-Myc expression, which was prevented by pretreatment with either with 5-oxo-ETE or an activator of PKC-ε, suggesting that the latter is required for the induction of c-Myc [92, 104]. A somewhat different mechanism appears to be responsible for the proliferative effect of the 5-oxo-ETE precursor 5-HETE on malignant pleural mesothelial cells, which appears to be mediated by induction of VEGF expression [100].

8.3 *Potential Involvement of 5-Oxo-ETE in Other Diseases*

Because of its stimulatory effects on neutrophils and monocytes 5-oxo-ETE could potentially play roles in other diseases in which these cells are involved. Although it is not as potent a chemoattractant for these cells as LTB_4, which is usually synthesized at the same time as 5-oxo-ETE, it is possible that it could play a role under certain circumstances. For example, neutrophils could become desensitized due to excessive exposure to LTB_4 [105], and in this situation 5-oxo-ETE might contribute to tissue infiltration of these cells. Furthermore, the synthesis of 5-oxo-ETE is regulated differently from other 5-LO products in that it is selectively enhanced under conditions of oxidative stress and cell death. Furthermore, the wide expression of 5-HEDH in both structural (e.g. both endothelial and epithelial) and inflammatory cells may be important for its transcellular biosynthesis at inflammatory sites. Thus 5-oxo-ETE could potentially be synthesized at necrotic sites following myocardial infarction and play a role in inducing neutrophil infiltration in these circumstances. It is also possible that 5-oxo-ETE could play a role in atherosclerosis, where it could be synthesized by infiltrating monocytes and macrophages subjected to oxidative stress.

It is also possible that 5-oxo-ETE could play a role in inflammation associated with pulmonary hypertension, as elevated levels were found in lung tissue from human subjects with severe pulmonary hypertension compared to control subjects [106]. In contrast, lower levels of LTB_4 were detected, which were not elevated in subjects with this condition. Interestingly, the levels of 5-oxo-ETE in lung tissue from hypertensive subjects who had been treated by intravenous infusion of prostacyclin were much lower than in untreated patients.

9 Conclusions

5-Oxo-ETE is a potent proinflammatory mediator produced by the 5-LO pathway. Its primary target appears to be the eosinophil and, apart from its less potent metabolites 5-oxo-15S-HETE and Fog_7, it is the only 5-LO product to have appreciable chemotactic activity for human eosinophils. 5-Oxo-ETE is formed by oxidation of 5S-HETE by the highly selective dehydrogenase 5-HEDH, which is widely distributed in both inflammatory and structural cells. Its synthesis requires conditions that promote the oxidation of intracellular NADPH to the obligatory 5-HEDH cofactor $NADP^+$, such as oxidative stress, the respiratory burst, and cell death. It can be formed both directly by inflammatory cells as well as by transcellular biosynthesis from inflammatory cell-derived 5S-HETE. The actions of 5-oxo-ETE are mediated by the G_i-coupled OXE receptor that is found in many species but not in rodents. This receptor is highly expressed on eosinophils and basophils, as well as neutrophils, monocytes/macrophages, and various cancer cell lines. Because of its potent actions on eosinophils 5-oxo-ETE may be an important proinflammatory mediator in asthma and other eosinophilic diseases. There is also evidence that 5-oxo-ETE may be involved in cancer, as it promotes the survival and proliferation of a number of cancer cell lines and rescues these cells from apoptosis induced by inhibitors of the 5-LO pathway. OXE receptor antagonists, which are currently under development, may be useful therapeutic agents in asthma and other diseases in which eosinophils play a prominent role and possibly also in cancer therapy.

Acknowledgements Work done in the authors' laboratories was supported by grants from the Canadian Institutes of Health Research (WSP, MOP-6254 and PPP-99490), the Quebec Heart and Stroke Foundation (WSP), the American Asthma Foundation (JR, 12–0049), and the National Heart, Lung, and Blood Institute (JR, R01HL081873). The Meakins-Christie Laboratories-MUHC-RI are supported in part by a Center grant from Le Fond de la Recherche en Santé du Québec as well as by the JT Costello Memorial Research Fund. JR also wishes to acknowledge the National Science Foundation for the AMX-360 (grant no. CHE-90-13145) and Bruker 400 MHz (grant no. CHE-03-42251) NMR instruments. The content is solely the responsibility of the authors and does not necessarily represent the official views of the National Heart, Lung, and Blood Institute or the National Institutes of Health.

References

1. Borgeat P, Hamberg M, Samuelsson B (1976) Transformation of arachidonic acid and homo-gamma-linolenic acid by rabbit polymorphonuclear leukocytes. Monohydroxy acids from novel lipoxygenases. J Biol Chem 251:7816–7820
2. Borgeat P, Samuelsson B (1979) Arachidonic acid metabolism in polymorphonuclear leukocytes: unstable intermediate in formation of dihydroxy acids. Proc Natl Acad Sci USA 76:3213–3217
3. Murphy RC, Hammarström S, Samuelsson B (1979) Leukotriene C: a slow-reacting substance from murine mastocytoma cells. Proc Natl Acad Sci USA 76:4275–4279
4. Serhan CN, Hamberg M, Samuelsson B (1984) Lipoxins: novel series of biologically active compounds formed from arachidonic acid in human leukocytes. Proc Natl Acad Sci USA 81 (17):5335–5339
5. Powell WS, Gravelle F (1988) Metabolism of 6-trans isomers of leukotriene B4 to dihydro products by human polymorphonuclear leukocytes. J Biol Chem 263:2170–2177
6. Powell WS, Gravelle F, Gravel S (1992) Metabolism of 5(S)-hydroxy-6,8,11,14-eicosatetraenoic acid and other 5(S)-hydroxyeicosanoids by a specific dehydrogenase in human polymorphonuclear leukocytes. J Biol Chem 267:19233–19241
7. O'Flaherty JT, Jacobson D, Redman J (1988) Mechanism involved in the mobilization of neutrophil calcium by 5-hydroxyeicosatetraenoate. J Immunol 140:4323–4328
8. Powell WS, Gravel S, MacLeod RJ, Mills E, Hashefi M (1993) Stimulation of human neutrophils by 5-oxo-6,8,11,14- eicosatetraenoic acid by a mechanism independent of the leukotriene B_4 receptor. J Biol Chem 268:9280–9286
9. O'Flaherty JT, Cordes J, Redman J, Thomas MJ (1993) 5-Oxo-eicosatetraenoate, a potent human neutrophil stimulus. Biochem Biophys Res Commun 192:129–134
10. Patel P, Cossette C, Anumolu JR, Erlemann KR, Grant GE, Rokach J, Powell WS (2009) Substrate selectivity of 5-hydroxyeicosanoid dehydrogenase and its inhibition by 5-hydroxy-delta(6)-long-chain fatty acids. J Pharmacol Exp Ther 329(1):335–341
11. Cossette C, Patel P, Anumolu JR, Sivendran S, Lee GJ, Gravel S, Graham FD, Lesimple A, Mamer OA, Rokach J, Powell WS (2008) Human neutrophils convert the sebum-derived polyunsaturated fatty acid sebaleic acid to a potent granulocyte chemoattractant. J Biol Chem 283(17):11234–11243
12. Powell WS, Gravel S, Gravelle F (1995) Formation of a 5-oxo metabolite of 5,8,11,14,17-eicosapentaenoic acid and its effects on human neutrophils and eosinophils. J Lipid Res 36 (12):2590–2598

13. Powell WS, Gravel S, Khanapure SP, Rokach J (1999) Biological inactivation of 5-oxo-6,8,11,14-eicosatetraenoic acid by human platelets. Blood 93(3):1086–1096
14. Erlemann KR, Cossette C, Grant GE, Lee GJ, Patel P, Rokach J, Powell WS (2007) Regulation of 5-hydroxyeicosanoid dehydrogenase activity in monocytic cells. Biochem J 403(1):157–165
15. Graham FD, Erlemann KR, Gravel S, Rokach J, Powell WS (2009) Oxidative stress-induced changes in pyridine nucleotides and chemoattractant 5-lipoxygenase products in aging neutrophils. Free Radic Biol Med 47(1):62–71
16. Yamazaki M, Sasaki M (1975) Formation of prostaglandin E1 from 15-ketoprostaglandin E1 by guniea pig lung 15-hydroxyprostaglandin dehydrogenase. Biochem Biophys Res Commun 66(1):255–261
17. Powell WS, Chung D, Gravel S (1995) 5-Oxo-6,8,11,14-eicosatetraenoic acid is a potent stimulator of human eosinophil migration. J Immunol 154:4123–4132
18. Zhang Y, Styhler A, Powell WS (1996) Synthesis of 5-oxo-6,8,11,14-eicosatetraenoic acid by human monocytes and lymphocytes. J Leukoc Biol 59(6):847–854
19. Zimpfer U, Dichmann S, Termeer CC, Simon JC, Schroder JM, Norgauer J (2000) Human dendritic cells are a physiological source of the chemotactic arachidonic acid metabolite 5-oxo-eicosatetraenoic acid. Inflamm Res 49(11):633–638
20. Grant GE, Gravel S, Guay J, Patel P, Mazer BD, Rokach J, Powell WS (2011) 5-oxo-ETE is a major oxidative stress-induced arachidonate metabolite in B lymphocytes. Free Radic Biol Med 50(10):1297–1304. doi:10.1016/j.freeradbiomed.2011.02.010, S0891-5849(11)00100-6 [pii]
21. Erlemann KR, Cossette C, Gravel S, Lesimple A, Lee GJ, Saha G, Rokach J, Powell WS (2007) Airway epithelial cells synthesize the lipid mediator 5-oxo-ETE in response to oxidative stress. Free Radic Biol Med 42(5):654–664
22. Erlemann KR, Cossette C, Gravel S, Stamatiou PB, Lee GJ, Rokach J, Powell WS (2006) Metabolism of 5-hydroxy-6,8,11,14-eicosatetraenoic acid by human endothelial cells. Biochem Biophys Res Commun 350(1):151–156
23. Cossette C, Walsh SE, Kim S, Lee GJ, Lawson JA, Bellone S, Rokach J, Powell WS (2007) Agonist and antagonist effects of 15R-prostaglandin (PG) D2 and 11-methylene-PGD2 on human eosinophils and basophils. J Pharmacol Exp Ther 320(1):173–179
24. Grant GE, Rubino S, Gravel S, Wang X, Patel P, Rokach J, Powell WS (2011) Enhanced formation of 5-oxo-6,8,11,14-eicosatetraenoic acid by cancer cells in response to oxidative stress, docosahexaenoic acid and neutrophil-derived 5-hydroxy-6,8,11,14-eicosatetraenoic acid. Carcinogenesis 32(6):822–828. doi:10.1093/carcin/bgr044, bgr044 [pii]
25. Folco G, Murphy RC (2006) Eicosanoid transcellular biosynthesis: from cell-cell interactions to in vivo tissue responses. Pharmacol Rev 58(3):375–388
26. Schafer FQ, Buettner GR (2001) Redox environment of the cell as viewed through the redox state of the glutathione disulfide/glutathione couple. Free Radic Biol Med 30(11):1191–1212
27. Hamberg M (1975) Decomposition of unsaturated fatty acid hydroperoxides by hemoglobin: structures of major products of 13L-hydroperoxy-9,11-octadecadienoic acid. Lipids 10 (2):87–92
28. Hall LM, Murphy RC (1998) Electrospray mass spectrometric analysis of 5-hydroperoxy and 5-hydroxyeicosatetraenoic acids generated by lipid peroxidation of red blood cell ghost phospholipids. J Am Soc Mass Spectrom 9(5):527–532
29. Khaselev N, Murphy RC (2000) Peroxidation of arachidonate containing plasmenyl glycerophosphocholine: facile oxidation of esterified arachidonate at carbon-5. Free Radic Biol Med 29(7):620–632. doi:10.1016/S0891-5849(00)00361-0 [pii]
30. Zarini S, Murphy RC (2003) Biosynthesis of 5-oxo-6,8,11,14-eicosatetraenoic acid from 5-hydroperoxyeicosatetraenoic acid in the murine macrophage. J Biol Chem 278:11190–11196
31. Bryant RW, She HS, Ng KJ, Siegel MI (1986) Modulation of the 5-lipoxygenase activity of MC-9 mast cells: activation by hydroperoxides. Prostaglandins 32(4):615–627

32. Bui P, Imaizumi S, Beedanagari SR, Reddy ST, Hankinson O (2011) Human CYP2S1 metabolizes cyclooxygenase- and lipoxygenase-derived eicosanoids. Drug Metab Dispos 39(2):180–190. doi:10.1124/dmd.110.035121, dmd.110.035121 [pii]
33. Schwenk U, Schröder JM (1995) 5-Oxo-eicosanoids are potent eosinophil chemotactic factors—functional characterization and structural requirements. J Biol Chem 270:15029–15036
34. Powell WS, MacLeod RJ, Gravel S, Gravelle F, Bhakar A (1996) Metabolism and biologic effects of 5-oxoeicosanoids on human neutrophils. J Immunol 156(1):336–342
35. Hevko JM, Bowers RC, Murphy RC (2001) Synthesis of 5-oxo-6,8,11,14-eicosatetraenoic acid and identification of novel omega-oxidized metabolites in the mouse macrophage. J Pharmacol Exp Ther 296(2):293–305
36. Berhane K, Ray AA, Khanapure SP, Rokach J, Powell WS (1998) Calcium/calmodulin-dependent conversion of 5-oxoeicosanoids to 6, 7- dihydro metabolites by a cytosolic olefin reductase in human neutrophils. J Biol Chem 273(33):20951–20959
37. Hevko JM, Murphy RC (2002) Formation of murine macrophage-derived 5-oxo-7-glutathionyl-8,11,14- eicosatrienoic acid (FOG7) is catalyzed by leukotriene C4 synthase. J Biol Chem 277(9):7037–7043
38. Stenson WF, Parker CW (1979) Metabolism of arachidonic acid in ionophore-stimulated neutrophils. Esterification of a hydroxylated metabolite into phospholipids. J Clin Invest 64:1457–1465
39. O'Flaherty JT, Taylor JS, Thomas MJ (1998) Receptors for the 5-oxo class of eicosanoids in neutrophils. J Biol Chem 273(49):32535–32541
40. O'Flaherty JT, Kuroki M, Nixon AB, Wijkander J, Yee E, Lee SL, Smitherman PK, Wykle RL, Daniel LW (1996) 5-Oxo-eicosatetraenoate is a broadly active, eosinophil- selective stimulus for human granulocytes. J Immunol 157(1):336–342
41. Czech W, Barbisch M, Tenscher K, Schopf E, Schröder JM, Norgauer J (1997) Chemotactic 5-oxo-eicosatetraenoic acids induce oxygen radical production, Ca^{2+}-mobilization, and actin reorganization in human eosinophils via a pertussis toxin-sensitive G-protein. J Invest Dermatol 108(1):108–112
42. Jones CE, Holden S, Tenaillon L, Bhatia U, Seuwen K, Tranter P, Turner J, Kettle R, Bouhelal R, Charlton S, Nirmala NR, Jarai G, Finan P (2003) Expression and characterization of a 5-oxo-6E,8Z,11Z,14Z-eicosatetraenoic acid receptor highly expressed on human eosinophils and neutrophils. Mol Pharmacol 63(3):471–477
43. Norgauer J, Barbisch M, Czech W, Pareigis J, Schwenk U, Schröder JM (1996) Chemotactic 5-oxo-icosatetraenoic acids activate a unique pattern of neutrophil responses - analysis of phospholipid metabolism, intracellular Ca^{2+} transients, actin reorganization, superoxide-anion production and receptor up-regulation. Eur J Biochem 236(3):1003–1009
44. O'Flaherty JT, Kuroki M, Nixon AB, Wijkander J, Yee E, Lee SL, Smitherman PK, Wykle RL, Daniel LW (1996) 5-oxo-eicosanoids and hematopoietic cytokines cooperate in stimulating neutrophil function and the mitogen-activated protein kinase pathway. J Biol Chem 271(30):17821–17828
45. Powell WS, Gravel S, Halwani F, Hii CS, Huang ZH, Tan AM, Ferrante A (1997) Effects of 5-oxo-6,8,11,14-eicosatetraenoic acid on expression of CD11b, actin polymerization and adherence in human neutrophils. J Immunol 159:2952–2959
46. Powell WS, Gravel S, Halwani F (1999) 5-oxo-6,8,11,14-eicosatetraenoic acid is a potent stimulator of L- selectin shedding, surface expression of CD11b, actin polymerization, and calcium mobilization in human eosinophils. Am J Respir Cell Mol Biol 20(1):163–170
47. Sturm GJ, Schuligoi R, Sturm EM, Royer JF, Lang-Loidolt D, Stammberger H, Amann R, Peskar BA, Heinemann A (2005) 5-Oxo-6,8,11,14-eicosatetraenoic acid is a potent chemoattractant for human basophils. J Allergy Clin Immunol 116(5):1014–1019
48. Urasaki T, Takasaki J, Nagasawa T, Ninomiya H (2001) Pivotal role of 5-lipoxygenase in the activation of human eosinophils: platelet-activating factor and interleukin-5 induce CD69 on eosinophils through the 5-lipoxygenase pathway. J Leukoc Biol 69(1):105–112

49. Monneret G, Boumiza R, Gravel S, Cossette C, Bienvenu J, Rokach J, Powell WS (2005) Effects of prostaglandin D_2 and 5-lipoxygenase products on the expression of CD203c and CD11b by basophils. J Pharmacol Exp Ther 312(2):627–634
50. Sozzani S, Zhou D, Locati M, Bernasconi S, Luini W, Mantovani A, O'Flaherty JT (1996) Stimulating properties of 5-oxo-eicosanoids for human monocytes: synergism with monocyte chemotactic protein-1 and -3. J Immunol 157(10):4664–4671
51. Iikura M, Suzukawa M, Yamaguchi M, Sekiya T, Komiya A, Yoshimura-Uchiyama C, Nagase H, Matsushima K, Yamamoto K, Hirai K (2005) 5-Lipoxygenase products regulate basophil functions: 5-oxo-ETE elicits migration, and leukotriene B(4) induces degranulation. J Allergy Clin Immunol 116(3):578–585
52. O'Flaherty JT, Cordes JF, Lee SL, Samuel M, Thomas MJ (1994) Chemical and biological characterization of oxo-eicosatetraenoic acids. Biochim Biophys Acta 1201(3):505–515
53. Powell WS, Ahmed S, Gravel S, Rokach J (2001) Eotaxin and RANTES enhance 5-oxo-6,8,11,14-eicosatetraenoic acid- induced eosinophil chemotaxis. J Allergy Clin Immunol 107(2):272–278
54. Sun FF, Crittenden NJ, Czuk CI, Taylor BM, Stout BK, Johnson HG (1991) Biochemical and functional differences between eosinophils from animal species and man. J Leukoc Biol 50:140–150
55. Schwenk U, Morita E, Engel R, Schröder JM (1992) Identification of 5-oxo-15-hydroxy-6,8,11,13-eicosatetraenoic acid as a novel and potent human eosinophil chemotactic eicosanoid. J Biol Chem 267:12482–12488
56. Stamatiou PB, Chan CC, Monneret G, Ethier D, Rokach J, Powell WS (2004) 5-Oxo-6,8,11,14-eicosatetraenoic acid stimulates the release of the eosinophil survival factor granulocyte-macrophage colony stimulating factor from monocytes. J Biol Chem 279:28159–28164
57. Gore V, Gravel S, Cossette C, Patel P, Chourey S, Ye Q, Rokach J, Powell WS (2014) Inhibition of 5-oxo-6,8,11,14-eicosatetraenoic acid-induced activation of neutrophils and eosinophils by novel indole OXE receptor antagonists. J Med Chem 57(2):364–377. doi:10.1021/jm401292m
58. Hosoi T, Koguchi Y, Sugikawa E, Chikada A, Ogawa K, Tsuda N, Suto N, Tsunoda S, Taniguchi T, Ohnuki T (2002) Identification of a novel eicosanoid receptor coupled to $G_{i/o}$. J Biol Chem 277:31459–31465
59. Takeda S, Yamamoto A, Haga T (2002) Identification of a G protein-coupled receptor for 5-oxo-eicosatetraenoic acid. Biomed Res Tokyo 23(2):101–108
60. Jones OT, Jones SA, Hancock JT, Topley N (1993) Composition and organization of the NADPH oxidase of phagocytes and other cells. Biochem Soc Trans 21:343–346
61. Bäck M, Powell WS, Dahlén SE, Drazen JM, Evans JF, Serhan CN, Shimizu T, Yokomizo T, Rovati GE (2014) International Union of Basic and Clinical Pharmacology. Update on leukotriene, lipoxin and oxoeicosanoid receptors: IUPHAR review 7. Br J Pharmacol. doi:10.1111/bph.12665
62. Offermanns S, Colletti SL, Lovenberg TW, Semple G, Wise A, IJzerman AP (2011) International Union of Basic and Clinical Pharmacology. LXXXII: Nomenclature and classification of hydroxy-carboxylic acid receptors (GPR81, GPR109A, and GPR109B). Pharmacol Rev 63(2):269–290. doi:10.1124/pr.110.003301, pr.110.003301 [pii]
63. Guo Y, Zhang W, Giroux C, Cai Y, Ekambaram P, Dilly AK, Hsu A, Zhou S, Maddipati KR, Liu J, Joshi S, Tucker SC, Lee MJ, Honn KV (2011) Identification of the orphan G protein-coupled receptor GPR31 as a receptor for 12-(S)-hydroxyeicosatetraenoic acid. J Biol Chem 286(39):33832–33840. doi:10.1074/jbc.M110.216564, M110.216564 [pii]
64. O'Flaherty JT, Rogers LC, Paumi CM, Hantgan RR, Thomas LR, Clay CE, High K, Chen YQ, Willingham MC, Smitherman PK, Kute TE, Rao A, Cramer SD, Morrow CS (2005) 5-Oxo-ETE analogs and the proliferation of cancer cells. Biochim Biophys Acta 1736(3):228–236

65. Sundaram S, Ghosh J (2006) Expression of 5-oxoETE receptor in prostate cancer cells: critical role in survival. Biochem Biophys Res Commun 339(1):93–98
66. Cooke M, Di CH, Maloberti P, Cornejo MF (2013) Expression and function of OXE receptor, an eicosanoid receptor, in steroidogenic cells. Mol Cell Endocrinol 371(1–2):71–78. doi:10.1016/j.mce.2012.11.003, S0303-7207(12)00495-9 [pii]
67. Enyedi B, Kala S, Nikolich-Zugich T, Niethammer P (2013) Tissue damage detection by osmotic surveillance. Nat Cell Biol 15(9):1123–1130. doi:10.1038/ncb2818, ncb2818 [pii]
68. O'Flaherty JT, Taylor JS, Kuroki M (2000) The coupling of 5-oxo-eicosanoid receptors to heterotrimeric G proteins. J Immunol 164(6):3345–3352
69. Hosoi T, Sugikawa E, Chikada A, Koguchi Y, Ohnuki T (2005) TG1019/OXE, a galpha(i/o)-protein-coupled receptor, mediates 5-oxo-eicosatetraenoic acid-induced chemotaxis. Biochem Biophys Res Commun 334(4):987–995
70. Langlois A, Chouinard F, Flamand N, Ferland C, Rola-Pleszczynski M, Laviolette M (2009) Crucial implication of protein kinase C (PKC)-delta, PKC-zeta, ERK-1/2, and p38 MAPK in migration of human asthmatic eosinophils. J Leukoc Biol 85(4):656–663
71. Sarveswaran S, Ghosh J (2013) OXER1, a G protein-coupled oxoeicosatetraenoid receptor, mediates the survival-promoting effects of arachidonate 5-lipoxygenase in prostate cancer cells. Cancer Lett 336(1):185–195. doi:10.1016/j.canlet.2013.04.027, S0304-3835(13)00361-3 [pii]
72. Schratl P, Sturm EM, Royer JF, Sturm GJ, Lippe IT, Peskar BA, Heinemann A (2006) Hierarchy of eosinophil chemoattractants: role of p38 mitogen-activated protein kinase. Eur J Immunol 36(9):2401–2409
73. Blättermann S, Peters I, Ottersbach PA, Bock A, Konya V, Weaver CD, Gonzalez A, Schroder R, Tyagi R, Luschnig P, Gab J, Hennen S, Ulven T, Pardo L, Mohr K, Gutschow M, Heinemann A, Kostenis E (2012) A biased ligand for OXE-R uncouples Gα and Gβγ signaling within a heterotrimer. Nat Chem Biol 8(7):631–638. doi:10.1038/nchembio.962, nchembio.962 [pii]
74. Konya V, Blattermann S, Jandl K, Platzer W, Ottersbach PA, Marsche G, Gutschow M, Kostenis E, Heinemann A (2014) A biased non-galphai OXE-R antagonist demonstrates that galphai protein subunit is not directly involved in neutrophil, eosinophil, and monocyte activation by 5-Oxo-ETE. J Immunol 192(10):4774–4782. doi:10.4049/jimmunol.1302013, jimmunol.1302013 [pii]
75. Patel P, Cossette C, Anumolu JR, Gravel S, Lesimple A, Mamer OA, Rokach J, Powell WS (2008) Structural requirements for activation of the 5-oxo-6E,8Z, 11Z,14Z-eicosatetraenoic acid (5-oxo-ETE) receptor: identification of a mead acid metabolite with potent agonist activity. J Pharmacol Exp Ther 325(2):698–707
76. Evans JF, Nathaniel DJ, Zamboni RJ, Ford-Hutchinson AW (1985) Leukotriene A_3. A poor substrate but a potent inhibitor of rat and human neutrophil leukotriene A_4 hydrolase. J Biol Chem 260(20):10966–10970
77. Patel P, Anumolu JR, Powell WS, Rokach J (2011) 5-Oxo-15-HETE: total synthesis and bioactivity. Bioorg Med Chem Lett 21(6):1857–1860. doi:10.1016/j.bmcl.2011.01.032, S0960-894X(11)00045-X [pii]
78. Powell WS, Rokach J (2013) The eosinophil chemoattractant 5-oxo-ETE and the OXE receptor. Prog Lipid Res 52(4):651–665. doi:10.1016/j.plipres.2013.09.001, S0163-7827(13)00053-2 [pii]
79. Gore V, Patel P, Chang CT, Sivendran S, Kang N, Ouedraogo YP, Gravel S, Powell WS, Rokach J (2013) 5-Oxo-ETE receptor antagonists. J Med Chem 56(9):3725–3732. doi:10.1021/jm400480j
80. Patel P, Reddy CN, Gore V, Chourey S, Ye Q, Ouedraogo YP, Gravel S, Powell WS, Rokach J (2014) Two potent OXE-R antagonists: assignment of stereochemistry. ACS Med Chem Lett 5(7):815–819
81. Schroeder JT (2009) Basophils beyond effector cells of allergic inflammation. Adv Immunol 101:123–161. doi:10.1016/S0065-2776(08)01004-3, S0065-2776(08)01004-3 [pii]

82. Muro S, Hamid Q, Olivenstein R, Taha R, Rokach J, Powell WS (2003) 5-oxo-6,8,11,14-eicosatetraenoic acid induces the infiltration of granulocytes into human skin. J Allergy Clin Immunol 112(4):768–774
83. Dallaire MJ, Ferland C, Page N, Lavigne S, Davoine F, Laviolette M (2003) Endothelial cells modulate eosinophil surface markers and mediator release. Eur Respir J 21(6):918–924
84. Guilbert M, Ferland C, Bosse M, Flamand N, Lavigne S, Laviolette M (1999) 5-Oxo-6,8,11,14-eicosatetraenoic acid induces important eosinophil transmigration through basement membrane components: comparison of normal and asthmatic eosinophils. Am J Respir Cell Mol Biol 21(1):97–104
85. Langlois A, Ferland C, Tremblay GM, Laviolette M (2006) Montelukast regulates eosinophil protease activity through a leukotriene-independent mechanism. J Allergy Clin Immunol 118 (1):113–119
86. Almishri W, Cossette C, Rokach J, Martin JG, Hamid Q, Powell WS (2005) Effects of prostaglandin D_2, 15-deoxy-$\Delta^{12,14}$-prostaglandin J_2, and selective DP_1 and DP_2 receptor agonists on pulmonary infiltration of eosinophils in Brown Norway rats. J Pharmacol Exp Ther 313(1):64–69
87. Stamatiou P, Hamid Q, Taha R, Yu W, Issekutz TB, Rokach J, Khanapure SP, Powell WS (1998) 5-Oxo-ETE induces pulmonary eosinophilia in an integrin-dependent manner in Brown Norway rats. J Clin Invest 102(12):2165–2172
88. Cossette C, Gravel S, Reddy CN, Gore V, Chourey S, Ye Q, Snyder NW, Mesaros CA, Blair IA, Lavoie JP, Reinero CR, Rokach J, Powell WS (2015) Biosynthesis and actions of 5-oxoeicosatetraenoic acid (5-oxo-ETE) on feline granulocytes. Biochem Pharmacol. doi:10.1016/j.bcp.2015.05.009
89. Murray JJ, Tonnel AB, Brash AR, Roberts LJ, Gosset P, Workman R, Capron A, Oates JA (1986) Release of prostaglandin D_2 into human airways during acute antigen challenge. N Engl J Med 315(13):800–804
90. Ghosh J, Myers CE (1998) Inhibition of arachidonate 5-lipoxygenase triggers massive apoptosis in human prostate cancer cells. Proc Natl Acad Sci USA 95(22):13182–13187
91. Ghosh J, Myers CE (1997) Arachidonic acid stimulates prostate cancer cell growth: critical role of 5-lipoxygenase. Biochem Biophys Res Commun 235(2):418–423
92. Sarveswaran S, Ghosh R, Morisetty S, Ghosh J (2015) MK591, a second generation leukotriene biosynthesis inhibitor, prevents invasion and induces apoptosis in the bone-invading C4-2B human prostate cancer cells: implications for the treatment of castration-resistant, bone-metastatic prostate cancer. PLoS One 10(4), e0122805. doi:10.1371/journal.pone. 0122805
93. Avis I, Hong SH, Martinez A, Moody T, Choi YH, Trepel J, Das R, Jett M, Mulshine JL (2001) Five-lipoxygenase inhibitors can mediate apoptosis in human breast cancer cell lines through complex eicosanoid interactions. FASEB J 15(11):2007–2009
94. Edderkaoui M, Hong P, Vaquero EC, Lee JK, Fischer L, Friess H, Buchler MW, Lerch MM, Pandol SJ, Gukovskaya AS (2005) Extracellular matrix stimulates reactive oxygen species production and increases pancreatic cancer cell survival through 5-lipoxygenase and NADPH oxidase. Am J Physiol Gastrointest Liver Physiol 289(6):G1137–G1147
95. Hammamieh R, Sumaida D, Zhang X, Das R, Jett M (2007) Control of the growth of human breast cancer cells in culture by manipulation of arachidonate metabolism. BMC Cancer 7:138
96. Moody TW, Leyton J, Martinez A, Hong S, Malkinson A, Mulshine JL (1998) Lipoxygenase inhibitors prevent lung carcinogenesis and inhibit non-small cell lung cancer growth. Exp Lung Res 24(4):617–628
97. Datta K, Biswal SS, Kehrer JP (1999) The 5-lipoxygenase-activating protein (FLAP) inhibitor, MK886, induces apoptosis independently of FLAP. Biochem J 340(Pt 2):371–375
98. Fischer AS, Metzner J, Steinbrink SD, Ulrich S, Angioni C, Geisslinger G, Steinhilber D, Maier TJ (2010) 5-Lipoxygenase inhibitors induce potent anti-proliferative and cytotoxic

effects in human tumour cells independently of suppression of 5-lipoxygenase activity. Br J Pharmacol 161(4):936–949. doi:10.1111/j.1476-5381.2010.00915.x
99. Sarveswaran S, Thamilselvan V, Brodie C, Ghosh J (2011) Inhibition of 5-lipoxygenase triggers apoptosis in prostate cancer cells via down-regulation of protein kinase C-epsilon. Biochim Biophys Acta 1813(12):2108–2117. doi:10.1016/j.bbamcr.2011.07.015, S0167-4889(11)00214-X [pii]
100. Romano M, Catalano A, Nutini M, D'Urbano E, Crescenzi C, Claria J, Libner R, Davi G, Procopio A (2001) 5-lipoxygenase regulates malignant mesothelial cell survival: involvement of vascular endothelial growth factor. FASEB J 15(13):2326–2336. doi:10.1096/fj.01-0150com, 15/13/2326 [pii]
101. Gupta S, Srivastava M, Ahmad N, Sakamoto K, Bostwick DG, Mukhtar H (2001) Lipoxygenase-5 is overexpressed in prostate adenocarcinoma. Cancer 91(4):737–743
102. Hassan S, Carraway RE (2006) Involvement of arachidonic acid metabolism and EGF receptor in neurotensin-induced prostate cancer PC3 cell growth. Regul Pept 133 (1–3):105–114
103. Cormier SA, Taranova AG, Bedient C, Nguyen T, Protheroe C, Pero R, Dimina D, Ochkur SI, O'Neill K, Colbert D, Lombari TR, Constant S, McGarry MP, Lee JJ, Lee NA (2006) Pivotal advance: eosinophil infiltration of solid tumors is an early and persistent inflammatory host response. J Leukoc Biol 79(6):1131–1139
104. Sarveswaran S, Chakraborty D, Chitale D, Sears R, Ghosh J (2015) Inhibition of 5-lipoxygenase selectively triggers disruption of c-Myc signaling in prostate cancer cells. J Biol Chem 290(8):4994–5006. doi:10.1074/jbc.M114.599035
105. Marleau S, Fortin C, Poubelle PE, Borgeat P (1993) In vivo desensitization to leukotriene B_4 (LTB_4) in the rabbit. Inhibition of LTB_4-induced neutropenia during intravenous infusion of LTB_4. J Immunol 150:206–213
106. Bowers R, Cool C, Murphy RC, Tuder RM, Hopken MW, Flores SC, Voelkel NF (2004) Oxidative stress in severe pulmonary hypertension. Am J Respir Crit Care Med 169 (6):764–769
107. Dereeper A, Guignon V, Blanc G, Audic S, Buffet S, Chevenet F, Dufayard JF, Guindon S, Lefort V, Lescot M, Claverie JM, Gascuel O (2008) Phylogeny.fr: robust phylogenetic analysis for the non-specialist. Nucleic Acids Res 36 (Web Server issue):W465–W469. doi: 10.1093/nar/gkn180, gkn180 [pii]
108. Krishnamoorthy S, Recchiuti A, Chiang N, Yacoubian S, Lee CH, Yang R, Petasis NA, Serhan CN (2010) Resolvin D1 binds human phagocytes with evidence for proresolving receptors. Proc Natl Acad Sci USA 107(4):1660–1665. doi:10.1073/pnas.0907342107, 0907342107 [pii]
109. Kanaoka Y, Maekawa A, Austen KF (2013) Identification of GPR99 protein as a potential third cysteinyl leukotriene receptor with a preference for leukotriene E4 ligand. J Biol Chem 288(16):10967–10972. doi:10.1074/jbc.C113.453704, C113.453704 [pii]

Lipoxins, Resolvins, and the Resolution of Inflammation

Antonio Recchiuti, Eleonora Cianci, Felice Simiele, and Mario Romano

Abstract Resolution of acute inflammation is an active process, where endogenous specialized pro-resolving mediators (SPM), derived from polyunsaturated fatty acids via lipoxygenase (LO)-driven biochemical pathways, have pivotal roles in "turning off" pro-inflammatory signals, prompting timely resolution. SPM potent and stereospecific bioactions are of considerable interest in immunopharmacology since unresolved inflammation represents a key pathogenetic mechanism of several widespread diseases.

In recent years, biosynthetic routes, chemical structure, and specific G-protein coupled receptors for many members of the SPM genus have been uncovered. This knowledge, that emphasizes the complex pathobiological role of LO in resolution of inflammation, provides previously unforeseen opportunity and strategies for innovative pharmacology for a variety of human pathologies.

Here, we provide an overview of biosynthetic pathways, receptors, bioactions and underlying mechanisms as well as of preclinical and clinical evidence of efficacy of the best-studied SPM.

List of Abbreviations

AA	Arachidonic acid (5Z, 8Z, 11Z, 14Z -eicosatetraenoic acid)
AKI	Acute kidney injury

A. Recchiuti • F. Simiele • M. Romano (✉)
Department of Medical, Oral and Biotechnological Sciences, G. D'Annunzio University of Chieti-Pescara, Via Luigi Polacchi 11/13, 66013 Chieti, Italy

Center of Excellence on Aging, G. D'Annunzio University of Chieti-Pescara, Via Luigi Polacchi 11/13, 66013 Chieti, Italy
e-mail: mromano@unich.it

E. Cianci
Department of Medical, Oral and Biotechnological Sciences, G. D'Annunzio University of Chieti-Pescara, Via Luigi Polacchi 11/13, 66013 Chieti, Italy

Department of Medicine and Aging Sciences, G. D'Annunzio University of Chieti-Pescara, Via Luigi Polacchi 11/13, 66013, Chieti, Italy

Center of Excellence on Aging, G. D'Annunzio University of Chieti-Pescara, Via Luigi Polacchi 11/13, 66013 Chieti, Italy

ALI	Acute lung injury
ALX/FPR2	Lipoxin A_4 receptor/formyl peptide receptor 2
ATL	Aspirin triggered lipoxin
ATLa	Aspirin triggered lipoxin analog
CCR5	C-C Chemokine receptor type 5
COX	Cyclooxygenase (prostaglandin H_2 synthase)
DHA	Docosahexaenoic acid (4Z, 7Z, 10Z, 13Z, 16Z, 19Z -docosahexaenoic acid)
EPA	Eicosapentaenoic acid (5Z, 8Z, 11Z, 14Z, 17Z -eicosapentaenoic acid)
GPR32/DRV1	G-Protein coupled receptor 32/resolvin D1 receptor 1
HSV	Herpes simplex virus
IFN	Interferon
IL	Interleukin
LO	Lipoxygenase
LM	Lipid mediator
LT	Leukotriene
LX	Lipoxin
MaR	Maresin
MCP-1	Monocyte chemotactic protein 1
NO	Nitric oxide
PD	Protectin
PDGF	Platelet-derived growth factor
PG	Prostaglandin
PUFA	Polyunsaturated fatty
Rv	Resolvin
T2D	Type 2 diabetes
TMJ	Temporomandibular joint
TNF-α	Tumor necrosis factor-α
VSMC	Vascular smooth muscle cells

1 The Resolution of the Inflammatory Response

The perception that acute inflammation is a detrimental process causing pain, tissue swelling, and organ malfunctioning, which must be therefore blocked with anti-inflammatory drugs, is widely diffused in the general population. In reality, acute inflammation is a physiological innate response that occurs in vascularized tissues to protect the host from physical injuries or microbial infections [1]. Thus, acute inflammation exerts many homeostatic functions that are required for maintaining health. For instance, inflammation helps to restore the nutrients balance within the body and expedites the healing process [2]. Hence, whether inflammation is detrimental or favorable to the host does not depend on how often or extensively it begins, but on how effectively and quickly it resolves.

The classical cardinal signs of inflammation, described by the Roman physician Celsus (first century BC), i.e. *rubor* (redness), *tumor* (swelling), *calor* (heat), and *dolor* (pain) are macroscopic manifestation of cellular and molecular processes that take place at the interface between the microcirculation and tissues. These encompass: (a) increase in vascular permeability, plasma leakage, and edema formation; (b) rapid leukocyte recruitment at inflamed loci by stimulation of adhesion to activated vascular endothelial cells and *diapedesi*s (i.e., active transmigration across blood vessels into surrounding tissues). Polymorphonuclear neutrophils (PMN) are the first leukocyte population to reach the damaged tissue. Next, monocytes enter the inflammatory site and differentiate into macrophages (MΦ), which have pivotal roles in initiating the resolution of inflammation and the return to homeostasis [3, 4]. These events are summarized in Fig. 1A.

Thus, the resolution of inflammation is an active response, tightly regulated at the cellular and molecular level. In order to define resolution in an unbiased, quantitative manner, mathematical indices were introduced by Bannenberg et al. by determining the cellular changes in exudates following an acute inflammatory stimulus [5] (Fig. 1B). The introduction of resolution indices allowed to pin point the cellular events occurring during the natural return from acute inflammation to homeostasis; to evaluate the role of endogenous chemical mediators in resolution, as well as to establish the biological actions of chemical entities on this process in preclinical models of disease [7–10]. These indices could be useful in human clinical settings to monitor the evolution of the inflammatory response and to assess the effects of therapeutics.

2 Lipoxygenases and the Biosynthesis of Pro-Resolution Lipid Mediators

Pioneer work by Serhan and coworkers [11–14] demonstrated that resolution of inflammation is orchestrated by specific endogenous lipid mediators, biosynthesized from essential polyunsaturated fatty acids (PUFA) and termed *specialized pro-resolving lipid mediators* (SPM) [6] or *immunoresolvents* [15–17], reviewed in [18]. The SPM genus includes lipoxins (LX), resolvins (Rv), protectins (PD), and maresins (MaR). They derive from lipoxygenase (LO)-driven conversion of PUFA, i.e. arachidonic acid (AA), eicosapentaenoic acid (EPA) and docosahexaenoic acid (DHA) that rapidly appear in inflammatory exudates [19]. In humans, both resident and blood cells carry the enzymatic machinery for the biosynthesis of SPM. Lipidometabolomic studies demonstrated that physiologically human biological fluids/tissues contain detectable concentrations of SPM [20–22], suggesting that they may have homeostatic functions. Basal levels of SPM can increase during physical exercise [23, 24], inflammation [25], *Mycobacterium tuberculosis* infection [26], coronary angioplasty [27], carbon monoxide exposure [28], or surgery [29] confirming their role in human pathophysiology.

This chapter will focus on the biosynthesis, receptors, and bioactions of the main LO-derived SPM.

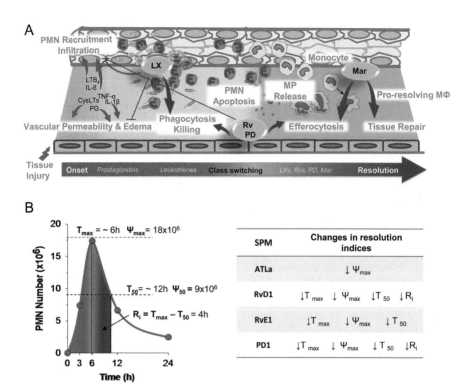

Fig. 1 Cellular and molecular mediators in acute inflammation and resolution. (**A**) Surgical intervention, tissue injuries, or microbial infections in vascularized tissues evoke an acute inflammatory response, characterized by edema, exudate formation, and leukocyte infiltration. Polymorphonuclear leukocytes (PMN) are among the first responders that fight microbes, followed by monocytes that differentiate locally into pro-resolving macrophages (MΦs). Efferocytosis of apoptotic PMN and microbes by pro-resolving MΦs and subsequent egress via lymphatics are hallmarks of tissue resolution. Activated leukocytes release microparticles (MPs) that can promote resolution. The ideal outcome of acute inflammation is resolution, leading to the restoration of homeostasis. Lipid autacoids prostaglandins (PGs) and leukotrienes (LTs) are classical mediators of the onset phase of inflammation, promoting edema, PMN recruitment, and pain. Specialized pro-resolving lipid mediators (SPM) are biosynthesized within resolving exudates by a class switching of the enzymatic activity. SPM, including lipoxins (LX), resolvins (Rv), protectins (PD) and maresins (MaR) drive the resolution of inflammation halting PMN infiltration, stimulating non-phlogistic MΦ efferocytosis, promoting tissue regeneration and controlling pain. (**B**) The resolution indices include: T_{max}, i.e., time point of maximum PMN infiltration (Ψ_{max}); T_{50}, time necessary to achieve 50 % reduction in PMN number (Ψ_{50}) from Ψ_{max}; the resolution interval ($R_i = T_{50} - T_{max}$), time interval between T_{max} and T_{50}. The introduction of resolution indices allowed to pin point the cellular events occurring during the natural return from acute inflammation to homeostasis, and also to evaluate the role of endogenous chemical mediators in resolution, as well as to establish the biological actions of chemical entities on this process in preclinical models of disease. The presence of ATLa lowers the Ψ_{max}; the presence of RvE1, RvD1 and PD1, also initiate the resolution at an earlier time point ($\downarrow T_{max}$ and $\downarrow T_{50}$); RvD1 and PD1 further shorten the resolution interval ($\downarrow R_i$) [5, 6]

2.1 Structure and Biosynthesis of LO-Driven LX

Lipoxins (from *lipoxygenase interaction products*) represent the first class of pro-resolution eicosanoids uncovered in the early 80' by Serhan and coworkers [14]. The leading compounds of this series were termed LXA_4 (5S,6R,15S-trihydroxy-7,9,13-*trans*-11-*cis*-eicosatetraenoic acid) and LXB_4 (5S,14R,15S-trihydroxy-6,10,12-*trans*-8-*cis*- eicosatetraenoic acid) [13].

LO play a pivotal role in LX biosynthesis. The sequential oxygenation of AA by 5- and 15-LO was the first route of LX biosynthesis to be elucidated [30]. In this case, AA can be initially attached by 5-LO to produce leukotriene (LT)A_4, which is converted to a 5S,6S,15S-epoxytetraene intermediate, subsequently transformed into LXA_4 and B_4. Alternatively, AA can be first taken by 15-LO, yielding 15S-hydroperoxy eicosatetreanoic acid (HpETE) and 15S-hydroxy eicosatetreanoic acid (HETE), which are converted by 5-LO into the 5S,6S,15S-epoxytetraene, which gives both LXA_4 and B_4. These pathways, first reported in neutrophils, do occur in a variety of cells, including eosinophils [31] and alveolar MΦs [32], or during cell/cell (tissue) interactions, for instance neutrophils/eosinophils [31] or neutrophils/lung tissue [25].

This route, confirmed in vivo by transfecting the human 15-LO gene into the rat kidney [33], may be prevalent in the airways [34]. An alternative, well-characterized route of LX biosynthesis involves 5- and 12-LO, during platelet/neutrophils interactions [35–37]. The metabolic pathway was conclusively elucidated using human megakaryocytes or 12-LO-transfected cells [38], platelets exposed to leukotriene A_4 (LTA_4) [39] and human platelet recombinant 12-LO incubated with LTA_4 in a cell-free system [40]. The chemistry of this reaction proceeds via 12-LO-catalyzed conversion of LTA_4 into a delocalized cation, which is attacked by water at carbon-6 to give LXA_4, and at carbon-14 to yield LXB_4 [38]. The LX synthase capability of 12-LO may be pathophysiologically relevant as 12-LO affinity for LTA_4 is comparable to that for AA [40]. Moreover, we have recently reported that platelets from cystic fibrosis patients exhibit reduced 12-LO catalytic activity, which impairs 12-HETE as well as LX generation [41]. This may have consequences on endogenous anti-inflammatory, pro-resolution circuits, since LX biosynthesis in vivo via the 5-/12-LO route does occur for instance during coronary angioplasty [27] and in healthy subjects undergoing strenuous physical exercise [23].

Five-LO is also involved in the biosynthesis of LX epimers, the 15-epi-LX also termed aspirin-triggered LX (ATL), since their formation requires the cooperation of aspirinated cyclooxygenase (COX)-2, which produces 15R-HETE. This is converted by 5-LO to an epoxide intermediate, which is in turn enzymatically transformed into 15-epi-LXA_4 (ATLa) and 15-epi-LXB_4. The biosynthetic pathways of LXA_4 (the best studied and most active among LX) and ATLa are illustrated in Fig. 2A. Cellular origins of ATL include mixed incubations of aspirin-treated human umbilical endothelial cells and neutrophils [42], interactions between neutrophils and A459 cells [43] or between hepatocytes and liver cells

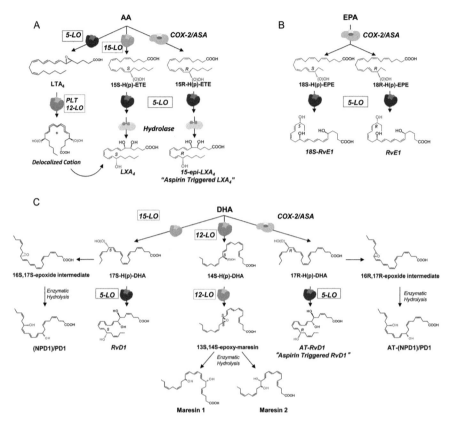

Fig. 2 Lipoxygenases and the biosynthesis of SPM. In self-limited inflammation, SPM are biosynthetized from lipoxygenase (LO)-driven conversion of PUFA, i.e. arachidonic acid (AA), eicosapentaenoic acid (EPA) and docosahexaenoic acid (DHA), which rapidly appear in inflammatory exudates. The main structures of key SPM and the biosynthetic pathways leading to lipoxins (**A**), E-series resolvins (**B**), D-series resolvins, protectins (**C**) and maresins (**D**), are illustrated. The complete stereochemistry of each of these SPM has been established, total organic synthesis achieved, and bioactions confirmed. See text and references for additional details

[44]. Liver from aspirin-treated rats also form ATL [44] as well as ex vivo stimulated human blood [45, 46]. ATLa formation can be observed in healthy volunteers taking aspirin [15, 47].

2.2 Structure and Biosynthesis of LO-Driven Rv

2.2.1 E-Series Resolvins

The important roles of EPA in the immune and cardiovascular systems are well documented [48–51]. Using an unbiased liquid chromatography-mass spectrometry

(LC-MS/MS)-based analysis, Serhan and coworkers found that EPA is enzymatically converted into novel potent lipid mediator (LM) that were named *E-series resolvins* (an acronym of *resolution phase interaction products*) because they: (*a*) were produced during cell-cell interactions in the resolution phase of acute inflammatory responses; (*b*) stopped excessive PMN infiltration in inflamed tissues (*c*) reduced exudate accumulation [5, 12, 52–54].

E-series Rv are endogenously biosynthesized in murine systems during resolution of inflammation and by isolated human cells (e.g., endothelial cells, leukocytes) and whole blood. The complete stereochemistry of the first member of this family, RvE1, has been established as $5S,12R,18R$-trihydroxy-$6Z,8E,10E,14Z,16E$-EPA [55]. In vascular endothelial cells, aspirin acetylated COX-2 converts EPA into $18R$-hydro (peroxy)-eicosapentaenoic acid (HEPE), which is rapidly taken up by activated leukocytes (e.g., PMN) and further metabolized into RvE1 (Fig. 2B). RvE1 is biosynthesized by human recombinant 5-LO and LTA_4 hydrolase from $18S$-HEPE [56] and by cytochrome P450 via oxygenation of EPA [12, 57]. RvE2 ($5S,18$-dihydroxy-EPE) is biosynthesized in resolving exudates and in human whole blood via reduction of $5S$-hydroperoxy,18-hydroxy-EPE, an intermediate in the biosynthetic pathway of RvE1 [56, 58, 59]. In addition, $18R$-RvE3 ($17R,18R$-dihydroxy-$5Z,8Z,11Z,13E,15E$-EPE) epimeric $17R$, $18S$-RvE3, and $14,20$-diHDHA, which are biosynthesized via 12/15-LO by eosinophils and bestow potent anti-inflammatory actions, have recently been identified [60–62].

2.2.2 D-Series Resolvins

Lipidometabolomic analyses of resolving exudates from mice given DHA and aspirin provided the first evidence of the formation of novel endogenous 17-hydroxy-containing mediators [63]. The biosynthetic pathways leading to this novel set of SMP was recapitulated using isolated human cells and recombinant enzymes. In particular, hypoxia-treated human endothelial cells convert DHA to 13-hydroxy-DHA that is switched to $17R$-HDHA by aspirin. This latter can be transformed by human PMN to di- and tri-hydroxy products termed *"aspirin triggered (AT)" D-series resolvins* [63]. In the absence of aspirin, D-series Rv carrying the $17S$-hydroxy group were identified in murine exudates and isolated human cells [52, 63]. The D-series Rv family also includes RvD2 ($7S$, $16R$, $17S$-trihydroxy-$4Z, 8E, 10Z, 12E, 14E, 19Z$-DHA) [64] and RvD3-RvD6. Each of these immunoresolvent is biosynthesized through specific enzymatic pathways, involving 15- and 5-LO (Fig. 2C), possesses distinct chemical structures, and exhibits potent bioactions [17, 65, 66].

2.2.3 (Neuro)protectins

DHA is also a precursor of distinct SPM termed *(Neuro)protectins*. The structure of the founding member of this family, PD1, was first disclosed by Serhan and Hong

[52, 63]. Several stereo- and positional isomers of PD1 were identified in human and mouse tissues [52, 63]. Also, aspirin promotes the biosynthesis of a 17R-epimeric form of PD1 (AT-PD1) [67]. The 15-LO is also involved in the biosynthesis of PD (Fig. 2C).

2.2.4 Maresins

Maresins, acronym of *MΦ mediator in resolving inflammation*, are a new family of SPM produced by MΦs [68]. A 12-LO-dependent biochemical pathway converts DHA into 14-hydroxydocosaexaenoic acid (HDHA), which is rapidly modified into 7,14-dihydroxydocosa-4Z,8E,10E,12Z,16Z,19Z-hexaenoic acid, named MaR1 [69–71]. In addition, 7S,14S-dihydroxydocosa-4Z,8E,10Z,12E,16Z,19Z-DHA (7S,14S-diHDHA), formed by consecutive lipoxygenation of 14-HDHA, and 13R,14S-diDHA, termed MaR2 is biosynthesized from DHA by 12-LO [72] (Fig. 2C).

3 Biological Actions of LX and Rv

3.1 *Lipoxins*

LX and ATL exert potent anti-inflammatory pro-resolution bioactions in vitro and in vivo. Leukocytes, expressing high level of ALX/FPR2, the LXA_4 and ATLa receptor (see below), are elective targets of these eicosanoids. For instance, LXA_4 and ATLa inhibit PMN and eosinophil chemotaxis [73, 74], PMN vascular adhesion, transendothelial and transepithelial migration, superoxide anion generation [30, 75–78], azurophilic degranulation [79], peroxynitrite formation and NF-κB activation [80]. On the other hand, LXA_4 and ATLa stimulate monocyte chemotaxis, MΦ efferocytosis and bacterial phagocytosis, reducing the release of inflammatory cytokines [81–84]. LXA_4 also downtones memory B-cell responses, thus modulating adaptive immunity [85]. Cellular LX bioactions are summarized in Table 1.

The organic synthesis of a number of LX analogs, more resistant to metabolic inactivation than native compounds [90, 131, 132] facilitated in vivo studies with a variety of animal models of disease (summarized in Table 2 and Fig. 3) and with human subjects to reduce bronco-constriction in asthmatic subjects [145] and more recently to treat infantile eczema [137]. Results from these studies are consistent with potent anti-inflammatory, pro-resolution properties of LXA_4 and ATLa, which can significantly improve disease outcome.

Table 1 Cellular bioactions of SPM and stable analogs

Target cell	Actions	References
LXA_4, ATL, and analogs		
PMN	↓ Chemotaxis, adhesion to/transmigration across endothelial and epithelial cells. ↓ ROS generation, CD11b/CD18 expression, pro-inflammatory cytokines	[75, 80, 86–90]
Monocytes/MΦs	↑ Non phlogistic chemotaxis and adhesion ↑ Phagocytic activity	[10, 81, 82, 91, 92]
Eosinophils	↓ Chemotaxis, IL-5, and eotaxin secretion	[73, 74, 93]
Platelets	↓ *Porphyromonas gingivalis*-induced aggregation	[86]
T lymphocytes	↓ TNF-α production ↑ CCR5 expression	[94, 95]
B lymphocytes	↓ IgM and IgG production by activated B cells and their proliferation through ALX/FPR2	[85]
NK cells	↓ Cytotoxicity ↑ Pro-resolution NK-mediated apoptosis of eosinophils and PMN	[96–98]
Endothelial cells	↓ ROS production, VEGF-induced proliferation adhesion molecules. ↑ Prostacyclin and NO production and HO-1 expression	[99–105]
Epithelial cells	↓ IL-8 release. ↑ epithelium repair through K channel activation and tight junction increase	[106–108]
VSMC	↓PDGF-induced migration. Regulates cell phenotype	[109, 110]
Fibroblasts	↑ Proliferation, pro-inflammatory cytokines, and MMP-3	[111, 112]
Mesangial cells	↑ Proliferation and pro-inflammatory cytokines	[113–115]
RvE1 and analogs		
PMN	↓ Transepithelial and endothelial migration ↓ TNF-α and LTB_4 signaling	[55, 116–118]
MΦs	↑ Efferocytosis	[10, 116]
Platelets	↓ Aggregation and ADP-$P2Y_{12}$ signaling	[118, 119]
Osteoclasts	↓ Bone resorption and cell fusion	[120, 121]
RvD1 and analogs		
PMN	↓ Transepithelial and endothelial migration ↓ Adhesion molecules, LTB_4 and IL-8 actions ↓ Actin remodeling and chemotaxis	[19, 122, 124]
MΦs	↑ Efferocytosis and killing of bacteria ↑ M2 phenotype and actions	[65, 123–125]
Microglial cells	↑ IL-1β	[63]
Endothelial cells	↓ PMN rolling, adhesion, and diapedesis	[126]
VSMC	↓ Proliferation, migration, leukocyte adhesion, and pro-inflammatory mediators	[127]
Gingival fibroblasts	↓ Cytokine-induced PGE_2 production ↑ LXA_4 and wound healing.	[128]
B lymphocytes	↑ IgG and IgM production	[129]

(continued)

Table 1 (continued)

Target cell	Actions	References
Lung fibroblasts and epithelial cells	↑ IL-6, IL-8, MCP-1, and PGE$_2$ production	[130]
RvD2		
PMN	↓L-Selectin shedding and CD18 expression ↑ Interactions with endothelial cells	[64]
MΦs	↑ Phagocytosis	[64]
Endothelial cells	↑ NO biosynthesis	[64]
VSMC	↓ Proliferation, migration, leukocyte adhesion, and pro-inflammatory mediators	[127]

Table 2 Bioactions of SPM in experimental diseases

Tissues organs	Diseases	SPM Tested	References
Skin	Inflammation and infection	LXA$_4$, ATL, RvE1, RvD1, RvD2, RvD5, PD1	[133, 134] [12, 52, 65, 135–137]
GI	Peritonitis and sepsis	LXA$_4$, RvD1, RvD2, RvD5, PD1	[5, 9, 10, 52, 64, 65, 123, 125, 138]
	Colitis	RvE1, RvD1, RvD2	[68, 139–141]
Airways	Asthma	LXA$_4$, RvE1, RvD1, AT-RvD1, MaR1	[7, 142–148]
	Cystic fibrosis	LXA$_4$	[149]
	Pleurisy	LXA$_4$	[73]
	Acute lung injury	LXA$_4$, RvE1, RvD1, AT-RvD1, MaR1	[150–154]
Periodontium	Periodontitis	LXA$_4$, ATL, RvE1, RvD1	[128, 155–157]
Eye	Retinopathy	RvE1, RvD1	[158, 159]
	Dry eye syndrome	RvE1	[160]
	HSV I infection	RvE1, PD1	[161, 162]
	Uveitis	RvD1	[163]
Joints	Arthritis and TMJ inflammation	LXA$_4$, RvD1	[164–166]
Others	I/R Injury	LXA$_4$, RvD1, RvE1	[19, 78, 167–170]
	Obesity and T2D	RvE1, RvD1	[171–175]
	Pain	LXA$_4$, ATL, RvD1, AT-RvD1, RvD2, RvE1	[165, 176–183]
	Allograft	LXA$_4$, RvE1	[184, 185]
	Parasite and microbe infection	LXA$_4$, RvD1, RvD2, RvD5, PD1	[64, 65, 138, 186–188]
	AKI and renal fibrosis	LXA$_4$, ATL, RvD1, RvD2, PD1	[169, 189]
	Vascular injury	ATL	[110]

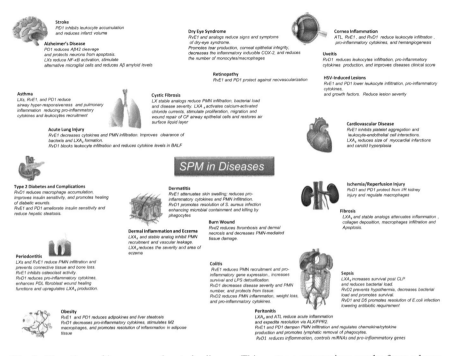

Fig. 3 Bioactions of immunoresolvents in disease. This cartoon summarizes results from a large body of preclinical and clinical studies with SPM

3.2 E-Series Resolvins

RvE1 displays potent stereoselective actions. For instance, at nanomolar concentrations, it reduces human PMN transendothelial diapedesis, dendritic cell migration and interleukin (IL)-12 production [53, 55] in vitro. Its most relevant cellular bioactions are listed in Table 1. In many pre-clinical models of disease, at nanogram to sub microgram doses, RvE1 carries counter-regulatory actions that prevent leukocyte-mediated tissue injury. These models are largely overlapping with those listed for LXA_4 and ATL (Table 2 and Fig. 3).

3.3 D-Series Resolvins

RvD1 and AT-RvD1 are potent regulators of inflammatory responses both in human and murine cells. For instance, they stop PMN transendothelial and transepithelial migration [122], and regulate endotoxin-induced cytokine production by MΦs [190].

In murine zymosan-induced acute peritonitis, RvD1 strikingly modifies the resolution indices, lowering the Ψ_{max} and shortening the R_i [9, 123, 125].

Recently, in vivo pharmacokinetics (PK) of RvD1 upon oral administration was established [9]. Oral gavage of a single bolus of RvD1 produced a rapid and transient increase in its plasma concentrations. Using classical PK mathematical models, it was calculated that the AUC_{0-last} of RvD1 is ~4.5 ng*h/mL. At these concentrations, RvD1 potently dampens PMN recruitment to vascular endothelial cells as well as endothelial transmigration both in mice and humans [122, 191], suggesting that RvD1 can exert regulatory functions on peripheral blood leukocyte diapedesis upon oral administration. The distribution volume of RvD1 was ~22 mL, which is much higher than the estimated mouse blood total volume [192], indicating that RvD1 may concentrate in peripheral depots (e.g., adipose tissue), in line with its lipophilic structure. RvD1 plasma clearance followed a 1st order kinetics, consistent with metabolic breakdown catalyzed by eicosanoid oxidoreductase as well as 11β-hydroxysteroid dehydrogenase [122, 193].

Using a pre-clinical model of peritoneal acute inflammation Recchiuti et al. also provided the first evidence of the pharmacological efficacy of orally administered RvD1, as oral RvD1 reduced total leukocyte infiltration into peritoneal cavities elicited by zymosan A in a dose and time dependent manner, sharply modifying select resolution indices, at doses as low as 0.05–5 µg/kg [9].

In relation to mechanisms of action, the observation that RvD1 reduces LTB_4 biosynthesis in MΦs by regulating 5-LO nuclear localization [194] is of particular interest since cytoplasmic 5-LO is capable of producing higher amounts of LXA_4. Thus, RvD1 by acting on regulatory mechanisms of 5-LO activation dampens the generation of pro-inflammatory mediators such as LTB_4, while promoting the biosynthesis of the pro-resolution LXA_4, showing a previously unappreciated interplay between the AA and the DHA pathways of inflammation and resolution.

RvD1 can also regulate 5-LO expression through microRNAs (miRNAs). Indeed, expression analyses of miRNAs from leukocytes collected during the onset and resolution phase of peritonitis demonstrated that RvD1 controls a specific set of pro-resolving miRNAs, namely miR-21, miR-146b, miR-208a, and miR-219, in vivo, in a time- and receptor-dependent manner, as part of its immunoresolving actions [123, 125]. Targets of the RvD1-regulated miRNAs include genes involved in the NF-kB activation pathway (e.g. IkB kinase and tumor necrosis factor receptor-associated factor 6), cytokines and chemokines (e.g., IL-8, 10, 12, interferon-α and β), programmed cell death 4, a tumor suppressor that acts as a translational repressor of IL-10 [195], and 5-LO. These results demonstrate for the first time the important roles of miRNAs in the resolution of inflammation. Along these lines, Fredman et al. reported that delayed resolution of acute peritonitis, triggered by high doses of zymosan A, dysregulates pro-resolving miRNA and lipid mediator profiles, decreasing miR-219 expression while increasing LTB_4 and reducing SPM production [196]. In addition, recent studies by Li et al. revealed that miR-466l is temporally regulated in murine exudate leukocytes and controls the biosynthesis of PG and immunoresolvents [197]. Finally, Brennan and coworkers found that LXA_4 attenuates the production of pro-fibrotic proteins (e.g., fibronectin,

N-cadherin, thrombospondin, and the notch ligand jagged-1) in cultured human proximal tubular epithelial cells via up-regulation of miRNA let-7c [198]. Collectively, these results demonstrate the involvement of miRNAs in the SPM protective and pro-resolution actions.

Additional information on RvD1 mechanisms of actions on murine MΦs during the resolution phase of acute peritonitis, was provided by recent studies from Recchiuti et al. These studies revealed that RvD1 selectively down-regulates chromatin remodeling enzymes involved in the control of genes related to immune responses [9].

Together, these results indicate that RvD1 acts on epigenetic mechanisms controlling transcriptional profiles in MΦs involved in resolution of inflammation.

Cellular actions and preclinical studies with resolvins are summarized in Tables 1 and 2 and in Fig. 3. It is evident that these compounds may have beneficial effects in different clinical settings, including cardiovascular, respiratory, renal, gastrointestinal, joint, neurological, ocular, dental, cutaneous, metabolic and infectious disease and that their pro-resolving actions are, in a variety of these settings, similar to those observed with LX. This may be in relation with overlapping mechanisms of actions and emphasize the relevance of the knowledge of receptors recognized by SPM and their signaling pathways.

4 Mechanisms of Action of LX and Rv: Receptor Identification, Cell Expression and Its Regulation, Genetic Variants, and Intracellular Signaling

4.1 ALX/FPR2, the Lipoxin A_4 Receptor

The first evidence of LXA_4 high affinity binding sites was obtained with PMN incubated with synthetic tritium labeled [11,12-^{3}H]-LXA_4 [199]. Later on, a systematic screening of cDNA clones obtained from PMN-differentiated promyelocytic HL60 cells led to the identification a cDNA clone encoding an orphan G protein coupled receptor (GPCR), belonging to the family of the formyl pepide receptor (FPR), and previously denominated formyl peptide receptor like (FPRL)-1 [200], which bound with high affinity to LXA_4 [201]. In 2009, the International Union of Basic and Clinical Pharmacology proposed ALX/FPR2 as official name for the LXA_4 receptor [202]. The ALX/FPR2 protein encompasses 351 amino acids arranged in seven putative transmembrane domains. It is highly expressed in myeloid cells and, at a lower extent, in a variety of other cell types (reviewed by Romano et al. 2007) [203]. Mouse and rat orthologues of human ALX/FPR2 have been identified [134, 204].

In addition to LXA_4, ATL and their stable analogs, this receptor is recognized by other pro-resolution mediators such as RvD1, AT-RvD1 [123, 124] and the glucocorticoid-induced protein, annexin (Anx) A1 and derived peptides

[53]. However, binding to ALX/FPR2 by a number of peptides carrying pro-inflammatory signalling has been reported (reviewed in [203, 205]). Thus, ALX/FPR2 signalling is cell and agonist specific, as it can be coupled to several G-protein and can activate different second messengers. In this respect, Cooray et al. recently reported that ALX/FPR2 undergoes homo- or hetero-dimerization with FPR1 and 3 in overexpressing HEK-293 cells, triggering distinct responses to its ligands [206]. These results provide further cues on ALX/FPR2 pharmacology.

Valuable information regarding the physiological role of this receptor in vivo has been provided by genetic manipulation of human ALX/FPR2 and its orthologue in mice. Myeloid-driven overexpression of human ALX/FPR2 in mice results not only in increased sensitivity to suboptimal doses of ATL, but also in reduced inflammatory responses in zymosan-induced peritonitis, in the absence of exogenously added ligands [123, 207]. On the other hand, ALX/FPR2 KO mice display an exacerbated inflammation and delayed resolution phenotype, do not respond to endogenous or administered pro-resolution ligands [208], and have a worse outcome of microbial sepsis [209]. Along these lines, ATL amounts and ALX/FPR2 expression levels dictate both the magnitude and duration of acute inflammation in humans [210]. Together, these data indicate that ALX/FPR2 expression level represents per se a key event in inflammation resolution, underlying the anti-inflammatory, pro-resolution function of ALX/FPR2 in humans. Thus, the knowledge of regulatory mechanisms of ALX/FPR2 expression is key for the designing of novel pro-resolution pharmacology.

In this respect, the identification of the ALX/FPR2 promoter [211, 212] and of a genetic variant of the promoter, associated with reduced transcriptional activity in a patient with metabolic syndrome and history of acute cardiovascular events and in his two daughters, both with arterial hypertension and increased body mass index [211], opens new perspectives for drug discovery and personalized medicine for inflammation-based disorders. More recently, Pierdomenico et al. reported regulation of ALX/FPR2 expression by miR-181b [213]. miR-181b down-regulates ALX/FPR2 expression in human MΦs and blunts pro-resolution responses of these cells. These results unravel novel regulatory mechanisms of ALX/FPR2 expression and agonist-triggered pro-resolutive responses that are paramount in resolution circuits.

4.2 E-Series Resolvin Receptors

Two GPCRs are involved in mediating RvE1 actions, namely ChemR23, also known as chemokine-like receptor 1, and LTB_4 receptor 1 (BLT1) [55, 116]. [^{3}H]-RvE1 binds to ChemR23 transfected cells with high stereoselectivity and affinity [55]. This binding is displaced by the synthetic peptide YHSFFFPGQFAFS, derived from human chemerin, suggesting that RvE1 and chemerin share recognition sites on ChemR23 [55, 156]. Notably, RvE1 binding sites on human PMN are pharmacologically distinct from ChemR23 [116]. In transfected cells, [^{3}H]-RvE1 also binds with high affinity to BLT1 but not to BLT2 and is displaced by unlabeled LTB$_4$. Together, these results demonstrate that RvE1 acts as a selective partial agonist on BLT1 attenuating LTB$_4$-induced pro-inflammatory responses [116]. This is another example of interplay between pro-inflammatory and pro-resolution pathways that involve LO-derived product and their receptors.

More recently, direct evidence for ligand-receptor interactions of RvE1 and its epimer 18S-RvE1 was provided using ChemR23 and BLT-1 β-arrestin cells [56]. Hence, RvE1 and 18S-RvE1 can share specific binding sites to human ChemR23 as well as to BLT-1.

RvE2 exerts potent and cell-specific bioactions on human leukocytes [56, 59]. Recently, experiments with tritium-labeled [^{3}H]-RvE2 and cells expressing ChemR23 and BLT-1 /β-arrestin revealing system, showed that RvE2 binds to human PMN and shares, at least in part, receptors with RvE1 [56].

4.3 D-Series Resolvin Receptors

Krishnamoorthy, Recchiuti et al. were the first to demonstrate specific high affinity binding of [^{3}H]-RvD1 to human peripheral blood PMN and monocytes. This binding is displaced by cold RvD1 (100 %) and LXA$_4$ (~60 %), but not by the AnxA1-derived Ac2-12 peptide [124], suggesting the existence of specific RvD1 recognition sites on human leukocyte membrane. Using a luciferase reporter system [124], these authors demonstrated that RvD1 reduces TNF-α-induced NF-κB activation in cells co-expressing NF-κB-responsive elements upstream the luciferase gene with either ALX/FPR2 or the orphan receptor, GPR32, but not with other GPCRs [124]. Moreover, using a GPCR-β-arrestin-coupled system based on enzyme fragment complementation, it was found that RvD1 concentration-dependently activates ALX/FPR2 and GPR32 [124].

Studies with genetically engineered mice and selective receptor antagonists or blocking antibodies confirmed the ALX/FPR2 and GPR32 dependency of RvD1 immunoresolving actions [107, 123–125, 144, 172, 174, 191, 214, 215], emphasizing the central role of ALX/FPR2 in inflammation resolution.

Human GPR32 has been identified in peripheral blood leukocytes and arterial and venous tissues using a cDNA array [123]. It is mostly abundant on PMN, monocytes and MΦs and is also present on vascular endothelial cells [124]. The murine orthologue of GPR32 is currently unknown, whereas it exists in monkeys, where LX are biosynthesized during acute pneumonia [216], further indicating that roles of SPM and their GPCR in immunity is evolutionarily conserved. Regulatory mechanisms of GPR32 are unknown.

Although specific receptors for RvD2, RvD3 and RvD4 have not yet been uncovered, the stereoselective actions of RvD2 are inhibited by pertussis toxin [64], implicating the involvement of GPCRs. More recently, Chiang et al. showed

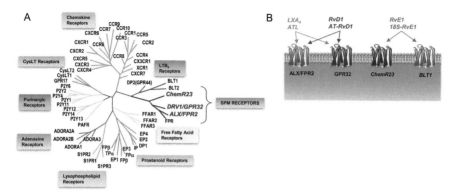

Fig. 4 SPM receptors. (**A**) The phylogenetic tree shows similarities in amino acid sequences among LXA$_4$, RvE1, and LTB$_4$ receptors deduced from the NCBI database. Cluster was generated with the ClustalW2 software (www.ebi.ac.uk/Tools/clustalw2). (**B**) Ligand recognition of ALX/FPR2, GPR32, ChemR23 and BLT1

that GPR32 is also recognized by RvD5 [65]. A phylogenetic tree and ligand recognition of SPM receptors are illustrated in Fig. 4.

Altogether, the identification of pro-resolution receptors and regulatory mechanisms of their expression and signaling represents a key step towards innovative pharmacology of inflammation resolution.

5 Conclusive Remarks

As our knowledge of lipoxygenases progresses, we appreciate the complexity and relevance of this family of enzymes. For instance, 5-LO has been historically regarded as a pro-inflammatory enzyme as it catalyzes the formation of LTB$_4$. Likewise, the pathophysiological role of 12- ad 15-LO is incompletely known. We now know that LO are also key enzymes in the biosynthesis of a genus of endogenous pro-resolution lipid mediators. This is a key concept for the development of mechanistic and temporal pharmacology of these enzymes that should be aimed to preserve their capability to catalyze reactions that lead to the biosynthesis of pro-resolution mediators, which, together with their receptors, are emerging as pivotal regulators of the host response and maintenance of homeostasis.

References

1. Majno G, Joris I (2004) Cells, tissues, and disease: principles of general pathology: principles of general pathology. Oxford University Press, USA
2. Medzhitov R (2008) Origin and physiological roles of inflammation. Nature 454 (7203):428–435

3. Gordon S (2007) The macrophage: past, present and future. Eur J Immunol 37(Suppl 1): S9–17
4. Serhan CN, Brain SD, Buckley CD, Gilroy DW, Haslett C, O'Neill LA, Perretti M, Rossi AG, Wallace JL (2007) Resolution of inflammation: state of the art, definitions and terms. FASEB J 21(2):325–332
5. Bannenberg GL, Chiang N, Ariel A, Arita M, Tjonahen E, Gotlinger KH, Hong S, Serhan CN (2005) Molecular circuits of resolution: formation and actions of resolvins and protectins. J Immunol 174(7):4345–4355
6. Serhan CN, Chiang N, Van Dyke TE (2008) Resolving inflammation: dual anti-inflammatory and pro-resolution lipid mediators. Nat Rev Immunol 8(5):349–361
7. Haworth O, Cernadas M, Yang R, Serhan CN, Levy BD (2008) Resolvin E1 regulates interleukin 23, interferon-gamma and lipoxin A4 to promote the resolution of allergic airway inflammation. Nat Immunol 9(8):873–879
8. Navarro-Xavier RA, Newson J, Silveira VL, Farrow SN, Gilroy DW, Bystrom J (2009) A new strategy for the identification of novel molecules with targeted proresolution of inflammation properties. J Immunol 184(3):1516–1525
9. Recchiuti A, Codagnone M, Pierdomenico AM, Rossi C, Mari VC, Cianci E, Simiele F, Gatta V, Romano M (2014) Immunoresolving actions of oral resolvin D1 include selective regulation of the transcription machinery in resolution-phase mouse macrophages. FASEB J 28(7):3090–3102
10. Schwab JM, Chiang N, Arita M, Serhan CN (2007) Resolvin E1 and protectin D1 activate inflammation-resolution programmes. Nature 447(7146):869–874
11. Levy BD, Clish CB, Schmidt B, Gronert K, Serhan CN (2001) Lipid mediator class switching during acute inflammation: signals in resolution. Nat Immunol 2(7):612–619
12. Serhan CN, Clish CB, Brannon J, Colgan SP, Chiang N, Gronert K (2000) Novel functional sets of lipid-derived mediators with antiinflammatory actions generated from omega-3 fatty acids via cyclooxygenase 2-nonsteroidal antiinflammatory drugs and transcellular processing. J Exp Med 192(8):1197–1204
13. Serhan CN, Hamberg M, Samuelsson B (1984) Lipoxins: Novel series of biologically active compounds formed from arachidonic acid in human leukocytes. Proc Natl Acad Sci USA 81(17):5335–5339
14. Serhan CN, Hamberg M, Samuelsson B (1984) Trihydroxytetraenes: a novel series of compounds formed from arachidonic acid in human leukocytes. Biochem Biophys Res Commun 118(3):943–949
15. Chiang N, Bermudez EA, Ridker PM, Hurwitz S, Serhan CN (2004) Aspirin triggers antiinflammatory 15-epi-lipoxin A4 and inhibits thromboxane in a randomized human trial. Proc Natl Acad Sci USA 101(42):15178–15183
16. Dalli J, Colas RA, Serhan CN (2013) Novel n-3 immunoresolvents: structures and actions. Sci Rep 3:1940
17. Dalli J, Winkler JW, Colas RA, Arnardottir H, Cheng CY, Chiang N, Petasis NA, Serhan CN (2013) Resolvin D3 and aspirin-triggered resolvin D3 are potent immunoresolvents. Chem Biol 20(2):188–201
18. Serhan CN, Chiang N (2013) Resolution phase lipid mediators of inflammation: agonists of resolution. Curr Opin Pharmacol 13(4):632–640
19. Kasuga K, Yang R, Porter TF, Agrawal N, Petasis NA, Irimia D, Toner M, Serhan CN (2008) Rapid appearance of resolvin precursors in inflammatory exudates: novel mechanisms in resolution. J Immunol 181(12):8677–8687
20. Colas RA, Shinohara M, Dalli J, Chiang N, Serhan CN (2014) Identification and signature profiles for pro-resolving and inflammatory lipid mediators in human tissue. Am J Physiol Cell Physiol 307(1):C39–C54
21. Psychogios N, Hau DD, Peng J, Guo AC, Mandal R, Bouatra S, Sinelnikov I, Krishnamurthy R, Eisner R, Gautam B, Young N, Xia J, Knox C, Dong E, Huang P,

Hollander Z, Pedersen TL, Smith SR, Bamforth F, Greiner R, McManus B, Newman JW, Goodfriend T, Wishart DS (2011) The human serum metabolome. PLoS One 6(2), e16957
22. Weiss GA, Troxler H, Klinke G, Rogler D, Braegger C, Hersberger M (2013) High levels of anti-inflammatory and pro-resolving lipid mediators lipoxins and resolvins and declining docosahexaenoic acid levels in human milk during the first month of lactation. Lipids Health Dis 12:89
23. Gangemi S, Luciotti G, D'Urbano E, Mallamace A, Santoro D, Bellinghieri G, Davi G, Romano M (2003) Physical exercise increases urinary excretion of lipoxin A4 and related compounds. J Appl Physiol (1985) 94(6):2237–2240
24. Markworth JF, Vella L, Lingard BS, Tull DL, Rupasinghe TW, Sinclair AJ, Maddipati KR, Cameron-Smith D (2013) Human inflammatory and resolving lipid mediator responses to resistance exercise and ibuprofen treatment. Am J Physiol Regul Integr Comp Physiol 305 (11):R1281–R1296
25. Edenius C, Kumlin M, Bjork T, Anggard A, Lindgren JA (1990) Lipoxin formation in human nasal polyps and bronchial tissue. FEBS Lett 272(1–2):25–28
26. Frediani JK, Jones DP, Tukvadze N, Uppal K, Sanikidze E, Kipiani M, Tran VT, Hebbar G, Walker DI, Kempker RR, Kurani SS, Colas RA, Dalli J, Tangpricha V, Serhan CN, Blumberg HM, Ziegler TR (2014) Plasma metabolomics in human pulmonary tuberculosis disease: a pilot study. PLoS One 9(10), e108854
27. Brezinski DA, Nesto RW, Serhan CN (1992) Angioplasty triggers intracoronary leukotrienes and lipoxin A4. Impact of aspirin therapy. Circulation 86(1):56–63
28. Chiang N, Shinohara M, Dalli J, Mirakaj V, Kibi M, Choi AMK, Serhan CN (2013) Inhaled carbon monoxide accelerates resolution of inflammation via unique proresolving mediator–heme oxygenase-1 circuits. J Immunol 190(12):6378–6388
29. Pillai PS, Leeson S, Porter TF, Owens CD, Kim JM, Conte MS, Serhan CN, Gelman S (2012) Chemical mediators of inflammation and resolution in post-operative abdominal aortic aneurysm patients. Inflammation 35(1):98–113
30. Serhan CN, Hamberg M, Samuelsson B, Morris J, Wishka DG (1986) On the stereochemistry and biosynthesis of lipoxin B. Proc Natl Acad Sci USA 83(7):1983–1987
31. Serhan CN, Hirsch U, Palmblad J, Samuelsson B (1987) Formation of lipoxin A by granulocytes from eosinophilic donors. FEBS Lett 217(2):242–246
32. Levy BD, Romano M, Chapman HA, Reilly JJ, Drazen J, Serhan CN (1993) Human alveolar macrophages have 15-lipoxygenase and generate 15(S)-hydroxy-5,8,11-cis-13-trans-eicosatetraenoic acid and lipoxins. J Clin Invest 92(3):1572–1579
33. Munger KA, Montero A, Fukunaga M, Uda S, Yura T, Imai E, Kaneda Y, Valdivielso JM, Badr KF (1999) Transfection of rat kidney with human 15-lipoxygenase suppresses inflammation and preserves function in experimental glomerulonephritis. Proc Natl Acad Sci USA 96(23):13375–13380
34. Planaguma A, Kazani S, Marigowda G, Haworth O, Mariani TJ, Israel E, Bleecker ER, Curran-Everett D, Erzurum SC, Calhoun WJ, Castro M, Chung KF, Gaston B, Jarjour NN, Busse WW, Wenzel SE, Levy BD (2008) Airway lipoxin A4 generation and lipoxin A4 receptor expression are decreased in severe asthma. Am J Respir Crit Care Med 178 (6):574–582
35. Edenius C, Haeggstrom J, Lindgren JA (1988) Transcellular conversion of endogenous arachidonic acid to lipoxins in mixed human platelet-granulocyte suspensions. Biochem Biophys Res Commun 157(2):801–807
36. Fiore S, Serhan CN (1990) Formation of lipoxins and leukotrienes during receptor-mediated interactions of human platelets and recombinant human granulocyte/macrophage colony-stimulating factor-primed neutrophils. J Exp Med 172(5):1451–1457
37. Serhan CN, Sheppard KA, Fiore S (1990) Lipoxin formation: evaluation of the role and actions of leukotriene A4. Adv Prostaglandin Thromboxane Leukot Res 20:54–62

38. Sheppard KA, Greenberg SM, Funk CD, Romano M, Serhan CN (1992) Lipoxin generation by human megakaryocyte-induced 12-lipoxygenase. Biochim Biophys Acta 1133(2):223–234
39. Romano M, Serhan CN (1992) Lipoxin generation by permeabilized human platelets. Biochemistry 31(35):8269–8277
40. Romano M, Chen XS, Takahashi Y, Yamamoto S, Funk CD, Serhan CN (1993) Lipoxin synthase activity of human platelet 12-lipoxygenase. Biochem J 296(Pt 1):127–133
41. Mattoscio D, Evangelista V, De Cristofaro R, Recchiuti A, Pandolfi A, Di Silvestre S, Manarini S, Martelli N, Rocca B, Petrucci G, Angelini DF, Battistini L, Robuffo I, Pensabene T, Pieroni L, Furnari ML, Pardo F, Quattrucci S, Lancellotti S, Davi G, Romano M (2010) Cystic fibrosis transmembrane conductance regulator (CFTR) expression in human platelets: impact on mediators and mechanisms of the inflammatory response. FASEB J 24(10):3970–3980
42. Claria J, Serhan CN (1995) Aspirin triggers previously undescribed bioactive eicosanoids by human endothelial cell-leukocyte interactions. Proc Natl Acad Sci USA 92(21):9475–9479
43. Claria J, Lee MH, Serhan CN (1996) Aspirin-triggered lipoxins (15-epi-LX) are generated by the human lung adenocarcinoma cell line (A549)-neutrophil interactions and are potent inhibitors of cell proliferation. Mol Med 2(5):583–596
44. Titos E, Chiang N, Serhan CN, Romano M, Gaya J, Pueyo G, Claria J (1999) Hepatocytes are a rich source of novel aspirin-triggered 15-epi-lipoxin A(4). Am J Physiol 277(5 Pt 1):C870–C877
45. Brezinski DA, Serhan CN (1991) Characterization of lipoxins by combined gas chromatography and electron-capture negative ion chemical ionization mass spectrometry: formation of lipoxin A4 by stimulated human whole blood. Biol Mass Spectrom 20(2):45–52
46. Sanak M, Levy BD, Clish CB, Chiang N, Gronert K, Mastalerz L, Serhan CN, Szczeklik A (2000) Aspirin-tolerant asthmatics generate more lipoxins than aspirin-intolerant asthmatics. Eur Respir J 16(1):44–49
47. Fiorucci S, Santucci L, Wallace JL, Sardina M, Romano M, del Soldato P, Morelli A (2003) Interaction of a selective cyclooxygenase-2 inhibitor with aspirin and NO-releasing aspirin in the human gastric mucosa. Proc Natl Acad Sci USA 100(19):10937–10941
48. Calder PC (2013) n-3 fatty acids, inflammation and immunity: new mechanisms to explain old actions. Proc Nutr Soc 72(3):326–336
49. De Caterina R (2011) n-3 fatty acids in cardiovascular disease. N Engl J Med 364(25):2439–2450
50. Dietary supplementation with n-3 polyunsaturated fatty acids and vitamin E after myocardial infarction: results of the GISSI-Prevenzione trial. Gruppo Italiano per lo Studio della Sopravvivenza nell'Infarto miocardico (1999). Lancet 354 (9177): 447–455
51. Iigo M, Nakagawa T, Ishikawa C, Iwahori Y, Asamoto M, Yazawa K, Araki E, Tsuda H (1997) Inhibitory effects of docosahexaenoic acid on colon carcinoma 26 metastasis to the lung. Br J Cancer 75(5):650–655
52. Hong S, Gronert K, Devchand PR, Moussignac RL, Serhan CN (2003) Novel docosatrienes and 17S-resolvins generated from docosahexaenoic acid in murine brain, human blood, and glial cells. Autacoids in anti-inflammation. J Biol Chem 278(17):14677–14687
53. Perretti M, Chiang N, La M, Fierro IM, Marullo S, Getting SJ, Solito E, Serhan CN (2002) Endogenous lipid- and peptide-derived anti-inflammatory pathways generated with glucocorticoid and aspirin treatment activate the lipoxin A4 receptor. Nat Med 8(11):1296–1302
54. Serhan CN, Gotlinger K, Hong S, Lu Y, Siegelman J, Baer T, Yang R, Colgan SP, Petasis NA (2006) Anti-inflammatory actions of neuroprotectin D1/protectin D1 and its natural stereoisomers: assignments of dihydroxy-containing docosatrienes. J Immunol 176(3):1848–1859
55. Arita M, Bianchini F, Aliberti J, Sher A, Chiang N, Hong S, Yang R, Petasis NA, Serhan CN (2005) Stereochemical assignment, antiinflammatory properties, and receptor for the omega-3 lipid mediator resolvin E1. J Exp Med 201(5):713–722

56. Tjonahen E, Oh SF, Siegelman J, Elangovan S, Percarpio KB, Hong S, Arita M, Serhan CN (2006) Resolvin E2: identification and anti-inflammatory actions: pivotal role of human 5-lipoxygenase in resolvin E series biosynthesis. Chem Biol 13(11):1193–1202
57. Haas-Stapleton EJ, Lu Y, Hong S, Arita M, Favoreto S, Nigam S, Serhan CN, Agabian N (2007) Candida albicans modulates host defense by biosynthesizing the pro-resolving mediator resolvin E1. PLoS One 2(12), e1316
58. Ogawa N, Kobayashi Y (2009) Total synthesis of resolvin E1. Tetrahedron Lett 50 (44):6079–6082
59. Oh SF, Dona M, Fredman G, Krishnamoorthy S, Irimia D, Serhan CN (2012) Resolvin e2 formation and impact in inflammation resolution. J Immunol 188(9):4527–4534
60. Isobe Y, Arita M, Iwamoto R, Urabe D, Todoroki H, Masuda K, Inoue M, Arai H (2013) Stereochemical assignment and anti-inflammatory properties of the omega-3 lipid mediator resolvin E3. J Biochem 153(4):355–360
61. Isobe Y, Arita M, Matsueda S, Iwamoto R, Fujihara T, Nakanishi H, Taguchi R, Masuda K, Sasaki K, Urabe D, Inoue M, Arai H (2012) Identification and structure determination of novel anti-inflammatory mediator resolvin E3, 17,18-dihydroxyeicosapentaenoic acid. J Biol Chem 287(13):10525–10534
62. Yokokura Y, Isobe Y, Matsueda S, Iwamoto R, Goto T, Yoshioka T, Urabe D, Inoue M, Arai H, Arita M (2014) Identification of 14,20-dihydroxy-docosahexaenoic acid as a novel anti-inflammatory metabolite. J Biochem 156(6):315–321
63. Serhan CN, Hong S, Gronert K, Colgan SP, Devchand PR, Mirick G, Moussignac R-L (2002) Resolvins: a family of bioactive products of omega-3 fatty acid transformation circuits initiated by aspirin treatment that counter proinflammation signals. J Exp Med 196 (8):1025–1037
64. Spite M, Norling LV, Summers L, Yang R, Cooper D, Petasis NA, Flower RJ, Perretti M, Serhan CN (2009) Resolvin D2 is a potent regulator of leukocytes and controls microbial sepsis. Nature 461(7268):1287–1291
65. Chiang N, Fredman G, Backhed F, Oh SF, Vickery T, Schmidt BA, Serhan CN (2012) Infection regulates pro-resolving mediators that lower antibiotic requirements. Nature 484 (7395):524–528
66. Winkler JW, Uddin J, Serhan CN, Petasis NA (2013) Stereocontrolled total synthesis of the potent anti-inflammatory and pro-resolving lipid mediator resolvin D3 and its aspirin-triggered 17R-epimer. Org Lett 15(7):1424–1427
67. Marcheselli VL, Hong S, Lukiw WJ, Tian XH, Gronert K, Musto A, Hardy M, Gimenez JM, Chiang N, Serhan CN, Bazan NG (2003) Novel docosanoids inhibit brain ischemia-reperfusion-mediated leukocyte infiltration and pro-inflammatory gene expression. J Biol Chem 278(44):43807–43817
68. Ishida T, Yoshida M, Arita M, Nishitani Y, Nishiumi S, Masuda A, Mizuno S, Takagawa T, Morita Y, Kutsumi H, Inokuchi H, Serhan CN, Blumberg RS, Azuma T (2009) Resolvin E1, an endogenous lipid mediator derived from eicosapentaenoic acid, prevents dextran sulfate sodium-induced colitis. Inflamm Bowel Dis 16(1):87–95
69. Dalli J, Zhu M, Vlasenko NA, Deng B, Haeggstrom JZ, Petasis NA, Serhan CN (2013) The novel 13S,14S-epoxy-maresin is converted by human macrophages to maresin 1 (MaR1), inhibits leukotriene A4 hydrolase (LTA4H), and shifts macrophage phenotype. FASEB J 27 (7):2573–2583
70. Serhan CN, Dalli J, Karamnov S, Choi A, Park CK, Xu ZZ, Ji RR, Zhu M, Petasis NA (2012) Macrophage proresolving mediator maresin 1 stimulates tissue regeneration and controls pain. FASEB J 26(4):1755–1765
71. Serhan CN, Yang R, Martinod K, Kasuga K, Pillai PS, Porter TF, Oh SF, Spite M (2009) Maresins: novel macrophage mediators with potent antiinflammatory and proresolving actions. J Exp Med 206(1):15–23

72. Deng B, Wang CW, Arnardottir HH, Li Y, Cheng CY, Dalli J, Serhan CN (2014) Maresin biosynthesis and identification of maresin 2, a new anti-inflammatory and pro-resolving mediator from human macrophages. PLoS One 9(7), e102362
73. Bandeira-Melo C, Serra MF, Diaz BL, Cordeiro RS, Silva PM, Lenzi HL, Bakhle YS, Serhan CN, Martins MA (2000) Cyclooxygenase-2-derived prostaglandin E2 and lipoxin A4 accelerate resolution of allergic edema in Angiostrongylus costaricensis-infected rats: relationship with concurrent eosinophilia. J Immunol 164(2):1029–1036
74. Soyombo O, Spur BW, Lee TH (1994) Effects of lipoxin A4 on chemotaxis and degranulation of human eosinophils stimulated by platelet-activating factor and N-formyl-L-methionyl-L-leucyl-L-phenylalanine. Allergy 49(4):230–234
75. Carlo T, Kalwa H, Levy BD (2013) 15-Epi-lipoxin A4 inhibits human neutrophil superoxide anion generation by regulating polyisoprenyl diphosphate phosphatase 1. FASEB J 27 (7):2733–2741
76. Colgan SP, Serhan CN, Parkos CA, Delp-Archer C, Madara JL (1993) Lipoxin A4 modulates transmigration of human neutrophils across intestinal epithelial monolayers. J Clin Invest 92 (1):75–82
77. Fierro IM, Colgan SP, Bernasconi G, Petasis NA, Clish CB, Arita M, Serhan CN (2003) Lipoxin A4 and aspirin-triggered 15-epi-lipoxin A4 inhibit human neutrophil migration: comparisons between synthetic 15 epimers in chemotaxis and transmigration with microvessel endothelial cells and epithelial cells. J Immunol 170(5):2688–2694
78. Scalia R, Gefen J, Petasis NA, Serhan CN, Lefer AM (1997) Lipoxin A4 stable analogs inhibit leukocyte rolling and adherence in the rat mesenteric microvasculature: role of P-selectin. Proc Natl Acad Sci USA 94(18):9967–9972
79. Gewirtz AT, Fokin VV, Petasis NA, Serhan CN, Madara JL (1999) LXA4, aspirin-triggered 15-epi-LXA4, and their analogs selectively downregulate PMN azurophilic degranulation. Am J Physiol 276(4 Pt 1):C988–C994
80. Jozsef L, Zouki C, Petasis NA, Serhan CN, Filep JG (2002) Lipoxin A4 and aspirin-triggered 15-epi-lipoxin A4 inhibit peroxynitrite formation, NF-kappa B and AP-1 activation, and IL-8 gene expression in human leukocytes. Proc Natl Acad Sci USA 99(20):13266–13271
81. Godson C, Mitchell S, Harvey K, Petasis NA, Hogg N, Brady HR (2000) Cutting edge: lipoxins rapidly stimulate nonphlogistic phagocytosis of apoptotic neutrophils by monocyte-derived macrophages. J Immunol 164(4):1663–1667
82. Maddox JF, Serhan CN (1996) Lipoxin A4 and B4 are potent stimuli for human monocyte migration and adhesion: selective inactivation by dehydrogenation and reduction. J Exp Med 183(1):137–146
83. Prescott D, McKay DM (2011) Aspirin-triggered lipoxin enhances macrophage phagocytosis of bacteria while inhibiting inflammatory cytokine production. Am J Physiol Gastrointest Liver Physiol 301(3):G487–G497
84. Romano M, Maddox JF, Serhan CN (1996) Activation of human monocytes and the acute monocytic leukemia cell line (THP-1) by lipoxins involves unique signaling pathways for lipoxin A4 versus lipoxin B4: evidence for differential Ca2+ mobilization. J Immunol 157 (5):2149–2154
85. Ramon S, Bancos S, Serhan CN, Phipps RP (2013) Lipoxin A modulates adaptive immunity by decreasing memory B-cell responses via an ALX/FPR2-dependent mechanism. Eur J Immunol 44(2):357–369
86. Borgeson E, Lonn J, Bergstrom I, Brodin VP, Ramstrom S, Nayeri F, Sarndahl E, Bengtsson T (2011) Lipoxin A(4) inhibits porphyromonas gingivalis-induced aggregation and reactive oxygen species production by modulating neutrophil-platelet interaction and CD11b expression. Infect Immun 79(4):1489–1497
87. Fiore S, Serhan CN (1995) Lipoxin A4 receptor activation is distinct from that of the formyl peptide receptor in myeloid cells: inhibition of CD11/18 expression by lipoxin A4-lipoxin A4 receptor interaction. Biochemistry 34(51):16678–16686

88. Levy BD, Fokin VV, Clark JM, Wakelam MJ, Petasis NA, Serhan CN (1999) Polyisoprenyl phosphate (PIPP) signaling regulates phospholipase D activity: a 'stop' signaling switch for aspirin-triggered lipoxin A4. FASEB J 13(8):903–911
89. Papayianni A, Serhan CN, Brady HR (1996) Lipoxin A4 and B4 inhibit leukotriene-stimulated interactions of human neutrophils and endothelial cells. J Immunol 156(6):2264–2272
90. Serhan CN, Maddox JF, Petasis NA, Akritopoulou-Zanze I, Papayianni A, Brady HR, Colgan SP, Madara JL (1995) Design of lipoxin A4 stable analogs that block transmigration and adhesion of human neutrophils. Biochemistry 34(44):14609–14615
91. Maddox JF, Hachicha M, Takano T, Petasis NA, Fokin VV, Serhan CN (1997) Lipoxin A4 stable analogs are potent mimetics that stimulate human monocytes and THP-1 cells via a G-protein-linked lipoxin A4 receptor. J Biol Chem 272(11):6972–6978
92. Mitchell D, O'Meara SJ, Gaffney A, Crean JKG, Kinsella BT, Godson C (2007) The lipoxin A4 receptor is coupled to SHP-2 activation: implications for regulation of receptor tyrosine kinases. J Biol Chem 282(21):15606–15618
93. Levy BD, De Sanctis GT, Devchand PR, Kim E, Ackerman K, Schmidt BA, Szczeklik W, Drazen JM, Serhan CN (2002) Multi-pronged inhibition of airway hyper-responsiveness and inflammation by lipoxin A(4). Nat Med 8(9):1018–1023
94. Ariel A, Chiang N, Arita M, Petasis NA, Serhan CN (2003) Aspirin-triggered lipoxin A4 and B4 analogs block extracellular signal-regulated kinase-dependent TNF-alpha secretion from human T cells. J Immunol 170(12):6266–6272
95. Ariel A, Fredman G, Sun Y-P, Kantarci A, Van Dyke TE, Luster AD, Serhan CN (2006) Apoptotic neutrophils and T cells sequester chemokines during immune response resolution through modulation of CCR5 expression. Nat Immunol 7(11):1209–1216
96. Barnig C, Cernadas M, Dutile S, Liu X, Perrella MA, Kazani S, Wechsler ME, Israel E, Levy BD (2013) Lipoxin A4 regulates natural killer cell and type 2 innate lymphoid cell activation in asthma. Sci Transl Med 5(174):174ra126
97. Ramstedt U, Ng J, Wigzell H, Serhan CN, Samuelsson B (1985) Action of novel eicosanoids lipoxin A and B on human natural killer cell cytotoxicity: effects on intracellular cAMP and target cell binding. J Immunol 135(5):3434–3438
98. Ramstedt U, Serhan CN, Nicolaou KC, Webber SE, Wigzell H, Samuelsson B (1987) Lipoxin A-induced inhibition of human natural killer cell cytotoxicity: studies on stereospecificity of inhibition and mode of action. J Immunol 138(1):266–270
99. Brezinski ME, Gimbrone MA Jr, Nicolaou KC, Serhan CN (1989) Lipoxins stimulate prostacyclin generation by human endothelial cells. FEBS Lett 245(1–2):167–172
100. Cezar-de-Mello PF, Nascimento-Silva V, Villela CG, Fierro IM (2006) Aspirin-triggered Lipoxin A4 inhibition of VEGF-induced endothelial cell migration involves actin polymerization and focal adhesion assembly. Oncogene 25(1):122–129
101. Cezar-de-Mello PF, Vieira AM, Nascimento-Silva V, Villela CG, Barja-Fidalgo C, Fierro IM (2008) ATL-1, an analogue of aspirin-triggered lipoxin A4, is a potent inhibitor of several steps in angiogenesis induced by vascular endothelial growth factor. Br J Pharmacol 153(5):956–965
102. Fierro IM, Kutok JL, Serhan CN (2002) Novel lipid mediator regulators of endothelial cell proliferation and migration: aspirin-triggered-15R-lipoxin A(4) and lipoxin A(4). J Pharmacol Exp Ther 300(2):385–392
103. Morris T, Stables M, Hobbs A, de Souza P, Colville-Nash P, Warner T, Newson J, Bellingan G, Gilroy DW (2009) Effects of low-dose aspirin on acute inflammatory responses in humans. J Immunol 183(3):2089–2096
104. Nascimento-Silva V, Arruda MA, Barja-Fidalgo C, Fierro IM (2007) Aspirin-triggered lipoxin A4 blocks reactive oxygen species generation in endothelial cells: a novel antioxidative mechanism. Thromb Haemost 97(1):88–98

105. Nascimento-Silva V, Arruda MA, Barja-Fidalgo C, Villela CG, Fierro IM (2005) Novel lipid mediator aspirin-triggered lipoxin A4 induces heme oxygenase-1 in endothelial cells. Am J Physiol Cell Physiol 289(3):C557–C563
106. Bonnans C, Fukunaga K, Levy MA, Levy BD (2006) Lipoxin A(4) regulates bronchial epithelial cell responses to acid injury. Am J Pathol 168(4):1064–1072
107. Buchanan PJ, McNally P, Harvey BJ, Urbach V (2013) Lipoxin A4-mediated KATP potassium channel activation results in cystic fibrosis airway epithelial repair. Am J Physiol Lung Cell Mol Physiol 305(2):L193–L201
108. Grumbach Y, Quynh NV, Chiron R, Urbach V (2009) LXA4 stimulates ZO-1 expression and transepithelial electrical resistance in human airway epithelial (16HBE14o-) cells. Am J Physiol Lung Cell Mol Physiol 296(1):L101–L108
109. Ho KJ, Spite M, Owens CD, Lancero H, Kroemer AH, Pande R, Creager MA, Serhan CN, Conte MS (2010) Aspirin-triggered lipoxin and resolvin E1 modulate vascular smooth muscle phenotype and correlate with peripheral atherosclerosis. Am J Pathol 177(4):2116–2123
110. Petri MH, Laguna-Fernandez A, Tseng CN, Hedin U, Perretti M, Back M (2014) Aspirin-triggered 15-epi-lipoxin A4 signals through FPR2/ALX in vascular smooth muscle cells and protects against intimal hyperplasia after carotid ligation. Int J Cardiol 179:370–372
111. Sodin-Semrl S, Taddeo B, Tseng D, Varga J, Fiore S (2000) Lipoxin A4 inhibits IL-1 beta-induced IL-6, IL-8, and matrix metalloproteinase-3 production in human synovial fibroblasts and enhances synthesis of tissue inhibitors of metalloproteinases. J Immunol 164(5):2660–2666
112. Wu SH, Wu XH, Lu C, Dong L, Chen ZQ (2006) Lipoxin A4 inhibits proliferation of human lung fibroblasts induced by connective tissue growth factor. Am J Respir Cell Mol Biol 34(1):65–72
113. McMahon B, Mitchell D, Shattock R, Martin F, Brady HR, Godson C (2002) Lipoxin, leukotriene, and PDGF receptors cross-talk to regulate mesangial cell proliferation. FASEB J 16(13):1817–1819
114. Mitchell D, Rodgers K, Hanly J, McMahon B, Brady HR, Martin F, Godson C (2004) Lipoxins inhibit Akt/PKB activation and cell cycle progression in human mesangial cells. Am J Pathol 164(3):937–946
115. Rodgers K, McMahon B, Mitchell D, Sadlier D, Godson C (2005) Lipoxin A4 modifies platelet-derived growth factor-induced pro-fibrotic gene expression in human renal mesangial cells. Am J Pathol 167(3):683–694
116. Arita M, Ohira T, Sun Y-P, Elangovan S, Chiang N, Serhan CN (2007) Resolvin E1 selectively interacts with leukotriene B4 receptor BLT1 and ChemR23 to regulate inflammation. J Immunol 178(6):3912–3917
117. Campbell EL, Louis NA, Tomassetti SE, Canny GO, Arita M, Serhan CN, Colgan SP (2007) Resolvin E1 promotes mucosal surface clearance of neutrophils: a new paradigm for inflammatory resolution. FASEB J 21(12):3162–3170
118. Dona M, Fredman G, Schwab JM, Chiang N, Arita M, Goodarzi A, Cheng G, von Andrian UH, Serhan CN (2008) Resolvin E1, an EPA-derived mediator in whole blood, selectively counterregulates leukocytes and platelets. Blood 112(3):848–855
119. Fredman G, Van Dyke TE, Serhan CN (2010) Resolvin E1 regulates adenosine diphosphate activation of human platelets. Arterioscler Thromb Vasc Biol 30(10):2005–2013
120. Herrera BS, Ohira T, Gao L, Omori K, Yang R, Zhu M, Muscara MN, Serhan CN, Van Dyke TE, Gyurko R (2008) An endogenous regulator of inflammation, resolvin E1, modulates osteoclast differentiation and bone resorption. Br J Pharmacol 155(8):1214–1223
121. Zhu M, Van Dyke TE, Gyurko R (2013) Resolvin E1 regulates osteoclast fusion via DC-STAMP and NFATc1. FASEB J 27(8):3344–3353
122. Sun YP, Oh SF, Uddin J, Yang R, Gotlinger K, Campbell E, Colgan SP, Petasis NA, Serhan CN (2007) Resolvin D1 and its aspirin-triggered 17R epimer. Stereochemical assignments, anti-inflammatory properties, and enzymatic inactivation. J Biol Chem 282(13):9323–9334

123. Krishnamoorthy S, Recchiuti A, Chiang N, Fredman G, Serhan CN (2012) Resolvin D1 receptor stereoselectivity and regulation of inflammation and proresolving microRNAs. Am J Pathol 180(5):2018–2027
124. Krishnamoorthy S, Recchiuti A, Chiang N, Yacoubian S, Lee CH, Yang R, Petasis NA, Serhan CN (2010) Resolvin D1 binds human phagocytes with evidence for proresolving receptors. Proc Natl Acad Sci USA 107(4):1660–1665
125. Recchiuti A, Krishnamoorthy S, Fredman G, Chiang N, Serhan CN (2011) MicroRNAs in resolution of acute inflammation: identification of novel resolvin D1-miRNA circuits. FASEB J 25(2):544–560
126. Norling LV, Dalli J, Flower RJ, Serhan CN, Perretti M (2012) Resolvin D1 limits polymorphonuclear leukocytes recruitment to inflammatory loci: receptor-dependent actions. Arterioscler Thromb Vasc Biol 32(8):1970–1978
127. Miyahara T, Runge S, Chatterjee A, Chen M, Mottola G, Fitzgerald JM, Serhan CN, Conte MS (2013) D-series resolvin attenuates vascular smooth muscle cell activation and neointimal hyperplasia following vascular injury. FASEB J 27(6):2220–2232
128. Mustafa M, Zarrough A, Bolstad AI, Lygre H, Mustafa K, Hasturk H, Serhan C, Kantarci A, Van Dyke TE (2013) Resolvin D1 protects periodontal ligament. Am J Physiol Cell Physiol 305(6):C673–C679
129. Ramon S, Gao F, Serhan CN, Phipps RP (2012) Specialized proresolving mediators enhance human B cell differentiation to antibody-secreting cells. J Immunol 189(2):1036–1042
130. Hsiao HM, Sapinoro RE, Thatcher TH, Croasdell A, Levy EP, Fulton RA, Olsen KC, Pollock SJ, Serhan CN, Phipps RP, Sime PJ (2013) A novel anti-inflammatory and pro-resolving role for resolvin D1 in acute cigarette smoke-induced lung inflammation. PLoS One 8(3), e58258
131. Clish CB, Levy BD, Chiang N, Tai HH, Serhan CN (2000) Oxidoreductases in lipoxin A4 metabolic inactivation: a novel role for 15-onoprostaglandin 13-reductase/leukotriene B4 12-hydroxydehydrogenase in inflammation. J Biol Chem 275(33):25372–25380
132. Sun YP, Tjonahen E, Keledjian R, Zhu M, Yang R, Recchiuti A, Pillai PS, Petasis NA, Serhan CN (2009) Anti-inflammatory and pro-resolving properties of benzo-lipoxin A(4) analogs. Prostaglandins Leukot Essent Fatty Acids 81(5–6):357–366
133. Clish CB, O'Brien JA, Gronert K, Stahl GL, Petasis NA, Serhan CN (1999) Local and systemic delivery of a stable aspirin-triggered lipoxin prevents neutrophil recruitment in vivo. Proc Natl Acad Sci USA 96(14):8247–8252
134. Takano T, Fiore S, Maddox JF, Brady HR, Petasis NA, Serhan CN (1997) Aspirin-triggered 15-epi-lipoxin A4 (LXA4) and LXA4 stable analogues are potent inhibitors of acute inflammation: evidence for anti-inflammatory receptors. J Exp Med 185(9):1693–1704
135. Bohr S, Patel SJ, Sarin D, Irimia D, Yarmush ML, Berthiaume F (2012) Resolvin D2 prevents secondary thrombosis and necrosis in a mouse burn wound model. Wound Repair Regen 21 (1):35–43
136. Kim TH, Kim GD, Jin YH, Park YS, Park CS (2012) Omega-3 fatty acid-derived mediator, resolvin E1, ameliorates 2,4-dinitrofluorobenzene-induced atopic dermatitis in NC/Nga mice. Int Immunopharmacol 14(4):384–391
137. Wu SH, Chen XQ, Liu B, Wu HJ, Dong L (2012) Efficacy and safety of 15(R/S)-methyl-lipoxin A(4) in topical treatment of infantile eczema. Br J Dermatol 168(1):172–178
138. Walker J, Dichter E, Lacorte G, Kerner D, Spur B, Rodriguez A, Yin K (2011) Lipoxin a4 increases survival by decreasing systemic inflammation and bacterial load in sepsis. Shock 36 (4):410–416
139. Arita M, Yoshida M, Hong S, Tjonahen E, Glickman JN, Petasis NA, Blumberg RS, Serhan CN (2005) Resolvin E1, an endogenous lipid mediator derived from omega-3 eicosapentaenoic acid, protects against 2,4,6-trinitrobenzene sulfonic acid-induced colitis. Proc Natl Acad Sci USA 102(21):7671–7676
140. Bento AF, Claudino RF, Dutra RC, Marcon R, Calixto JB (2011) Omega-3 fatty acid-derived mediators 17(R)-hydroxy docosahexaenoic acid, aspirin-triggered resolvin D1 and resolvin D2 prevent experimental colitis in mice. J Immunol 187(4):1957–1969

141. Campbell EL, MacManus CF, Kominsky DJ, Keely S, Glover LE, Bowers BE, Scully M, Bruyninckx WJ, Colgan SP (2010) Resolvin E1-induced intestinal alkaline phosphatase promotes resolution of inflammation through LPS detoxification. Proc Natl Acad Sci USA 107(32):14298–14303
142. Aoki H, Hisada T, Ishizuka T, Utsugi M, Kawata T, Shimizu Y, Okajima F, Dobashi K, Mori M (2008) Resolvin E1 dampens airway inflammation and hyperresponsiveness in a murine model of asthma. Biochem Biophys Res Commun 367(2):509–515
143. Aoki H, Hisada T, Ishizuka T, Utsugi M, Ono A, Koga Y, Sunaga N, Nakakura T, Okajima F, Dobashi K, Mori M (2010) Protective effect of resolvin E1 on the development of asthmatic airway inflammation. Biochem Biophys Res Commun 400(1):128–133
144. Barnig C, Cernadas M, Dutile S, Liu X, Perrella MA, Kazani S, Wechsler ME, Israel E, Levy BD (2013) Lipoxin A4 regulates natural killer cell and type 2 innate lymphoid cell activation in asthma. Sci Transl Med 5(174):174ra126
145. Christie PE, Spur BW, Lee TH (1992) The effects of lipoxin A4 on airway responses in asthmatic subjects. Am Rev Respir Dis 145(6):1281–1284
146. Krishnamoorthy N, Burkett PR, Dalli J, Abdulnour RE, Colas R, Ramon S, Phipps RP, Petasis NA, Kuchroo VK, Serhan CN, Levy BD (2015) Cutting edge: maresin-1 engages regulatory T cells to limit type 2 innate lymphoid cell activation and promote resolution of lung inflammation. J Immunol 194(3):863–867
147. Levy BD, De Sanctis GT, Devchand PR, Kim E, Ackerman K, Schmidt BA, Szczeklik W, Drazen JM, Serhan CN (2002) Multi-pronged inhibition of airway hyper-responsiveness and inflammation by lipoxin A(4). Nat Med 8(9):1018–1023
148. Rogerio AP, Haworth O, Croze R, Oh SF, Uddin M, Carlo T, Pfeffer MA, Priluck R, Serhan CN, Levy BD (2012) Resolvin D1 and aspirin-triggered resolvin D1 promote resolution of allergic airways responses. J Immunol 189(4):1983–1991
149. Karp CL, Flick LM, Park KW, Softic S, Greer TM, Keledjian R, Yang R, Uddin J, Guggino WB, Atabani SF, Belkaid Y, Xu Y, Whitsett JA, Accurso FJ, Wills-Karp M, Petasis NA (2004) Defective lipoxin-mediated anti-inflammatory activity in the cystic fibrosis airway. Nat Immunol 5(4):388–392
150. Abdulnour RE, Dalli J, Colby JK, Krishnamoorthy N, Timmons JY, Tan SH, Colas RA, Petasis NA, Serhan CN, Levy BD (2014) Maresin 1 biosynthesis during platelet-neutrophil interactions is organ-protective. Proc Natl Acad Sci USA 111(46):16526–16531
151. El Kebir D, Gjorstrup P, Filep JG (2012) Resolvin E1 promotes phagocytosis-induced neutrophil apoptosis and accelerates resolution of pulmonary inflammation. Proc Natl Acad Sci USA 109(37):14983–14988
152. El Kebir D, Jozsef L, Pan W, Wang L, Petasis NA, Serhan CN, Filep JG (2009) 15-epi-lipoxin A4 inhibits myeloperoxidase signaling and enhances resolution of acute lung injury. Am J Respir Crit Care Med 180(4):311–319
153. Tang H, Liu Y, Yan C, Petasis NA, Serhan CN, Gao H (2014) Protective actions of aspirin-triggered (17R) resolvin D1 and its analogue, 17R-hydroxy-19-para-fluorophenoxy-resolvin D1 methyl ester, in C5a-dependent IgG immune complex-induced inflammation and lung injury. J Immunol 193(7):3769–3778
154. Wang B, Gong X, Wan JY, Zhang L, Zhang Z, Li HZ, Min S (2011) Resolvin D1 protects mice from LPS-induced acute lung injury. Pulm Pharmacol Ther 24(4):434–441
155. Hasturk H, Kantarci A, Goguet-Surmenian E, Blackwood A, Andry C, Serhan CN, Van Dyke TE (2007) Resolvin E1 regulates inflammation at the cellular and tissue level and restores tissue homeostasis in vivo. J Immunol 179(10):7021–7029
156. Hasturk H, Kantarci A, Ohira T, Arita M, Ebrahimi N, Chiang N, Petasis NA, Levy BD, Serhan CN, Van Dyke TE (2006) RvE1 protects from local inflammation and osteoclast-mediated bone destruction in periodontitis. FASEB J 20(2):401–403
157. Serhan CN, Jain A, Marleau S, Clish C, Kantarci A, Behbehani B, Colgan SP, Stahl GL, Merched A, Petasis NA, Chan L, Van Dyke TE (2003) Reduced inflammation and tissue

damage in transgenic rabbits overexpressing 15-lipoxygenase and endogenous anti-inflammatory lipid mediators. J Immunol 171(12):6856–6865
158. Connor KM, SanGiovanni JP, Lofqvist C, Aderman CM, Chen J, Higuchi A, Hong S, Pravda EA, Majchrzak S, Carper D, Hellstrom A, Kang JX, Chew EY, Salem N, Serhan CN, Smith LEH (2007) Increased dietary intake of omega-3-polyunsaturated fatty acids reduces pathological retinal angiogenesis. Nat Med 13(7):868–873
159. Jin Y, Arita M, Zhang Q, Saban DR, Chauhan SK, Chiang N, Serhan CN, Dana R (2009) Anti-angiogenesis effect of the novel anti-inflammatory and pro-resolving lipid mediators. Invest Ophthalmol Vis Sci 50(10):4743–4752
160. Li N, He J, Schwartz CE, Gjorstrup P, Bazan HEP (2010) Resolvin E1 improves tear production and decreases inflammation in a dry eye mouse model. J Ocul Pharmacol Ther 26(5):431–439
161. Rajasagi NK, Reddy PB, Mulik S, Gjorstrup P, Rouse BT (2013) Neuroprotectin D1 reduces the severity of herpes simplex virus-induced corneal immunopathology. Invest Ophthalmol Vis Sci 54(9):6269–6279
162. Rajasagi NK, Reddy PBJ, Suryawanshi A, Mulik S, Gjorstrup P, Rouse BT (2011) Controlling herpes simplex virus-induced ocular inflammatory lesions with the lipid-derived mediator resolvin E1. J Immunol 186(3):1735–1746
163. Settimio R, Clara DF, Franca F, Francesca S, Michele D (2012) Resolvin D1 reduces the immunoinflammatory response of the rat eye following uveitis. Mediators Inflamm 2012:318621
164. Conte FP, Menezes-de-Lima O Jr, Verri WA Jr, Cunha FQ, Penido C, Henriques MG (2010) Lipoxin A(4) attenuates zymosan-induced arthritis by modulating endothelin-1 and its effects. Br J Pharmacol 161(4):911–924
165. Lima-Garcia JF, Dutra RC, da Silva K, Motta EM, Campos MM, Calixto JB (2011) The precursor of resolvin D series and aspirin-triggered resolvin D1 display anti-hyperalgesic properties in adjuvant-induced arthritis in rats. Br J Pharmacol 164(2):278–293
166. Norling LV, Spite M, Yang R, Flower RJ, Perretti M, Serhan CN (2011) Cutting edge: Humanized nano-proresolving medicines mimic inflammation-resolution and enhance wound healing. J Immunol 186(10):5543–5547
167. Brancaleone V, Gobbetti T, Cenac N, le Faouder P, Colom B, Flower RJ, Vergnolle N, Nourshargh S, Perretti M (2013) A vasculo-protective circuit centered on lipoxin A4 and aspirin-triggered 15-epi-lipoxin A4 operative in murine microcirculation. Blood 122(4):608–617
168. Chiang N, Gronert K, Clish CB, O'Brien JA, Freeman MW, Serhan CN (1999) Leukotriene B4 receptor transgenic mice reveal novel protective roles for lipoxins and aspirin-triggered lipoxins in reperfusion. J Clin Invest 104(3):309–316
169. Duffield JS, Hong S, Vaidya VS, Lu Y, Fredman G, Serhan CN, Bonventre JV (2006) Resolvin D series and protectin D1 mitigate acute kidney injury. J Immunol 177(9):5902–5911
170. Keyes KT, Ye Y, Lin Y, Zhang C, Perez-Polo JR, Gjorstrup P, Birnbaum Y (2010) Resolvin E1 protects the rat heart against reperfusion injury. Am J Physiol Heart Circ Physiol 299(1): H153–H164
171. Gonzalez-Periz A, Horrillo R, Ferre N, Gronert K, Dong B, Moran-Salvador E, Titos E, Martinez-Clemente M, Lopez-Parra M, Arroyo V, Claria J (2009) Obesity-induced insulin resistance and hepatic steatosis are alleviated by omega-3 fatty acids: a role for resolvins and protectins. FASEB J 23(6):1946–1957
172. Hellmann J, Tang Y, Kosuri M, Bhatnagar A, Spite M (2011) Resolvin D1 decreases adipose tissue macrophage accumulation and improves insulin sensitivity in obese-diabetic mice. FASEB J 25(7):2399–2407
173. Rius B, Titos E, Moran-Salvador E, Lopez-Vicario C, Garcia-Alonso V, Gonzalez-Periz A, Arroyo V, Claria J (2014) Resolvin D1 primes the resolution process initiated by calorie restriction in obesity-induced steatohepatitis. FASEB J 28(2):836–848

174. Tang Y, Zhang MJ, Hellmann J, Kosuri M, Bhatnagar A, Spite M (2013) Proresolution therapy for the treatment of delayed healing of diabetic wounds. Diabetes 62(2):37–41
175. Titos E, Rius B, Gonzalez-Periz A, Lopez-Vicario C, Moran-Salvador E, Martinez-Clemente M, Arroyo V, Claria J (2011) Resolvin D1 and its precursor docosahexaenoic acid promote resolution of adipose tissue inflammation by eliciting macrophage polarization toward an M2-like phenotype. J Immunol 187(10):5408–5418
176. Abdelmoaty S, Wigerblad G, Bas DB, Codeluppi S, Fernandez-Zafra T, el El-Awady S, Moustafa Y, Abdelhamid Ael D, Brodin E, Svensson CI (2013) Spinal actions of lipoxin A4 and 17(R)-resolvin D1 attenuate inflammation-induced mechanical hypersensitivity and spinal TNF release. PLoS One 8(9), e75543
177. Bang S, Yoo S, Yang TJ, Cho H, Hwang SW (2011) 17(R)-resolvin D1 specifically inhibits transient receptor potential ion channel vanilloid 3 leading to peripheral antinociception. Br J Pharmacol 165(3):683–692
178. Huang L, Wang CF, Serhan CN, Strichartz G (2011) Enduring prevention and transient reduction of postoperative pain by intrathecal resolvin D1. Pain 152(3):557–565
179. Park CK, Xu ZZ, Liu T, Lu N, Serhan CN, Ji RR (2011) Resolvin D2 is a potent endogenous inhibitor for transient receptor potential subtype V1/A1, inflammatory pain, and spinal cord synaptic plasticity in mice: distinct roles of resolvin D1, D2, and E1. J Neurosci 31 (50):18433–18438
180. Quan-Xin F, Fan F, Xiang-Ying F, Shu-Jun L, Shi-Qi W, Zhao-Xu L, Xu-Jie Z, Qing-Chuan Z, Wei W (2012) Resolvin D1 reverses chronic pancreatitis-induced mechanical allodynia, phosphorylation of NMDA receptors, and cytokines expression in the thoracic spinal dorsal horn. BMC Gastroenterol 12(1):148
181. Svensson CI, Zattoni M, Serhan CN (2007) Lipoxins and aspirin-triggered lipoxin inhibit inflammatory pain processing. J Exp Med 204(2):245–252
182. Xu Z-Z, Zhang L, Liu T, Park JY, Berta T, Yang R, Serhan CN, Ji R-R (2010) Resolvins RvE1 and RvD1 attenuate inflammatory pain via central and peripheral actions. Nat Med 16 (5):592–597
183. Xu ZZ, Berta T, Ji RR (2012) Resolvin E1 inhibits neuropathic pain and spinal cord microglial activation following peripheral nerve injury. J Neuroimmune Pharmacol 8 (1):37–41
184. Devchand PR, Schmidt BA, Primo VC, Q-y Z, Arnaout MA, Serhan CN, Nikolic B (2005) A synthetic eicosanoid LX-mimetic unravels host-donor interactions in allogeneic BMT-induced GvHD to reveal an early protective role for host neutrophils. FASEB J 19 (2):203–210
185. Levy BD, Zhang QY, Bonnans C, Primo V, Reilly JJ, Perkins DL, Liang Y, Amin Arnaout M, Nikolic B, Serhan CN (2011) The endogenous pro-resolving mediators lipoxin A4 and resolvin E1 preserve organ function in allograft rejection. Prostaglandins Leukot Essent Fatty Acids 84(1–2):43–50
186. Molina-Berrios A, Campos-Estrada C, Henriquez N, Faundez M, Torres G, Castillo C, Escanilla S, Kemmerling U, Morello A, Lopez-Munoz RA, Maya JD (2013) Protective role of acetylsalicylic acid in experimental trypanosoma cruzi infection: evidence of a 15-epi-lipoxin A(4)-mediated effect. PLoS Negl Trop Dis 7(4), e2173
187. Shryock N, McBerry C, Salazar Gonzalez RM, Janes S, Costa FT, Aliberti J (2013) Lipoxin A (4) and 15-epi-lipoxin A(4) protect against experimental cerebral malaria by inhibiting IL-12/IFN-gamma in the brain. PLoS One 8(4), e61882
188. Tobin DM, Roca FJ, Oh SF, McFarland R, Vickery TW, Ray JP, Ko DC, Zou Y, Bang ND, Chau TT, Vary JC, Hawn TR, Dunstan SJ, Farrar JJ, Thwaites GE, King MC, Serhan CN, Ramakrishnan L (2012) Host genotype-specific therapies can optimize the inflammatory response to mycobacterial infections. Cell 148(3):434–446
189. Borgeson E, Docherty NG, Murphy M, Rodgers K, Ryan A, O'Sullivan TP, Guiry PJ, Goldschmeding R, Higgins DF, Godson C (2011) Lipoxin A(4) and benzo-lipoxin A(4) attenuate experimental renal fibrosis. FASEB J 25(9):2967–2979

190. Merched AJ, Ko K, Gotlinger KH, Serhan CN, Chan L (2008) Atherosclerosis: evidence for impairment of resolution of vascular inflammation governed by specific lipid mediators. FASEB J 22(10):3595–3606
191. Norling LV, Dalli J, Flower RJ, Serhan CN, Perretti M (2012) Resolvin D1 limits polymorphonuclear leukocyte recruitment to inflammatory loci: receptor-dependent actions. Arterioscler Thromb Vasc Biol 32(8):1970–1978
192. Green EL (1966) Biology of the laboratory mouse, 2nd edn. Dover Publications, New York
193. Claria J, Dalli J, Yacoubian S, Gao F, Serhan CN (2012) Resolvin D1 and resolvin D2 govern local inflammatory tone in obese fat. J Immunol 189(5):2597–2605
194. Fredman G, Ozcan L, Spolitu S, Hellmann J, Spite M, Backs J, Tabas I (2014) Resolvin D1 limits 5-lipoxygenase nuclear localization and leukotriene B4 synthesis by inhibiting a calcium-activated kinase pathway. Proc Natl Acad Sci USA 111(40):14530–14535
195. Sheedy FJ, Palsson-McDermott E, Hennessy EJ, Martin C, O'Leary JJ, Ruan Q, Johnson DS, Chen Y, O'Neill LAJ (2010) Negative regulation of TLR4 via targeting of the proinflammatory tumor suppressor PDCD4 by the microRNA miR-21. Nat Immunol 11(2):141–147
196. Fredman G, Li Y, Dalli J, Chiang N, Serhan CN (2012) Self-limited versus delayed resolution of acute inflammation: temporal regulation of pro-resolving mediators and microRNA. Sci Rep 2:639
197. Li Y, Dalli J, Chiang N, Baron RM, Quintana C, Serhan CN (2013) Plasticity of leukocytic exudates in resolving acute inflammation is regulated by microRNA and proresolving mediators. Immunity 39(5):885–898
198. Brennan EP, Nolan KA, Borgeson E, Gough OS, McEvoy CM, Docherty NG, Higgins DF, Murphy M, Sadlier DM, Ali-Shah ST, Guiry PJ, Savage DA, Maxwell AP, Martin F, Godson C (2013) Lipoxins attenuate renal fibrosis by inducing let-7c and suppressing TGFbetaR1. J Am Soc Nephrol 24(4):627–637
199. Fiore S, Ryeom SW, Weller PF, Serhan CN (1992) Lipoxin recognition sites. Specific binding of labeled lipoxin A4 with human neutrophils. J Biol Chem 267(23):16168–16176
200. Murphy PM, Ozcelik T, Kenney RT, Tiffany HL, McDermott D, Francke U (1992) A structural homologue of the N-formyl peptide receptor. Characterization and chromosome mapping of a peptide chemoattractant receptor family. J Biol Chem 267(11):7637–7643
201. Fiore S, Maddox JF, Perez HD, Serhan CN (1994) Identification of a human cDNA encoding a functional high affinity lipoxin A4 receptor. J Exp Med 180(1):253–260
202. Ye RD, Boulay F, Wang JM, Dahlgren C, Gerard C, Parmentier M, Serhan CN, Murphy PM (2009) International union of basic and clinical pharmacology. LXXIII. Nomenclature for the formyl peptide receptor (FPR) family. Pharmacol Rev 61(2):119–161
203. Romano M, Recchia I, Recchiuti A (2007) Lipoxin receptors. TheScientificWorldJournal 7:1393–1412
204. Chiang N, Takano T, Arita M, Watanabe S, Serhan CN (2003) A novel rat lipoxin A4 receptor that is conserved in structure and function. Br J Pharmacol 139(1):89–98
205. Chiang N, Serhan CN, Dahlen SE, Drazen JM, Hay DW, Rovati GE, Shimizu T, Yomizo T, Brink C (2006) The lipoxin receptor ALX: potent ligand-specific and stereoselective actions in vivo. Pharmacol Rev 58(3):463–487
206. Cooray SN, Gobbetti T, Montero-Melendez T, McArthur S, Thompson D, Clark AJL, Flower RJ, Perretti M (2013) Ligand-specific conformational change of the G-protein-coupled receptor ALX/FPR2 determines proresolving functional responses. Proc Natl Acad Sci 110(45):18232–18237
207. Devchand PR, Arita M, Hong S, Bannenberg G, Moussignac RL, Gronert K, Serhan CN (2003) Human ALX receptor regulates neutrophil recruitment in transgenic mice: roles in inflammation and host defense. FASEB J 17(6):652–659
208. Dufton N, Hannon R, Brancaleone V, Dalli J, Patel HB, Gray M, D'Acquisto F, Buckingham JC, Perretti M, Flower RJ (2010) Anti-inflammatory role of the murine formyl-peptide

receptor 2: ligand-specific effects on leukocyte responses and experimental inflammation. J Immunol 184(5):2611–2619
209. Gobbetti T, Coldewey SM, Chen J, McArthur S, le Faouder P, Cenac N, Flower RJ, Thiemermann C, Perretti M (2014) Nonredundant protective properties of FPR2/ALX in polymicrobial murine sepsis. Proc Natl Acad Sci USA 111(52):18685–18690
210. Morris T, Stables M, Colville-Nash P, Newson J, Bellingan G, de Souza PM, Gilroy DW (2010) Dichotomy in duration and severity of acute inflammatory responses in humans arising from differentially expressed proresolution pathways. Proc Natl Acad Sci USA 107 (19):8842–8847
211. Simiele F, Recchiuti A, Mattoscio D, De Luca A, Cianci E, Franchi S, Gatta V, Parolari A, Werba JP, Camera M, Favaloro B, Romano M (2012) Transcriptional regulation of the human FPR2/ALX gene: evidence of a heritable genetic variant that impairs promoter activity. FASEB J 26(3):1323–1333
212. Waechter V, Schmid M, Herova M, Weber A, Gunther V, Marti-Jaun J, Wust S, Rosinger M, Gemperle C, Hersberger M (2012) Characterization of the promoter and the transcriptional regulation of the lipoxin A4 receptor (FPR2/ALX) gene in human monocytes and macrophages. J Immunol 188(4):1856–1867
213. Pierdomenico AM, Recchiuti A, Simiele F, Codagnone M, Mari VC, Davi G, Romano M (2014) MicroRNA-181b regulates ALX/FPR2 expression and proresolution signaling in human macrophages. J Biol Chem 290(6):3592–3600
214. Gavins FN, Hughes EL, Buss NA, Holloway PM, Getting SJ, Buckingham JC (2011) Leukocyte recruitment in the brain in sepsis: involvement of the annexin 1-FPR2/ALX anti-inflammatory system. FASEB J 26(12):4977–4989
215. Lee HN, Surh YJ (2013) Resolvin D1-mediated NOX2 inactivation rescues macrophages undertaking efferocytosis from oxidative stress-induced apoptosis. Biochem Pharmacol 86 (6):759–769
216. Dalli J, Kraft BD, Colas RA, Shinohara M, Fredenburgh LE, Hess DR, Chiang N, Welty-Wolf KE, Choi AM, Piantadosi CA, Serhan CN (2015) Proresolving lipid mediator profiles in baboon pneumonia are regulated by inhaled carbon monoxide. Am J Respir Cell Mol Biol 53 (3):314–325

Index

A
airway smooth muscle cells (ASMC), 90
ALX/FPR2 receptor, 223
arachidonate 15-lipoxygenase-1 orthologs (ALOX15)
    bioactivities of, 50
    catalytic multiplicity
        leukotriene synthase activity, 71–72
        lipohydroperoxidase activity, 70–71
        lipoxygenase activity, 65–70
    degree of homology, of mammalian, 62
    enzymology
        human, 57–59
        murine, 60–61
        nonhuman primates, 59–60
        porcine, 56
        rabbit, 54–55
    multiplicity, 49
    perspectives, 72–73
    structural biology
        pig, 64
        rabbit, 62–64
arachidonic acid (AA)
    leukotrienes, 31, 32
    lipoxygenation, 102
    metabolism, 102
    oxygenation, 61
    oxygen insertion, 1, 2
    p12-LOX, 166
aspirin induced asthma (AIA), 42
asthma, 200
atherosclerosis
    animal models
        omega-3 supplementation, 115
        targeting 5-LO/FLAP, 111
        targeting 12/15-LO, 113
        targeting LT receptors, 113
        targeting LXA_4 receptor, 114
    effects of 5-LO knock-out, 112
    effects of 12/15-LO knock-out, 114
    effects of omega-3 supplementation, 116
    E-series resolvin signaling, 111
    expression and metabolism, 108
    LT signaling, 109
    LXs signaling, 109, 110
    pathophysiology and clinical context, 107
atopic dermatitis, 173
ATP, 15

B
broken helix, 10

C
cancer
    breast cancer, 134
    colorectal cancer, 138
    hemato-oncological malignancies, 141
    inhibitors and mitogenic effects, 145
    lacking therapeutic efficacy, 146
    pancreatic cancer, 133
    prostate cancer, 137
chemokine-like receptor 1, 224
chronic inflammatory diseases, 90
coactosin-like protein (CLP), 12
cysteinyl-leukotrienes (cysLTs)
    bioactions, 38
    classical bioactions of, 39

cysteinyl-leukotrienes (cysLTs) (*cont.*)
   elicit, 38
   LTC4S, 41

D
DDM. *See* Dodecylmaltoside (DDM)
diacylglycerides, 16
Dicer, 12
diHETEs, 188
DNA methylation, 19
docosahexaenoic acid (DHA), 85
dodecylmaltoside (DDM), 40
D-series resolvins
   LO-driven, 217
   LX, 221–223
   receptors, 225
   RvD1 and AT-RvD1, 221–223

E
electron paramagnetic resonance (EPR), 10
endogenous PUFA, 198
eosinophil chemoattractants, 193
eosinophilic diseases, 200
epidermis-type LOX-3 (eLOX-3), 168
E-series resolvins
   LO-driven, 216–217
   LX, 221
   receptors, 225–226
   RvE1, 219–221

F
FLAP. *See* 5-lipoxygenase-activating protein (FLAP)

G
GC-boxes, 20, 21, 56
glutathione peroxidases, 92
granulocytes, 192
GSH molecule, 40

H
helicobacter pylori, 90
hemostasis, 86, 89
12(S)-HETE, 87
human 15-lipoxygenase-2, 167
human ALOX15, 57
human GPR32, 225

5-hydroxyeicosanoid dehydrogenase (5-HEDH), 187, 189
12-hydroxyeicosatetrenoate (12-HETE), 5, 87

I
inflammation
   classical cardinal signs, 213
   injury and wound healing, 170
   oxylipins on, 89
   resolution response, 212

L
leukotriene, 5
   biosynthesis, 8, 14
   formation, 18
leukotriene A_4 (LTA$_4$), 186
leukotriene A4 hydrolase (LTA4H)
   aminopeptidase activity, 33
   catalytic mechanism, 36
   cellular location, 33
   crystal structure, 32, 34
   drug target, 37
leukotriene C4 synthase (LTC4S)
   drug target, 41
   MAPEG members, 39
   regulation, 39
   topology, 38
leukotriene synthase activity, 71, 72
lipohydroperoxidase
   activity, 70
   mechanistic scheme, 66
lipoxin A4 receptor, 223
lipoxins (LX)
   biological actions
      D-series resolvins, 221–223
      E-series resolvins, 221
      lipoxins, 218
   mechanisms of action
      D-series resolvin receptors, 225–226
      E-series resolvin receptors, 225–226
      receptor identification and cell expression, 223–224
5-lipoxygenase (5-LO), 162
   discovery, 7
   domains, 10
   gene expression, 19
      DNA methylation determines expression, 19
      effect of trichostatin A, 20
      possible regulation by alternative splicing and by miRNA, 22

Index

upregulation by TGFß and 1,25(OH)
  2D3, 21
vitamin D responsive elements, in
  promotor, 21
interacting proteins
  coactosin, 12
  dicer, 12
  FLAP, 11
in leukotriene biosynthesis, 8
regulation, of enzyme activity, 13
  ATP, 15
  Ca^{2+}, 14
  cell membranes and PC, 15
  diacylglycerides, 16
  leukotriene formation, gender-related
    differences, 18
  phosphorylation, 17
  redox regulation, 16
  subcellular localization, 18
pathway, 186
structure of, 10
5-lipoxygenase activating protein (FLAP), 11,
  31
12-lipoxygenase (12(S)-LOX), 85
regulation, oxylipin
  exposure, 92, 93
  signal outcome, 93
lipoxygenases (LOXs)
activity, 65, 68
  biomembranes, 69
  cholesterol esters, 69
  fatty acids, 65
  free fatty acids, 68
  lipoproteins, 69
  oxygenation of phospholipids, 69
arachidonic acid
  metabolism of, 102, 103
  omega 3 fatty acids, 104
atherosclerosis
  animal models, 111, 113–115
  E-series resolvin signaling
    implications, 111
  expression and metabolism, 108
  LT signaling, 109
  LXs signaling
    implications, 109, 110
  pathophysiology and clinical context,
    107
catalytic property, 1
ChemR23, 107
clinical implications
  anti-LTs, 120
  omega 3 fatty acids, 121
epidermis-type lipoxygenase 3, 168
family, 84
FPR2/ALX, 106
future direction, 147
GPR32, 106
human 15-lipoxygenase-1, 165
human 15-lipoxygenase-2, 167
inflammatory skin diseases, 172
  atopic dermatitis, 173
  psoriasis, 172
  systemic sclerosis, 173
intimal hyperplasia, 117
  LT receptors, 118
  LX-signaling in smooth muscle cells,
    119
  pathophysiology and clinical context,
    117
  resolvin-signaling in smooth muscle
    cells, 119
LT receptors, 104
mechanistic scheme, 65
metabolism of omega-3 fatty acids, 105
mouse 8-lipoxygenase, 167
multiplicity, 49
  bacterial genes and LOX-isoforms, 53
  human genes and LOX-isoforms, 49
  mouse genes and LOX-isoforms, 50
  zebrafish genes and LOX-isoforms, 51
oxylipin effect, 86
oxylipins in inflammation, 89
physiological and pathophysiological
  roles, 5
in platelets, 84
platelet-type 12-LOX, 166
regulation, 2
12R-lipoxygenase, 168
skin
  cancer, 174
  inflammation, injury and wound
    healing, 170
  metabolites in human and
    mouse, 162
  12R-LOX/eLOX-3 role, 169
  structural and metabolic functions,
    160
tumorigenesis role
  breast cancer, 134
  colorectal cancer, 138
  hemato-oncological malignancies, 141
  inhibitors and mitogenic effects, 145
  lacking therapeutic efficacy, 146
  pancreatic cancer, 133
  prostate cancer, 137

LO-driven LX, 215–216
LO-driven Rv
   D-series resolvins, 217
   E-series resolvins, 216–217
   maresins, 218
   (Neuro)protectins, 217, 218

M
Maresins, 218
membrane-Associated Proteins in Eicosanoid and Glutathione Metabolism (MAPEG), 39
mouse 8-lipoxygenase, 167
murine ALOX15 orthologs, 60

N
(neuro)protectins, 217
neutrophils, 11, 192

O
omega-3 fatty acids, 93
oxoeicosanoid (OXE) receptor
   antagonists, 199
   expression, 196
   signaling mechanisms, 196
   structural requirements for activation, 197, 198
5-Oxo-6E,8Z,11Z,14Z-eicosatetraenoic acid (5-Oxo-ETE)
   biological activities, 198
   cellular synthesis regulation, 190
   5-HEDH
      distribution, 189
      properties, 188
      substrate selectivity, 187
      transcellular biosynthesis, 189
   leukocytes effects
      granulocytes, 192
      monocytes, 194
   metabolism, 191
   OXE receptor, 194, 195
      antagonists, 199
      expression, 196
      signaling mechanisms, 196
      structure-activity relationships, 197
   potential roles, 202
      asthma, 200
      cancer, 201
      eosinophilic diseases, 200
   synthesis pathways, 190
Oxylipins, 88, 91

P
phosphorylation, 17
platelet-activating factor (PAF), 33
platelets, 86, 89
   hemostatic and immune functions of, 84
   hemostatic role, 83
   influencing matters, for 12(S)-LOX signal outcome, 93
   lipoxygenases
      oxylipin effect, 86
      inflammation, 89
   12(S)-LOX, 89
   mechanism for regulation, 88
   original characterization of, 86
   pro-inflammatory role, 84
   regulation, of 12(S)-LOX oxylipin exposure, 92
   trans-cellular regulation, 90
platelet-type 12-LOX (p12-LOX), 166
polymorphonuclear leukocytes (PMNs), 22, 32, 214
polymorphonuclear neutrophils (PMN), 111, 170, 213
polyunsaturated fatty acids (PUFA), 1, 8, 84, 165, 168, 190, 213
porcine ALOX15 (pigALOX15), 56, 64, 69
Pro-Gly-Pro (PGP), 33, 34
prothrombotic observations, 87
psoriasis, 172
PUFAs. *See* Polyunsaturated fatty acids (PUFA)

R
rabbit ALOX15 (rabALOX15)
   arachidonic acid oxygenation, 55, 72
   calcium concentrations, 55
   fatty acid oxygenation, 55, 68
   reaction kinetics, 54
   structural biology
      crystal structure, 62
      solution structure, 63
redox cycle, 16
resolvins (Rv)
   biological actions
      E-series resolvins, 221–223
      lipoxins, 218
   mechanisms of action
      D-series resolvin receptors, 225–226
      E-series resolvin receptors, 225–226
      receptor identification and cell expression, 223–224

S

skin
- cancer, 174
- inflammation, injury and wound healing, 170
- metabolites in human and mouse, 162
- 12R-LOX/eLOX-3 role, 169
- structural and metabolic functions, 160

specialized pro-resolving mediators (SPM)
- inflammatory resolution, 212
- lipoxygenases, 213
- pro-resolution lipid mediators, biosynthesis of, 213

systemic sclerosis, 173

T

thrombosis, 86
trichostatin A, 20
trimeric functional unit, 40
tumorigenesis
- breast cancer, 134
- colorectal cancer, 138
- hemato-oncological malignancies, 141
- inhibitors and mitogenic effects, 145
- lacking therapeutic efficacy, 146
- pancreatic cancer, 133
- prostate cancer, 137

Z

zebrafish genome, 52

PGMO 06/14/2018